Next-Generation Food Crops for Human Health

As the global population surges, the challenge of feeding the world becomes increasingly urgent. Meeting this demand requires doubling crop yields and delivering safe, nutritious, and affordable food, but current agricultural practices fall short in combating malnutrition, climate change impacts, and the decline in food diversity.

Next-Generation Food Crops for Human Health delves into the advances in genetic and genomic research that are revolutionizing the development of productive, nutrient-dense food crops. It presents insights into tools including high-throughput phenomics, DNA sequencing, and genomic selection, which enable scientists to discover functionally characterized genes and enhance staple crops, such as grains, legumes, fruits, vegetables, and oil crops. By leveraging these technologies, researchers are creating a new generation of foods that optimize essential nutrients, from complex carbohydrates and proteins to vitamins and bioactive compounds.

A volume in the *Nextgen Agriculture: Novel Concepts and Innovative Strategies* series, this book is valuable to graduate and postgraduate students, postdoctoral researchers, and policymakers working toward improving nutritional security worldwide.

NextGen Agriculture: Novel Concepts and Innovative Strategies

Series Editor: Chittaranjan Kole

Allele Mining for Genomic Designing of Fruit Crops
Chittaranjan Kole, Kenta Shirasawa, and Anil Kumar Singh

Allele Mining for Genomic Designing of Cereal Crops
Chittaranjan Kole, Sharat Kumar Pradhan, and Vijay K Tiwari

Allele Mining for Genomic Designing of Vegetable Crops
Chittaranjan Kole, Tusar Kanti Behera, and Prashant Kaushik

Allele Mining for Genomic Designing of Oilseed Crops
Chittaranjan Kole, Manish Kumar Pandey, and Naveen Puppala

Sustainable Urban Agriculture
New Frontiers
Kheir Al-Kodmany, Madhav Govind, Sharmin Khan, and Chittaranjan Kole

Allele Mining for Genomic Designing of Grain Legume Crops
Chittaranjan Kole, C. Bharadwaj, and Abhimanyu Sarkar

The Vertical Farm
Scientific Advances and Technological Developments
Kheir Al-Kodmany, Andrew Keong Ng, Abel Tablada, and Chittaranjan Kole

Next-Generation Food Crops for Human Health
Edited by Sangam L. Dwivedi, Rodomiro Ortiz, and Chittaranjan Kole

For more information, please visit our website: https://www.routledge.com/Nextgen-Agriculture/book-series/NGA

Next-Generation Food Crops for Human Health

Edited by Sangam L. Dwivedi, Rodomiro Ortiz and Chittaranjan Kole

CRC Press
Taylor & Francis Group
Boca Raton London New York

CRC Press is an imprint of the Taylor & Francis Group, an **informa** business

Designed cover image: Shutterstock

First edition published 2026
by CRC Press
2385 NW Executive Center Drive, Suite 320, Boca Raton FL 33431

and by CRC Press
4 Park Square, Milton Park, Abingdon, Oxon, OX14 4RN

CRC Press is an imprint of Taylor & Francis Group, LLC

ISBN: 978-1-032-73908-3 (hbk)
ISBN: 978-1-032-74032-4 (pbk)
ISBN: 978-1-003-46733-5 (ebk)

DOI: 10.1201/9781003467335

Typeset in Times
by Apex CoVantage, LLC

Dedicated to Dr. Norman Ernest Borlaug, the Father of Green Revolution, Nobel Peace Prize winner and the Founder of the World Food Prize,

who inspired my research and academic works with his generous appreciations and advice.

Preface to the Series

Agriculture is going through a phase of transformation now. It has both challenges and opportunities. One of the major challenges it is facing is obviously food security. A population of about 9.3 billion projected by 2050 necessitates an increase of crop production by 60–70%. Another 10–15% increase would be needed to counter the reduction of crop production due to global warming and climate change. Nutrition security is another challenge with more than 800 million people suffering from malnutrition. Depleted fossil fuels and increasing dependence on biofuels from crop residues are other challenges. Loss of biodiversity and land degradation also add to the list of challenges. The cultivable area is also decreasing fast due to housing, transport, and industrialization. Above all, climate change and global warming have a huge impact on agriculture, health, and environment.

Several novel concepts and strategies are emerging to address the above challenges. For example, the practice of precision farming will lead to precise and timely use of inputs, resulting in higher yield and better quality of produce on the one hand and reduction of loss of inputs on the other. Similarly, urban agriculture and vertical agriculture will add a lot of cultivable areas in the future. Protected farming will facilitate growing crops under controlled conditions, leading to a reduction in biotic and abiotic stresses and precise utilization of inputs.

Agriculture requires maintenance of the philosophy that farmers have been following since its inception over thousands of years, which are being sold today in new packages with sophisticatedly coined labels such as traditional agriculture, natural agriculture, sustainable agriculture, conservation agriculture, integrated agriculture, regenerative agriculture, organic agriculture, and so forth.

In the changing scenarios, next-generation agriculture will need designed crop varieties and breeds produced through designed practices. This book series aims at deliberations on several novel concepts and innovative strategies developed for next-generation agriculture and will hopefully benefit agricultural education, research, and outreach.

Kolkata Chittaranjan Kole
January 10, 2024

Special Acknowledgement to Phullara Kole, for her outstanding assistance in editing this book series

Contents

Preface

The global population is projected to reach 9.8 billion by 2050 and 11.2 billion by 2100. The crop yields must roughly double, and so too must the supply of safe and nutritious food for the growing world population. The current yield levels achieved through plant breeding are not enough to meet the growing demand for nutritious food worldwide. The average income is on the rise in developing economies, which means there is increased demand for affordable, safer, and more nutritious foods, especially from those who are economically disadvantaged in the Global South.

Malnutrition, unhealthy diets, and lifestyle changes are the major risk factors for overweight and obesity-linked chronic diseases in humans, adversely impacting the achievement of the Sustainable Development Goals. The food produced today is less nutritious due to climate change across the world's food-basket regions. The adoption of Green Revolution technologies, and the predominant cultivation of rice and wheat, which have replaced many traditional crops grown on marginal lands, mainly millets and pulses, have reduced the diversity in food, resulting in widespread malnutrition in the developing world. In the late 1990s, there were efforts to increase and improve the oil and protein quality of staple crops, again ignoring critical elements, such as fibers, minerals, vitamins, and bioactive compounds, which are very critical for the health and overall well-being of humans.

Advances in high-throughput phenomics (for characterizing grains for proximate composition, metabolomics, and capturing consumer-preferred traits) and DNA-sequencing technologies to carry out high-throughput genotyping, functional genomics aided gene discovery, high-resolution trait mapping, haplotype discovery, and genomic selection tools in a variety of crops are enabling researchers to identify functionally characterized genes and DNA markers to find variations in gene pools and apply such knowledge in plant breeding.

A concerted effort is now directed toward improving these traits in key cereals, legumes, fruits, and vegetables (including roots and tubers), as well as in oil crops. A new generation of foods and ingredients with improved intrinsic nutritional quality and preferred attributes must be developed through advanced breeding technologies to improve public health by increasing nutritional density and optimizing the quality of complex carbohydrates, proteins, lipids, minerals, and vitamins. There is also growing awareness among chefs and diners regarding food that meets the body's nutrient requirements.

This book contains 20 chapters dealing with the advances in the development of genetic and genomic resources for developing productive, safe, and nutritious staple food crops (grains, oil, fruits, and vegetables). The target audience are graduate and postgraduate students, postdoctoral fellows, and younger-generation researchers who want to improve the nutritional quality of food crops. Those involved in formulating government policies that support public health will benefit from this book, especially in the Global South. It will also serve as a guide for chefs engaged in providing nutritionally enhanced ingredients.

Sangam L. Dwivedi, Rodomiro Ortiz, and Chittaranjan Kole

About the Editors

Sangam L. Dwivedi, MSc, PhD (genetics and plant breeding), is an Indian national who worked for ICRISAT as a senior scientist (groundnut breeding) and as a consultant in various capacities (plant genetic resources, breeding and genetics, biotechnology, HarvestPlus) at ICRISAT or in other countries (Myanmar, Sri Lanka, the USA, and Vietnam) for almost 35 years, and then retired. They contributed to the release of 25 peanut cultivars (Cyprus, Ghana, India, Korea, the Philippines, Sri Lanka, Thailand, and Vietnam) and registered several elite peanut germplasms with specific attributes for the benefit of the peanut researchers globally. Published 150 journal articles, including research articles and invited reviews (biofortification, climate change, crop wild relatives, deleterious mutations, drought tolerance, flavonoids, genetic gains, genetic resources, genetic tradeoffs, haploids, heirloom cultivars, landraces, molecular breeding, nutrition, reproductive trait biology, root plasticity, seed genomics, seed phytate, transgene) in prestigious journals.

Rodomiro Ortiz (Lima, Perú; 1958) is a faculty professor, the chair of genetics and plant breeding at the Swedish University of Agricultural Sciences (SLU), and the SLU Breeding Network chairman. He holds a PhD in plant breeding and genetics from the University of Wisconsin-Madison, and worked as a young researcher at UNALM (Perú), Centro Internacional de la Papa (CIP, Perú), and Rutgers University. He was a scientist in and a director of various CGIAR centers (CIMMYT. ICRISAT, IITA), and held a Nordic professorship on plant genetic resources at the then Royal Veterinary and Agricultural University of Denmark (merged thereafter with the University of Copenhagen). Professor Ortiz has written around 1000 reports, of which about 50% are journal articles or edited book chapters with an h-index of 75 and cited by 22,700 (per Google Scholar on June 9, 2023). The CGIAR awarded to IITA the prestigious 1994 King Baudouin Award for the multidisciplinary research of the team working in plantain and banana improvement, in which he was program leader. In 2012, *Plant Breeding Reviews* dedicated its volume 36 to him. Professor Ortiz was the principal investigator of SLU/ICARDA-led research development partnership project *Adapting durum wheat varieties to the Senegal Basin for food security*, which won the 2017 Olam Prize for Innovation in Food Security. He was given the Bertebos Prize 2022 for mastering the use of new molecular scientific methods to be a driving force in the breeding of crops of great importance as food in Africa, the Americas, and Europe. Professor Ortiz was a member of CGIAR Independent Science and Partnership Council (ISPC, 2015–2019) and of the Commission on Sustainable Agriculture Intensification (CoSAI, 2020–2021) and has been an international fellow of the Royal Swedish Academy of Agriculture and Forestry (KSLA) since 2019, a member of the Royal Physiographic Society at Lund (KFS) from 2021 onward, and a fellow of the African Academy of Sciences (FAAS) since 2021.

Chittaranjan Kole is an internationally reputed scientist with an illustrious professional career spanning over 37 years and original contributions in the fields of plant genomics and biotechnology, leading to the publication of more than 150 research articles in globally leading journals. He has edited over 180 books for leading publishers of the world, including Springer Nature, Wiley-Blackwell, and Taylor & Francis Group. His works and editing acumen have been appreciated by seven Nobel Laureates, including Profs. Norman Borlaug, Arthur Kornberg, Werner Arber, Phillip Sharp, Günter Blobel, Lee Hartwell, and Roger Kornberg. Previously, he was a visiting professor at Pennsylvania State University and Clemson University. He also served as the vice chancellor of the BC Agricultural University in India. He is also the president of the International Climate-Resilient Crop Genomics Consortium (ICRCGC), International Phytomedomics and Nutriomics Consortium (IPNC), and Genome India International (GII). Presently, he is working as the chief advisor of the Neotia University, Kolkata, India, and chairman of the Prof. Chittaranjan Kole Foundation for Science and Society.

Contributors

Sandip K. Bera
ICAR-Directorate of Groundnut Research
Junagadh, India

Deekshitha Bomireddy
International Crops Research Institute for the Semi-Arid Tropics (ICRISAT)
Patancheru, India

Sangam L. Dwivedi
Independent Researcher
Hyderabad, Telangana, India

Sunil S. Gangurde
International Crops Research Institute for the Semi-Arid Tropics (ICRISAT)
Patancheru, India

Baozhu Guo
Crop Protection and Management Research Unit
USDA-ARS, Tifton, GA, USA

Chittaranjan Kole
Chittaranjan Kole Foundation for Science and Society
Kolkata, India

Boshou Liao
Oil Crops Research Institute (OCRI)
Chinese Academy of Agricultural Sciences (CAAS)
Wuhan, China

Khaja Moinuddin
International Crops Research Institute for the Semi-Arid Tropics (ICRISAT)
Patancheru, India
and
University of Agricultural Sciences
Raichur, India

Yogesh Dashrath Naik
Rajendra Prasad Central Agricultural University (RPCAU)
Samastipur, India

Rodomiro Ortiz
Swedish University of Agricultural Sciences
Uppsala Sweeden

Manish K. Pandey
International Crops Research Institute for the Semi-Arid Tropics (ICRISAT)
Patancheru, India

U. Nikhil Sagar
International Crops Research Institute for the Semi-Arid Tropics (ICRISAT)
Patancheru, India
and
Acharya NG Ranga Agricultural University
Guntur, Andhra Pradesh, India

Priya Shah
International Crops Research Institute for the Semi-Arid Tropics (ICRISAT)
Patancheru, India

D.S. Shailaja
University of Agricultural Sciences
Dharwad, 580005
Karnataka, India

Hari Kishan Sudini
International Crops Research Institute for the Semi-Arid Tropics (ICRISAT)
Patancheru, India

H.V. Veerendrakumar
International Crops Research Institute for the Semi-Arid Tropics (ICRISAT)
Patancheru
India and Acharya NG Ranga Agricultural University
Guntur, Andhra Pradesh, India

Xinping Wang
Department of Microbiology and Cell Science
Institute of Food and Agricultural Sciences
University of Florida, Gainesville, FL, USA.

Chuanzhi Zhao
Institute of Crop Germplasm Resources (Biotechnology Research Center), Shandong Academy of Agricultural Sciences; Shandong Provincial Key Laboratory of Crop Genetic Improvement
Ecology and Physiology
Jinan, Shandong, PR China

Introduction

Malnutrition, unhealthy diet, and lifestyle changes are the major risk factors for overweight and obesity-linked chronic diseases (cardiovascular diseases, stroke, certain types of cancers, hypertension, type 2 diabetes, fatty liver, respiratory diseases, and more), adversely impacting the Sustainable Development Goals. The food produced today is less nutritious due to climate change across the world's food-basket regions. Having a healthy diet of colored grains, nuts, fruits, and vegetables, together with changes in lifestyle, could minimize the risk of these diseases in humans.

Advances in genomic science are enabling researchers to identify genes and DNA markers to unfold variations in gene pools and apply knowledge to develop nutrient-dense crops. A new generation of nutrient-dense crops, for example, those rich in fiber (soluble and non-soluble fiber), bioactive compounds (carotenoids, flavonoids), and micronutrients, with appropriate balance in proteins and lipids, is being developed to cater to the needs of various consumers. There is a growing awareness among the public at large, unlike in the past, regarding consuming nutritious foods and a willingness to pay more for a healthy diet. Chefs are looking for new ingredients (such as colored grains with unique culinary, sensory, and textural traits; micronutrient-dense grains with reduced seed phytate; and fruits and vegetables rich in bioactive compounds) to prepare unique dishes with specific aromas, textures, and nutritional properties. The evidence suggests a growing demand for food that meets consumers' tastes and nutritional requirements. The 20 chapters contained in this book synthesize recent advances in developing genetic and genomic resources, which may be deployed to create a new generation of food crops that are climate-resilient, higher-yielding, and nutrient-dense.

Dwivedi and Kole, in Chapter 1, discuss multiple negative effects of climate change on agricultural production as well as on global public health. Adapting for smart agriculture, crop and farming system diversification, optimizing synthetic fertilizers and agrochemicals use, regenerative and/or organic agriculture, and harnessing plant-microbe interaction are expected to enhance food and nutritional security and ecosystem sustainability.

Ortiz and Dwivedi, in Chapter 2, discuss the impact of Green Revolution technologies that brought significant advancements in agricultural research, development, and technology transfer, aimed at increasing food production to meet the growing demands of a rapidly expanding global population. While many regions achieved self-reliance in food production, adoption of Green Revolution technologies caused widespread malnutrition and ecosystem degradation, which necessitated the need to diversify crops and cropping systems for dietary diversity in the Global South.

Most nutritional traits are highly influenced by genotype × environment interaction and genotype × management interaction, necessitating the need for multi-environment testing to identify lines with significant advantage in nutritional traits, as highlighted by Dwivedi and Mattoo in Chapter 3.

Mattoo and Dwivedi, in Chapter 4, discuss the combined use of lexicons (a set of standardized vocabularies developed by highly trained human panels to define sensory attributes) together with volatile compounds and physiochemical attributes for sensory evaluation of edible produce using AI-based machine learning and statistical models to identify the best models for applying in crop-breeding programs, and document significant breeding milestones of the uptake of genetic and genomic resources for the improvement of sensory traits via crop breeding.

Ortiz and Dwivedi, in Chapter 5, discuss chefs' and consumers' preferences as a guide for developing new cultivars with unique nutritional and health-promoting characteristics. They recommend an interactive collaboration with participatory plant breeding and participatory varietal selection, involving farmers, diners, chefs, plant breeders, and biodiversity specialists, including seed savers, to address the various needs of stakeholders involved in agri-food business with diverse tastes and health-conscious choices in the modern world.

Ortiz and Dwivedi, in Chapter 6, discuss the role of crop wild relatives, landraces, and heirloom germplasms/cultivars with unique tastes and flavors. Integrating such germplasms and knowledge into plant-breeding programs could lead to the development of food crops like fruits and vegetables with unique sensory, textural, and nutritional value and climate resilience, which will ensure the preservation, in perpetuity, of such priceless genetic resources for the benefit of humanity.

Dwivedi and Mattoo, in Chapter 7, discuss functional allelic variation associated with culinary, sensory, and textural traits in cereals and food legumes. Accessions with unique culinary and sensory characteristics are available in many crops, with their genetic architecture unlocked. The functionally characterized genes and DNA markers could be deployed in breeding programs to develop next-generation food crops, combining yield, nutrition, and climate resilience with better culinary and sensory traits.

Dwivedi and Mattoo, in Chapter 8, discuss exploiting natural and CRISPR/Cas9-induced variation in betaine aldehyde dehydrogenase (BADH) to develop fragrant crops. For example, grain's aroma trait has been widely exploited to develop aromatic rice that is marketed as basmati or jasmine rice. BADH has also been discovered in mung beans, vegetable soybeans, and lablab beans, as well as in fruit crops, such as cucumbers, sponge gourds, and winter melons. The CRISPR/Cas9 system is equally effective in creating allelic variation associated with fragrance in grain crops. Consumers will soon have aromatic maize- and sorghum-derived food products on their food plates.

Ortiz and Dwivedi, in Chapter 9, emphasize achieving significant milestones in agricultural biofortification. Crop cultivars with specific biofortified traits include fragrant basmati and jasmine rice; flavonoid-rich barley, rice, sorghum, and wheat; iron- and zinc-dense beans, rice, and wheat; low-seed-phytate soybean; flavonoid-enriched potatoes; β-carotene-rich cassava, rice, and sweet potato; and flavonoid-rich tomatoes. Millions of households across the Global South are cultivating these crops. Combining fragrance, flavonoids, micronutrients, and β-carotenoids with plant-based seed proteins will offer nutritionally balanced diets and sensory pleasure to diners.

Veerendrakumar et al., in Chapter 10, provide an overview on the types of mycotoxins reported in maize and peanuts, their effects on animal and human health, management strategies, and the ways to minimize the risk of toxin contamination through crossbreeding and biotechnological tools. Multifunctional and multidimensional approaches to plant breeding and biotechnology should be integrated to minimize the risk of toxin contamination in food/feed crops.

Sagar et al., in Chapter 11, discuss the progress to date achieved in developing high-oleate and/or hypoallergenic peanuts. Enhancing the shelf-life of peanut products and minimizing risk of peanut allergy are two major issues limiting peanut consumption. A ratio of oleic (O)/linoleic (L) fatty acid determines the shelf-life of peanut products: the higher the ratio, the longer the shelf-life. Numerous high-oleate cultivars, in all botanical and market types, are currently being commercially grown to saturate peanut products with the high-oleate trait. Consumers prefer peanuts with firm and crispy texture over those with soft and mushy texture. Significant progress has been achieved to incorporate these traits into peanut cultivars. Efforts are underway to develop hypoallergenic (low-allergen) peanuts using transgenic and genome-editing tools.

Dwivedi and Mattoo, in Chapter 12, summarize the progress to date achieved in reducing bitterness in Cucurbitaceae and glucosinolates (GLSs) in Brassicaceae. *Bl* and *Bt* regulate bitterness in Cucurbitaceae. KASP markers were deployed to differentiate bitter-free segregants in cucumber. Maintaining desired levels of seed GLSs without disrupting GLSs in vegetative tissues is a significant breeding challenge. CRISPR/Cas9-based editing of the glucosinolate transporter (GT) lowered the seed GLSs at globally acceptable levels without disrupting the synthesis in vegetative tissues to protect plants from pest damage. Applying divergent mass selection effectively altered total and individual GLS contents.

Dwivedi and Mattoo, in Chapter 13, discuss genes associated with fruit morphotypes and the nutritional quality in Cucurbitaceae vegetables. Genes regulating cell division patterning are universal shape regulators in horticultural crops. Assessing and exploiting such variation in the gene pool, via gene-based markers that perfectly co-segregate with characteristic fruit morphotypes and aroma, and newly emergent information on key genes and proteins regulating variation in edible organ size and shape may provide additional opportunities to modify the form and contour of vegetable crops.

Dwivedi and Mattoo, in Chapter 14, discuss the genetic and molecular basis of horticultural and sensory traits of melons (*Cucumis melo* and *Citrullus lunatus*). The functionally characterized genes and markers are already in use in melon breeding. The pangenomes and doubled haploid technology provide additional opportunities to tailor and accelerate the development of inbred lines and hybrid cultivars with specific fruit qualities.

Dwivedi and Kole, in Chapter 15, synthesize the genetic and molecular basis of bioactive compounds in the bitter melon (or bitter gourd). Bioactive compounds from bitter melons provide health benefits to humans. Sufficient natural variation for bioactive compounds is reported in gene pools. Bitter melon genomes contain many genes associated with stress tolerance, carotenoids, cucurbitacin, terpenes, and antioxidant activity. A major gene cluster (*Bi*) is conserved in cucumbers, melons, and watermelons and is involved in biosynthesis of cucurbitane triterpenoids contributing to the bitter taste. Combining morphological characters with nutrient and nutraceutical contents has great potential to develop functionally rich bitter melons germplasm/cultivars.

Ortiz and Dwivedi, in Chapter 16, discuss the nutritional quality of the tomato heirloom germplasm. Heirloom tomatoes are prized for their long shelf-life, juiciness, unique texture, flavors, shapes, colors, and taste. Variation in phenolics, flavonoids, carotenoids, and volatile compounds contribute to their unique flavor and taste. Key alleles associated with unique tomato flavors were identified, providing unique opportunities to restore some of the tomato flavors lost through domestication or genetic improvement. Genetics, environment, and agronomic practices contribute to unique tomato taste and flavor. Heirloom germplasm should be profiled for metabolites and textural traits to preserve unique landraces for enhancing the flavor and taste of modern tomatoes. CRISPR/Cas9 systems hold promise to accelerate the development of tomatoes with unique flavor and taste.

Mattoo and Dwivedi, in Chapter 17, discuss the progress realized in developing genetic and chemical toolboxes to restore the flavor and taste of modern tomatoes. A few wild tomato species closely related to cultivated tomatoes possess an intense fruit flavor. ILs in the genetic background of modern tomatoes with superior flavor and taste are developed, and many genes and SNPs associated with improved fruit flavor are unlocked. DAMs and DEGs during fruit ripening may provide an additional opportunity to unlock candidate metabolites and new genes related to improved flavor. Tomato pan- and super-pangenomes contain many gene variants that are absent in reference genomes, which may be used to unlock genetic factor(s) underlying new genes associated with flavor and taste. Genetic transformation and genome-editing tools are equally effective for enhancing tomato flavor and taste without any adverse impact on agronomic performance.

Dwivedi and Mattoo, in Chapter 18, discuss profiling of *Brassica* leafy germplasm for variation in glucosinolates (GLSs) and their hydrolyzed compounds. GLSs and their compounds are effective in cancer treatments, enhancing product shelf-life, and inhibiting microbial growth without any adverse impact on the organoleptic properties of the product. *Brassica* leafy germplasm shows abundant variability in leaf morphotypes, bioactive compounds, and minerals. Complex genetic networks regulate cell proliferation and cell expansion of leaves. Multiple genes and QTLs control variation in leaf morphotypes. Applying divergent mass selection resulted in lines with distinct GLS profiles that yielded at par or excelled better than the original population. ILs with diverse morphotypes should be exploited in *Brassica* leafy breeding programs. Targeted editing has effectively reduced seed GLS content while maintaining high GLSs in other plant parts without adversely impacting growth or defense parameters.

Ortiz and Dwivedi, in Chapter 19, provide molecular insights regulating GLS metabolites in the *Brassica* germplasm. The gene network coordinates the stepwise conversion of amino acids into the synthesis of glucosinolate structures. Over 120 different structures of GLS and 101 Brassicaceae genes associated with GLS biosynthesis are unfolded. Mutants involving GTS significantly reduced seed GLS levels. A complex interplay between genetics and metabolism or between genetic factors and environmental cues shapes plant secondary metabolite profiles in *Brassica* species, adapted to ecological niches. This knowledge can be used to develop cultivars with diverse characteristics, enhanced climate resilience, and better nutritional profiles.

Dwivedi and Mattoo, in Chapter 20, summarize the progress achieved to develop flavonoid-enriched fruits and vegetables using genome-editing tools. Flavonoid content of grapes, pomegranates, melons, tomatoes, Chinese cabbages, and lettuces has been improved, with no adverse impact on yield or quality. The public responds more positively to genome-edited crops than GMOs. Inappropriate government regulatory policies across countries/regions are probably the biggest obstacle in the acceptance of genome-edited crops.

Abbreviations

AFs	aflatoxins
AFB1	aflatoxin B1
AI	artificial intelligence
Al	alanine
ANN	artificial neural network
ARMS-PCR	amplification refractory mutation-system polymerase chain reaction
As	arsenic
AS-PCR	allele-specific polymerase chain reaction
BMFE	bitter melon fruit extract
BST	breeders' sensory test
C	carbon
CA	citric acid
Ca	calcium
CAPS	cleaved amplified polymorphic sequence
CATA	check all that apply
Cd	cadmium
C:N ratio	carbon-nitrogen ratio
CNVs	copy number variations
Co	cobalt
Cr	chromium
C:N:P ratio	carbon-nitrogen-phosphorus ratio
C:P ratio	carbon-phosphorus ratio
CTGs	cucurbitane-type triterpene glycosides
Cu	copper
CRU	controlled release of urea
CT	cooking time
CWR	crop wild relatives
DA	descriptive analysis
DAMs	differentially accumulate metabolites
DNA	deoxyribonucleic acid
DEGs	differentially expressed genes
DH	double haploid
DMS	divergence mass selection
DON	deoxynivalenol
eCO_2	elevated carbon dioxide
EMS	ethyl methanesulfonate
eQTL	expression quantitative trait loci
ELISA	enzyme-linked immunosorbent assay
pQTL	*trans*-protein quantitative trait loci
E	environment
FA	fatal allergens

FACE	free air carbon dioxide enrichment facility
Fe	iron
FER	*Fusarium*
FL	free listing
FT-NIR	Fourier-transform near infrared
FTIR-ATR	Fourier-transform infrared spectroscopy with attenuated total reflection
G	genotype
GEI	genotype × environment × interaction
GEBV	genomic estimated breeding value
GER	*Gibberella* ear rot
GI	geographical identification tag
GMI	genotype × management × interaction
GHG	greenhouse gas
GLSs	glucosinolates
GMOs	genetically modified organisms
GS	genomic selection
GWAS	genome-wide association studies
Hg	mercury
HS-GC-MS	headspace–gas chromatography–mass spectrometry
HS-SPME-GC-MS	headspace–solid-phase microextraction–gas chromatography–mass spectrometry
HS-SPME-GC×GC-TOFMS	headspace–solid-phase microextraction–full two-dimensional gas chromatography–time-of-flight mass spectrometry
HIV/AIDS	human immunodeficiency virus/acquired immunodeficiency syndrome
IBBs	iron-biofortified beans
ICGVs	ICRISAT groundnut varieties
ILs	introgression lines
IMMTs	indicators, metrics, methods, and tools
K	potash
KASP	kompetitive allele-specific PCR
LDA	linear discriminant analysis
LSL	long shelf-life
MA	malic acid
MAB	marker-assisted breeding
MABB	marker-assisted back-crossbreeding
MAGIC	multiparent advanced generation intercross
MAS	marker-assisted selection
MEL	monitoring, evaluation, and learning
MFCW	mature fruit cuticular wax
Mg	magnesium
ML	machine learning
ML	malformed leaves

Mn	manganese
mGWAS	meta genome-wide association studies
mQTL	meta quantitative trait loci
MRP	Maillard reaction products
N	nitrogen
Na	sodium
NAM	nested association mapping
NCDs	noncommunicable diseases
NGS	next-generation sequencing
Ni	nickel
NILs	near isogenic lines
NIR	near infrared
NO3-N	nitrate-nitrogen leaching
NORG	nonorganic
N:P:K	nitrogen-phosphorus-potash ratio
N:P ratio	nitrogen-phosphorus ratio
O_3	ozone
O/L ratio	oleic/linoleic fatty acid ratio
ORG	organic
OTGs	oleanane-type triterpene glycosides
P	phosphorus
PA	proanthocyanidins
Pb	lead
PGPR	plant-growth-promoting rhizobacteria
PGR	plant genetic resources
PLS-DA	partial least square discriminant analysis
QTL	quantitative trait loci
RATA	rate all that apply
RGB	red, green, blue
RILs	recombinant inbred lines
RNAi	RNA interference
RTB	root, tuber, banana
RV	regressor vector
RWCS	rice wheat cropping system
Se	selenium
SDG2	Sustainable Development Goal 2
SKC	Seed to Kitchen Collaborative
SKCS	single-kernel characterization system
SSL	short shelf-life
Sn	tin
SNPs	single-nucleotide polymorphisms
SSRs	simple sequence repeats
SVs	structural variations
TEs	transposon elements
TGRC	tomato genetic resource center
TNC	total nonstructural carbohydrate

TSC	nonstructural carbohydrate
TSSC	total soluble solid content
TTA	total titrable acids
UNSDGs	United Nations Sustainable Development Goals
UV-B	ultraviolet irradiation
VOCs	volatile organic compounds
WGS	whole-genome sequencing
WT	wild type
Zn	zinc

Glossary

agrobiodiversity: The entire diversity of genes, species, cultivars, and ecosystems in agriculture, including taxonomic, ecological, morphological, and molecular diversity.

bioactive compounds: A substance with biological activity and affecting a living organism. The effect could be positive or negative depending on the substance, dose, and bioavailability. **If consumed in sufficient quantities, it provides health benefits to humans and plays a critical role in plant growth, development, and stress tolerance.**

biofortification: Developing micronutrient-dense cultivars through crossbreeding or genetic engineering.

conservation agriculture: An approach for managing agro-ecosystems for improved and sustained productivity, increased profits, and food security while preserving and enhancing the resource base and the environment.

crop diversification: Growing more than one crop in an area, which is accomplished by adding a new crop species or different cultivars or by changing the cropping system currently in use. It enhances soil health, breaks the insect-disease cycle, reduces weeds and soil erosion, and ensures food and nutritional security from climate shocks.

crop rotation: A practice that involves alternating the types of crops planted in a specific area over time to prevent the spread of viruses and diseases and reduce weeds.

genome selection (GS): A method to predict the genetic value of selection candidates based on the genomic estimated breeding value (GEBV) predicted from high-density markers positioned throughout the genome. GEBV is based on all markers, including both minor and major marker effects, capturing greater genetic variation for a given trait (or a combination of traits) under selection.

geographical identification (GI) tag: A product signifying its origin from a specific geographic location having characteristic quality attributes. It is covered as an element of intellectual property rights under the Paris Convention for the Protection of Industrial Property.

greenhouse gas (GHG): Carbon dioxide (CO_2), methane (CH_4), nitrous oxide (N_2O), and ozone (O_3) in an atmosphere that absorbs and emits radiation within the thermal infrared range.

Global South: A group of countries in Asia, Africa, Latin America, and Oceania that are less developed than those in the so-called Global North (most of Europe, the USA, Canada, Australia, and so forth).

genetically modified organism (GMO): An organism (plant, human, or microbe) in which one or more changes have been effected into the genome using another organism's DNA via genetic engineering.

glucosinolates (GLSs): A group of sulfur-containing compounds isolated from plants known for their strong odor and volatility, protecting plants from herbivores, pathogens, and insects and providing health benefits (anticancer, anti-inflammatory, and antioxidant properties) to humans.

Green Revolution: A suite of agricultural technologies, including irrigation, synthetic nitrogen fertilizer, pesticides, and improved cultivars revolutionizing maize, rice, and wheat production in many parts of the world.

genome-wide association study (GWAS): Assessing marker-trait association (MTA) by involving a large number of germplasms and single-nucleotide polymorphisms (SNPs/InDels/SVs/PAVs) markers. It exploits historic linkage disequilibrium to link phenotypes with genotypes to find MTAs underlying complex traits.

heirloom germplasm: Traditional or older cultivars that are open-pollinated, passed down from gardener to gardener or handed down in families, and often not used in large-scale agricultural enterprises.

introgression lines (ILs): A set of lines developed and constructed by using a combination of backcrossing and selfing to replace chromosome fragments of the recipient genome with chromosome fragments of the donor genome. They are powerful pre-breeding stocks to broaden the genetic base of crop cultivars.

landrace: A dynamic population of a cultivated species with prehistoric origin and distinct identity that was locally adapted in association with traditional farming systems.

MAGIC: Multiparent advanced generation intercross (MAGIC) recombinant inbred lines (RILs) that are genetically diverse and have undergone many generations of recombination. They display emerging features over experimental biparental and germplasm populations in combining significant levels of genetic recombination, a lack of genetic structure, and high genetic and phenotypic diversity.

malnutrition: A condition that results from eating a diet whose nutrients are either insufficient or are in excess such that the diet causes health problems.

marker-assisted selection: An indirect selection process where a trait of interest is selected based on a marker (morphological, biochemical, or DNA/RNA variation) linked to a trait of interest.

mGWAS: A research method that combines metabolomics (small molecules within an organism) data with genetic information from a genome-wide association study (GWAS) to identify genetic variants associated with specific metabolite levels, providing a powerful link between genotypes and phenotypes.

mQTL: A specific region on the genome (chromosome) that is associated with variations in the levels of metabolites within an organism.

nested association mapping (NAM): A large panel of RILs that share a common recurrent parent(s) in the form of line × tester mating system.

next-generation sequencing (NGS): A modern high-throughput method that sequences millions of DNA fragments simultaneously, revolutionizing genomic research.

noncommunicable diseases (NCDs): A group of long-lasting chronic diseases (e.g., cancer and cardiovascular diseases) in humans, not caused by infectious agents.

pangenome: An entire set of genes within a species, including core (containing sequences shared between all individuals of the species) and dispensable (a set of genes within a species that are present in some individuals but not in others) genomes. A pangenome is more useful than a single-reference genome for unlocking the genomic basis of complex traits.

polymerase chain reaction (PCR): A laboratory technique to denature and renature short segments of DNA.

quantitative trait loci (QTL): DNA segments carrying genes controlling quantitative traits.

recombinant inbred lines (RILs): A product of biparental crossing followed by numerous generations of selfing and single-seed descent—producing homozygous lines that are a mosaic of chromosomal regions from the founder parents. They are limited to a maximum of two alleles at any one locus, which may exclude the majority of variation, caused by additive and epistatic interactions, contributing to phenotype in natural populations.

root microbiomes: A community of microbes (bacteria, fungi, and archaea) living in association with a plant's roots, and contributing to plant fitness, reproduction, and nutritional quality of edible produce.

structural variants (or copy number variants, CNVs): Genomic polymorphisms (DNA segments >1 kb in length), such as inversions, insertions, duplications, or deletions, resulting from gains or losses of DNA segments or chromosome rearrangements.

Sustainable Development Goals (SDGs): A set of 17 international goals (https://www.undp.org/sustainable-development-goals) aimed to end poverty, protect the planet, and create a sustainable world by 2030.

whole-genome sequencing (WGS): A comprehensive method (https://sapac.illumina.com/techniques/sequencing/dna-sequencing/whole-genome-sequencing.html) for analyzing entire genomes of a given species, including bacteria, viruses, fungi, plants, and animals, using NGS technology.

1 Importance of Health and Nutrition Security

Origin and Evolution of the Concept

Sangam L. Dwivedi and Chittaranjan Kole

1.1 INTRODUCTION

Today's lifestyle is largely different from that of our ancestors. They practiced food gathering and hunting as the basis of nutrition until humans began domesticating plants and animals. Lifestyle changes (from active to sedentary) and dietary shift from hunting and gathering of wild animals and plants to dependency on domesticated animals and crops, with greater emphasis to meat-based diets, may have brought surge in noncommunicable diseases in modern-day humans. Our culture, rather than the millions of years of ancestry, has driven changes in our lifestyle and diet from the earliest primates to hunter-gatherers, and from the early agriculture and industry to present-day agriculture. Increasingly, more people around the world suffer from various diseases, ranging from diet-related intolerance (e.g., allergy) to widespread occurrence of noncommunicable diseases, despite the wide range of foods available. The calories supplied from processed foods and drinks have a high energy density but provide hardly any fiber and micronutrients (Alt et al., 2022).

Six of the nine planetary boundaries (climate change, biosphere integrity, land system change, freshwater change, biogeochemical flows, ocean acidification, atmospheric aerosol loading, stratospheric ozone depletion, novel entities) are already transgressed, reflecting unprecedented human disruption of the earth's system. Earth is now well beyond being a safe operating space for humanity. Ocean acidification is close to being breached, aerosol loading regionally exceeds the boundary, while stratospheric ozone levels have slightly recovered. The transgression level has increased for all boundaries earlier identified as overstretched. The production of agricultural goods drives the earth's system's functional biosphere integrity, which is also transgressed. The land system change boundary, therefore, is the most powerful means that humanity has at its disposal to combat climate change. Restoring global forest cover back to the levels of the late 20th century, for example, would provide a substantial cumulative sink for atmospheric CO_2 in 2100 (Richardson et al., 2023). Transition toward food and agricultural systems that include both food security and planetary health should be built upon for economically feasible solutions through new innovations (Hofman-Bergholm, 2022).

DOI: 10.1201/9781003467335-1

'Food security here refers to when all people at all times have physical and economic access to sufficient and nutritious food for an active and healthy life' (http://www.fao.org/wfs/index_en.htm). About 800 million individuals do not have access to enough food, and >2 billion people experience key micronutrient deficiencies, of which 60% of individuals are food-insecure in low-income countries. Food insecurity adversely impacts human capital at family, community, region, and country levels. It has serious repercussions in achieving the United Nations Sustainable Development Goals, or SDGs (file:///C:/Users/sanga/Downloads/The-Sustainable-Development-Goals-Report–2022.pdf; accessed on May 3, 2024). A sound, equitable, and sustainable food system is essential for countries to achieve SDGs (Pérez-Escamilla, 2017).

Sustainable Development Goal 2 (SDG2) aims to end all forms of hunger and malnutrition by 2030 to ensure that all people, including children, have sufficient and nutritious food all year round. Promoting sustainable agriculture, supporting small-scale farmers, equal access to land, technology, and markets, and minimizing food waste may realize 100% SDG2 targets. Achieving long-term solutions to avoid food waste necessitates involvement of all players in the food supply chain, namely, farmers, manufacturers, distributors, and consumers. Traditional food waste reduction strategies are ineffective in lowering GHG emissions and creating suitable food waste treatment. Building an inventory of theoretical and practical methods of prevention and minimization of food waste and losses are needed to minimize tradeoffs involving food/nutritional security and environmental sustainability (Wani et al., 2024).

In this chapter, we list some of the most pressing issues adversely impacting agricultural production worldwide and, more particularly, greater risks to food and nutritional security in the Global South.

1.2 CLIMATE CHANGE

Climate change has multiple negative effects on agricultural production and global public health. The changes in soil fertility, crop yield and composition, and bioavailability of nutrients in foods increase the risk of malnutrition (i.e., undernutrition or obesity) and diet-related noncommunicable diseases (i.e., cancer, type 2 diabetes, cardiovascular diseases) in humans, particularly in the Global South (Dwivedi et al., 2023; Owino et al., 2022). In addition, food systems substantially contribute to greenhouse gas emissions, i.e., ~16 CO_2 eq yr^{-1} in 2018, or one-third of the global anthropogenic total (Tubiello et al., 2021). Food security and adequate nutrition for the global population can be achieved using climate-smart, sustainable agricultural practices, while reducing the negative environmental impacts of agriculture, including greenhouse gas emissions (Owino et al., 2022; Wijerathna-Yapa and Pathirana, 2022). Countries/regions need to make changes in dietary shift to minimize the risk of malnutrition and age-old diseases in humans and for ecosystem sustainability. For example, an increased intake of nutrient-dense foods to address malnutrition in sub-Saharan Africa and South Asia, while for Europe and North America, a shift toward more plant-based diets would be healthier and simultaneously reduce the per capita environmental footprints (Chen et al., 2022).

1.3 CONSERVATION AGRICULTURE

Regenerative soil-building farming practices (i.e., no-till, cover crops, and diverse rotations) as compared to conventional (synthetically fertilized and herbicide-treated) farming enhance the nutritional profiles of crops and livestock (Montgomery et al., 2022). Adopting an organic agriculture system, including the use of microbial inoculants, compost, and mulch to build soil organic matter, improves crop yields, soil properties, and produce quality (Gopinath et al., 2023; Manzeke-Kangara et al., 2023; Montgomery et al., 2022). The organically grown produce relative to conventional farming is healthier because of the higher levels of phytochemicals it contains. Higher phytochemicals exhibit health-protective antioxidants and anti-inflammatory properties. Reducing our dependence on synthetic fertilizers and agrochemicals will make the production systems more sustainable and viable in changing global climate (Wang et al., 2022).

1.4 CROPS AND FARMING SYSTEMS DIVERSIFICATION

Crop and farming system diversification, instead of a monocropping system, has a greater potential to improve dietary diversity to enhance food and nutritional security and ecosystem sustainability (Dwivedi et al., 2017; Nicholson et al., 2021). Crop diversification refers to a shift from regional dominance of one crop to regional production of many crops, while in farming system diversification, the farming practices and landscapes that increase functional biodiversity at multiple spatial and/or temporal scales are included, providing critical inputs to agriculture (Kremen et al., 2012; Rautaray and Sucharita, 2024). Designing of cropping systems with minimal or lesser impact on environment and use of genetics to tailor new crop cultivars are probably the best options to optimize productivity and nutritional quality in most challenging environments (Roberts and Mattoo, 2019).

1.5 RHIZOSPHERE MICROBIOME INTERACTION

Ample evidence suggests that host (plant)-root microbiome interactions minimize the adverse impact of climate change on agricultural production, reduce soil-borne diseases, and improve the nutritional quality of food crops. Over millennia, plants have evolved intricate relationships with a diverse array of microorganisms inhabiting their roots, which should be further researched to identify beneficial microorganisms forming a dynamic and interconnected network within and around the plant to promote plant growth and development through nutrient acquisition. Microbes solubilize essential minerals, fix atmospheric nitrogen, and convert organic matter into accessible forms, thereby augmenting the nutrient pool available to the plant (Mukherjee et al., 2024).

1.6 ADOPTING POLLINATOR-FRIENDLY PRACTICES

Pollination services are essential for food and nutritional security. Many crops, including fruits and vegetables, depend on pollinators for reproduction (van der

Sluijs and Vaage, 2016). Reports suggest global decline in pollinators (e.g., insects, birds, bats) due to climate change, changes in vegetation and land use patterns, pollution, and pesticides. The decline in pollination services in agricultural landscape endangers food and nutritional security (Tscharntke, 2021). Understanding plant-pollinator interactions and adopting for pollinator-friendly production systems may provide opportunities to minimize the risk associated with loss of pollinators on agricultural production (Potts et al., 2016).

1.7 PROMOTING THE CULTIVATION OF PERENNIAL GRAIN CROPS

Cultivating perennial grains, instead of row crops (annual grain crops), offer sustainable food production system to minimize the impact of climate change in agricultural production. It also maintains ecosystem sustainability and provides nutritional security (Chapman et al., 2022; Soto-Gömez and Pérez-Rodríguez, 2022). The release of perennial cultivars, perennial rice, and intermediate wheatgrass (Kenza) (Bajgain et al., 2020; Huang et al., 2018) demonstrates the potential of this approach to minimize the adverse impact of climate change on food and nutritional security. The cultivation of perennial grain crops is expected to improve, or at least maintain, their soil quality because of least soil disturbance post-seeding (Cattani, 2024).

1.8 ADOPTING BIOFORTIFICATION (AGRONOMIC AND GENETIC) PRACTICES

Understanding the physiological and molecular responses of nutrient acquisition, transport, and distribution in plants, and advances in agronomic- and genetic-based biofortification approaches for optimizing yield and nutritional quality (including protein, minerals, and phytochemicals) under varying environments is the need of the hour to enhance food and nutritional security in the Global South (Dwivedi et al., 2023; Khan et al., 2023). Colored fruits, vegetables, and grains are the source of bioactive compounds. The accumulated evidence suggests that regular consumption of such kinds of foods minimizes the risk of noncommunicable diseases in humans. A paradigm shift in crop improvement programs is therefore recommended to enhance the availability of such food crops for the benefit of humanity. Significant progress has been achieved to tailor new germplasm with required characteristics to harness nutritional benefits of bioactive compounds in the colored edible produce (Dwivedi et al., 2022; Mattoo et al., 2022).

1.9 RESOURCE-USE-EFFICIENT CROPS

Feeding safe and nutritious food to the growing world population requires continuous investment of research and development to develop and apply resource-use-efficient technologies to ease pressure on limited land and freshwater resources. The root-soil-microbe interactions are complex in nature, involving physical, chemical, and biological processes that influence nutrient-use efficiency, plant growth, and health. Unlocking these rhizosphere processes may provide new opportunities to harness such interactions to ensure crop productivity and nutrient use efficiency. Further, integrating in-depth knowledge about above- and belowground plant traits, including plant-soil-microbe interactions, in crop improvement programs may help us to ensure crop resilience in stressful environments to produce sufficient, safe, and nutritious food in a more sustainable way (Oburger et al., 2022).

1.10 CONCLUSION

Adoption of climate-smart farming, together with production system diversification, with greater emphasis on production and consumption of colorful fruits, vegetables, and grain crops (both annual and perennial), practicing regenerative and/or organic agriculture with judicial use of synthetic fertilizers and agrochemicals to maintain productivity, and harnessing variations in plant-microbe interaction, may provide an additional layer of safety to achieve food/nutritional security, environmental sustainability, and ultimately achieving the SDGs in the Global South.

REFERENCES

Alt, K.W., Al-Ahmad, A., and Woelber, J.P. (2022). Nutrition and health in human evolution–Past to present. *Nutrients* 14, 3594. https://doi.org/10.3390/nu14173594

Bajgain, P., Zhang, X., Jungers, J.M., DeHaan, L.R., Heim, B., Sheaffer, C.C., et al. (2020). 'MN-Clearwater', the first food-grade intermediate wheatgrass (Kernza perennial grain) cultivar. *J. Plant Regist.* 14, 288–297. https://doi.org/10.1002/plr2.20042

Cattani, D.J. (2024). Perennial grains and oilseeds: Current status and future prospects. *Agriculture* 14, 1035. https://doi.org/10.3390/agriculture14071035

Chapman, E.A., Thomsen, H.C., Tulloch, S., Correira, P.M.P., Luo, G., Najafi, J., et al. (2022). Perennials as future grain crops: Opportunities and challenges. *Front. Plant Sci.* 13, 898769. https://doi.org/10.3389/fpls.2022.898769

Chen, C., Chaudhary, A., and Mathys, A. (2022). Dietary change and global sustainable development goals. *Front. Sustain. Food Syst.* 6, 771041. https://doi.org/10.3389/fsufs.2022.771041

Dwivedi, S.L., Garcia-Oliveira, A.L., Govindraj, M., and Ortiz, R. (2023). Biofortification to avoid malnutrition in humans in a changing climate: Enhancing micronutrient bioavailability in seed, tuber, and storage roots. *Front. Plant Sci.* 14, 1119148. https://doi.org/10.3389/fpls.2023.1119148

Dwivedi, S.L., Mattoo, A.K., Garg, M., Dutt, S., Singh, B., and Ortiz, R. (2022). Developing germplasm and promoting consumption of anthocyanin-rich grains for health benefits. *Front. Sust. Food Syst.* 6, 867897. https://doi.org/10.3389/fsufs.2022.867897

Dwivedi, S.L., van Bueren, E.T.L., Ceccarelli, S., Grando, S., Upadhyaya, H.D., and Ortiz, R. (2017). Diversifying food systems in the pursuit of sustainable food production and healthy diets. *Trends Plant Sci.* 22, 842–856. https://doi.org/10.1016/j.tplants.2017.06.011

Gopinath, K.A., Venkatesh, G., Manjunath, M., Jayalakshmi, M., Prasad, T.V., Rajkumar, B., et al. (2023). Impact of organic and integrated production systems on yield and seed quality of rainfed crops and on soil properties. *Front. Nutr.* 10, 1127970. https://doi.org/10.3389/fnut.2023.1127970

Hofman-Bergholm, M. (2022). A transition towards a food and agricultural system that includes both food security and planetary health. *Foods* 12, 12. https://doi.org/10.3390/foods12010012

Huang, G., Qin, S., Zhang, S., Cai, X., Wu, S., Dao, J., et al. (2018). Performance, economics and potential impact of perennial rice PR23 relative to annual rice cultivars at multiple locations in Yunnan province of China. *Sustainability* 10, 1086. https://doi.org/10.3390/su10041086

Khan, M.I.R., Nazir, F., Maheshwari, C., Chopra, P., Chhillar, H., and Sreenivasulu, N. (2023). Mineral nutrients in plants under changing environments: A road to future food and nutrition security. *Plant Genome* 16, e20362. https://doi.org/10.1002/tpg2.20362

Kremen, C., Iles, A., and Bacon, C. (2012). Diversified farming systems: An agroecological, systems-based alternative to modern industrial agriculture. *Ecol. Soc.* 17, 44. http://doi.org/10.5751/ES-05103-170444

Manzeke-Kangara, M.C., Joy, E.J.M., Lark, R.M., Redfern, S., Eilander, A., and Broadley, M.R. (2023). Do agronomic approaches aligned to regenerative agriculture improve the micronutrient concentrations of edible portions of crops? A scoping review of evidence. *Front. Nutr.* 10, 1078667. https://doi.org/10.3389/fnut.2023.1078667

Mattoo, A.K., Dwivedi, S.L., Dutt, S., Singh, B., Garg, M., and Ortiz, R. (2022). Anthocyanin-rich vegetables for human consumption—focus on potato, sweet potato, and tomato. *Int. J. Mol. Sci.* 23, 2634. https://doi.org/10.3390/ijms23052634

Montgomery, D.R., Biklé, A., Archuleta, R., Brown, P., and Jordan, J. (2022). Soil health and nutrient density: Preliminary comparison of regenerative and conventional farming. *Peer J.* 10, e12848. https://doi.org/10.7717/peerj.12848

Mukherjee, A., Singh, B.N., Kaur, S., Sharma, M., de Araújo, A.S.F., de Araújo, A.S.F., et al. (2024). Unearthing the power of microbes as plant microbiome for sustainable agriculture. *Microbial. Res.* 286, 127780. https://doi.org/10.1016/j.micres.2024.127780

Nicholson, C.C., Emery, B.F., and Niles, M.T. (2021). Global relationships between crop diversity and nutritional stability. *Nat. Commun.* 2, 5310. ttps://doi.org/10.1038/s41467-021-25615-2

Oburger, E., Schmidt, H., and Staudinger, C. (2022). Harnessing belowground processes for sustainable intensification of agricultural systems. *Plant Soil.* 478, 177–209. https://doi.org/10.1007/s11104-022-05508-z

Owino, V., Kumwenda, C., Ekesa, B., Parker, M.E., Ewoldt, L., Roos, N., et al. (2022). The impact of climate change on food systems, diet quality, nutrition, and health outcomes: A narrative review. *Front. Clim.* 4, 941842. https://doi.org/10.3389/fclim.2022.941842

Pérez-Escamilla, R.P. (2017). Food security and the 2015–2030 sustainable development goals: From human to planetary health. *Curr. Dev. Nutr.* 1, e000513. https://doi.org/10.3945/cdn.117.000513

Potts, S., Imperatriz-Fonseca, V., Ngo, H.T., Aizen, M.A., Biesmeijer, J.C., Breeze, T.D., et al. (2016). Safeguarding pollinators and their values to human well-being. *Nature* 540, 220–229. https://doi.org/10.1038/nature20588

Rautaray, S.K., and Sucharita, S. (2024). Crops diversification for sustainable agriculture. *Indian Farming* 74, 19–22.

Richardson, K., Steffen, W., Lucht, W., Bendtsen, J., Cornell, S.E., Donges, J.F., et al. (2023). Earth beyond six of nine planetary boundaries. *Sci. Adv.* 9, eadh2458. https://doi.org/10.1126/sciadv.adh2458

Roberts, D.P., and Mattoo, A.K. (2019). Sustainable crop production systems and human nutrition. *Front. Sustain. Food Syst.* 3, 72. https://doi.org/10.3389/fsufs.2019.00072

Soto-Gömez, D., and Pérez-Rodríguez, P. (2022). Sustainable agriculture through perennial grains: Wheat, rice, maize, and other species. A review. *Agric. Ecosyst. Environ.* 325, 107747. https://doi.org/10.1016/j.agee.2021.107747

Tscharntke, T. (2021). Disrupting plant-pollinator systems endangers food security. *One Earth* 4, 1217–1219. http://doi.org/10.1016/joneeart.2021.08.022

Tubiello, F.N., Rosenweig, C., Conchedda, G., Karl, K., Gütschow, J., Xueyao, P., et al. (2021). Greenhouse gas emissions from food systems: Building the evidence base. *Environ. Res. Lett.* 16, 065007. https://doi.org/10.1088/1748-9326/ac018e

Van der Sluijs, J.P., and Vaage, N.S. (2016). Pollinators and global food security: The need for holistic global stewardship. *Food Ethics* 1, 75–91. https://doi.org/10.1007/s41055-016-0003-z

Wang, L., Rengel, Z., Zhang, K., Jin, K., Lyu, Y., Zhang, L., et al. (2022). Ensuring future food security and resource sustainability: Insights into the rhizosphere. *iScience* 25, 104168. https://doi.org/10.1016/j.isci.2022.104168

Wani, N.R., Rather, R.A., Farooq, A., Padder, S.A., Baba, T.R., Sharma, S., et al. (2024). New insights in food security and environmental sustainability through waste food management. *Environ. Sci. Pollut. Res. Int.* 31, 17835–17857. https://doi.org/10.1007/s11356-023-26462-y

Wijerathna-Yapa, A., and Pathirana, R. (2022). Sustainable agro-food systems for addressing climate change and food security. *Agriculture* 12, 1554. https://doi.org/10.3390/agriculture12101554

2 The Green Revolution Impact on Nutritional Quality and Malnutrition in the Global South

Rodomiro Ortiz and Sangam L. Dwivedi

2.1 INTRODUCTION

The Green Revolution introduced a suite of innovations, including high-yielding cultivars, synthetic fertilizers, and pesticides, as noted by Pingali (2023). These technological advancements revolutionized agricultural practices and led to remarkable increases in crop yields. For instance, according to Pingali (2012), the introduction of high-yielding varieties of rice and wheat resulted in a tripling of production, while only requiring a 30% increase in land area. Overall, the Green Revolution represented a landmark period in agricultural history, significantly increasing food production and transforming agricultural systems worldwide. However, its legacy also includes debates over its environmental and social impacts, underscoring the importance of considering sustainability and equity in future agricultural innovations and interventions. While it achieved remarkable success in boosting rice and wheat production, particularly in the Global South, the unintended consequences of this technological revolution are now evident (Ortiz, 2011). This agricultural revolution had a profound impact on numerous nations in Asia and Latin America, significantly boosting food production (Hazell, 2009). Without the innovations introduced during this period, crop yields in the developing world would likely have been 20% lower, and food prices could have been 19% higher than they were in 2000. The Green Revolution played a pivotal role in averting a potential decline in calorie consumption by about 5%, preventing an increase in malnourished children by at least 2%. In fact, the health status of 32 to 42 million preschool children saw improvement due to the positive outcomes of this agricultural revolution.

As we navigate the 21st century, agriculture must play a pivotal role in fostering a healthy and prosperous society. This entails providing more nutritious and healthy food through a combination of genetic improvement and sound crop and resource management practices. Only through such comprehensive and sustainable approaches can agriculture effectively meet the pressing challenges of our time. This chapter explores how the Green Revolution's emphasis on mass production overlooked nutritional quality, contributing to the widespread issue of malnutrition in

DOI: 10.1201/9781003467335-2

the Global South. As we stand in the present day, the global demand for nutritious and high-quality food, feed, fiber, and fuel continues to rise. However, the challenges have evolved, and by 2030, there will be an additional 1.7 billion people to feed. Compounding this issue is a decline in the ratio of arable land by 40 to 55%, and many regions grappled with water scarcity, land erosion, intensified droughts, stagnant progress in crop productivity, dwindling groundwater aquifers, overgrazing of pastures, tropical deforestation, species extinction, overfishing, and the impacts of anthropogenic climate change. Addressing these contemporary challenges necessitates a shift toward a more intricate and integrated agricultural research-for-development agenda. This should emphasize livelihood improvement alongside the resilience, eco-efficiency, and sustainability of agro-ecosystems, moving beyond the sole focus on crop productivity gains. Sustainable intensification of agro-ecosystems becomes paramount, aiming to produce more food with fewer inputs while adapting agriculture to the realities of climate change. Moreover, the conservation of agrobiodiversity through its active utilization, adding value throughout the food chain, improving the nutritional quality of human diets, and ensuring that markets work for smallholders are crucial components of this broader agenda.

2.2 THE GREEN REVOLUTION'S TECHNOLOGICAL ADVANCEMENTS

The Green Revolution indeed brought about a profound transformation in agricultural practices, particularly through the introduction of innovative technologies. High-yielding semi-dwarf, photoperiod-insensitive cultivars of rice and wheat, developed through extensive plant-breeding efforts, played a central role in increasing crop yields and generating economic benefits for farmers. These new cultivars were specifically designed to produce higher yields under optimal growing conditions, contributing to significant increases in agricultural productivity. In addition to improved cultivars, the widespread adoption of synthetic fertilizers and pesticides further bolstered crop productivity by addressing nutrient deficiencies and minimizing losses caused by pathogens and pests. These chemical inputs helped farmers achieve more stable and predictable yields, reducing the risks associated with crop failure due to pest infestations or soil nutrient deficiencies. However, as emphasized by Fischer (2016), the success of the Green Revolution cannot be solely attributed to technological innovations alone. Political commitment and supportive policies played a crucial role in driving the widespread adoption and success of these technologies, particularly in regions like Asia where the Green Revolution had a profound impact. Governments in these regions implemented policies and provided resources to support farmers in adopting new technologies, including investments in infrastructure, extension services, and agricultural research and development.

Fischer's insights challenge the perception that Green Revolution technologies were inherently scale-neutral. Instead, they highlight the importance of political will and supportive policies in facilitating the adoption and dissemination of these technologies, especially in the context of developing countries where agricultural transformation was most needed. Overall, the Green Revolution represents a complex interplay of technological innovation, political commitment, and supportive policies that collectively transformed agricultural systems and significantly increased food production worldwide. However, its legacy also includes debates over its environmental and social impacts, underscoring the importance of balancing productivity with sustainability and equity in future agricultural development efforts.

2.3 MASS PRODUCTION AND ITS IMPACT ON NUTRITION

The Green Revolution undoubtedly achieved its primary objective of significantly boosting food production, but it did so with a heavy focus on quantity over nutritional quality. The cultivars developed during this period were primarily selected for their ability to produce higher yields per hectare rather than for their nutritional content. This resulted in a situation where the emphasis on a few high-yielding crops like rice and wheat led to monoculture practices, which further diminished the diversity of nutrients in local diets. As highlighted by Pingali (2019), this shift toward monoculture not only reduced the variety of nutrients available but also made agricultural systems more vulnerable to pathogens, pests, and environmental degradation. However, as noted by Davis et al. (2019), diversifying crop production by increasing the cultivation of millets and sorghum can address some of these issues. These coarse grains are not only rich in protein and iron but also offer environmental benefits such as reduced greenhouse gas emissions, lower irrigation water requirements, and decreased energy inputs compared to rice cultivation. The challenge lies, therefore, in reallocating agricultural land from rice cultivation, which dominates many regions due to its high grain yield, to the production of these more nutritious and environmentally sustainable crops. This shift would require policy support, technological innovation, and changes in farming practices to incentivize farmers to diversify their crop choices and promote the cultivation of millets and sorghum alongside traditional staples like rice and wheat.

2.4 MONOCULTURE AND DIETARY DIVERSITY

The practice of monoculture, which characterized the spread of the Green Revolution elsewhere, had profound effects on both production systems and local diets. While it streamlined agricultural production and enabled large-scale farming operations, it also led to a loss of biodiversity in diets at the local level. For example, intensive rice monoculture farming replaced wild leafy vegetables and fish in the diet of poor people in the Philippines (Cagauan, 1995). The prioritization of high-yield crops such as rice and wheat during the Green Revolution had significant implications for agricultural biodiversity and food security. This emphasis on boosting yields often led to the neglect of traditional and locally adapted cultivars of indigenous crops, as noted by Eliazer Nelson et al. (2019). Consequently, many communities became increasingly reliant on a narrow range of staple foods, which replaced distinct native crops that were once an integral part of their diets. This displacement of indigenous crops by high-yield cultivars not only diminished agricultural biodiversity but also posed risks to food security and resilience. Indigenous crops often possess unique genetic traits that make them well-suited to local environmental conditions and more resilient to pathogens, pests, and climate variability. By favoring high-yield cultivars over indigenous crops, there is always a risk of losing valuable genetic diversity that could be crucial for future agricultural sustainability and adaptation to changing environmental conditions. Furthermore, the extinction or decline of indigenous crops can have cultural and socioeconomic consequences for communities, as these crops may hold cultural significance and provide sources of income and livelihoods. Addressing these challenges requires recognizing the importance of preserving agricultural biodiversity and promoting the cultivation of indigenous crops alongside high-yield cultivars. Such an approach not only enhances dietary diversity and nutrition but also contributes to ecological resilience and supports the preservation of traditional knowledge and cultural heritage. It also underscores the need for more inclusive and sustainable agricultural practices that prioritize both productivity and biodiversity conservation.

The overreliance on a limited number of crops made communities more vulnerable to nutritional deficiencies, as they lacked the diverse array of nutrients provided by a varied diet. In line with these concerns, research by Hao et al. (2022) sheds light on the long-term effects of intensive crop-breeding practices on the nutritional content of crops. Their findings indicate significant declines in copper (Cu), nitrogen (N), and phosphorus (P) over an 80-year period of wheat breeding in China. This decline in essential nutrients could further exacerbate nutritional deficiencies in populations relying heavily on wheat as a staple food. Overall, the emphasis on monoculture during the Green Revolution had complex consequences, including diminished biodiversity in diets and potential declines in the nutritional quality of staple crops. Addressing these challenges requires a holistic approach that prioritizes sustainable agricultural practices, promotes crop diversification, and preserves traditional knowledge of locally adapted crops to ensure food security and nutrition for all.

The research by Headey and Hoddinott (2015) highlights the nuanced impacts of the Green Revolution in Bangladesh. While it led to weight gain among young children, it had limited effects on postnatal linear growth. This suggests that while there may have been improvements in nutritional intake, there were still challenges in ensuring comprehensive child development. Moreover, recent findings by Debnath et al. (2023) shed light on concerning trends related to the quality of rice and wheat, the primary crops of the Green Revolution in South Asia. The study indicates a worrying decline in the quality of these crops since the 1970s. Specifically, the concentration of zinc (Zn) in rice and wheat grains decreased significantly, potentially leading to a decrease in the mineral-diet quality index. This decline in mineral content could have far-reaching implications for public health, particularly in India. A reduction in essential minerals like zinc could contribute to the prevalence of noncommunicable diseases (NCDs) such as iron-deficiency anemia or respiratory, cardiovascular, and musculoskeletal disorders by 2040. These findings underscore the need for sustainable agricultural practices and efforts to ensure the nutritional adequacy of staple crops to mitigate potential health risks associated with declining crop quality.

2.5 LIMITED MICRONUTRIENT AVAILABILITY

The Green Revolution undoubtedly brought about significant increases in the production of staple crops like rice and wheat, thereby addressing issues of food scarcity in many regions. However, as highlighted by Gómez et al. (2013), the focus on boosting yields of macronutrient-rich crops often came at the expense of neglecting micronutrient content in diets. Micronutrients, such as iron, zinc, and vitamin A, are essential for human health, playing crucial roles in various physiological functions. The oversight of these micronutrients in agricultural interventions during the Green Revolution contributed to a decline in the overall nutritional quality of diets, particularly in regions of the Global South. Consequently, many populations in these regions have experienced deficiencies in key micronutrients, leading to a rise in malnutrition-related health issues, such as stunting, anemia, and impaired immune function. Addressing these challenges requires a shift toward a more holistic approach to agrifood systems that prioritize not only increasing food production but also improving the nutritional quality of diets. This new approach may involve promoting crop diversification to include more nutrient-rich foods, supporting the cultivation of traditional and locally adapted crops rich in micronutrients, and integrating nutrition-sensitive agricultural practices. By adopting such an approach, it may be possible to reduce both undernourishment and malnutrition, ultimately improving the overall health and well-being of populations in the Global South. This underscores the importance of considering not only quantity but also quality when designing agricultural interventions and agrifood systems.

2.6 IMPACT ON VULNERABLE POPULATIONS

The consequences of the Green Revolution were not uniform across all segments of society. Vulnerable populations, particularly those in rural areas with limited access to diverse and nutritious foods, bore the brunt of nutritional shortcomings. Women and children, in particular, faced higher risks of malnutrition due to their increased nutritional requirements and often limited access to a variety of foods.

Women, who often bear the primary responsibility for household food preparation and childcare, may have experienced increased workloads and stress due to changes in agricultural practices. Additionally, their nutritional needs, especially during pregnancy and lactation, may not have been adequately met, leading to higher susceptibility to malnutrition. Children, as a vulnerable demographic group, were also at an increased risk of malnutrition. The emphasis on high-yielding staple crops like rice and wheat may have led to dietary monotony, limiting children's access to a diverse range of nutrients essential for growth and development. Furthermore, factors such as poor sanitation and inadequate healthcare infrastructure in rural areas could have exacerbated the prevalence of malnutrition among children.

Overall, the Green Revolution's focus on increasing agricultural productivity often overlooked the need for ensuring equitable access to nutritious foods, particularly for vulnerable populations. Addressing the nutritional shortcomings resulting from the Green Revolution requires comprehensive strategies that prioritize the diverse dietary needs of all segments of society, including women and children in rural areas. This may involve promoting agricultural diversification, improving access to education and healthcare, and implementing targeted interventions to address the specific nutritional needs of vulnerable populations.

2.7 ECONOMIC CONSIDERATIONS AND MARKET FORCES

The Green Revolution undeniably achieved significant strides in increasing agricultural productivity and enhancing food security, often measured through economic metrics such as increased income for farmers and national-level food availability. However, the pursuit of market-driven forces and profit maximization sometimes came at the expense of neglecting the needs of small-scale farmers and marginalized communities.

As highlighted by Pingali (2012, 2019, 2023), the focus on economic indicators may have obscured the uneven distribution of benefits, with marginalized communities and small-scale farmers often being left behind. While increased production levels may have boosted overall food availability, these benefits did not always translate into improved nutritional outcomes for all segments of society. For example, small-scale farmers, who may lack access to resources and support systems, faced challenges in fully benefiting from the technological advancements of the Green Revolution. Limited access to credit, inputs, and extension services, coupled with market dynamics that favored larger, more commercialized agricultural operations, often marginalized smallholders, perpetuating inequalities within the agricultural sector. Furthermore, marginalized communities, including indigenous peoples and rural populations living in remote areas, may have been excluded from the benefits of increased production due to barriers such as limited infrastructure, land tenure issues, and social marginalization.

Addressing these disparities requires a more holistic approach to agricultural development that prioritizes inclusivity, equity, and sustainability. This may involve targeted interventions to support small-scale farmers, such as improved access to resources, training, and market linkages. Additionally, policies and programs should be designed with a focus on promoting dietary diversity, ensuring access to nutritious foods for all, and addressing the underlying determinants of malnutrition, including poverty and social inequality. By recognizing the importance of inclusive and equitable agricultural development, it becomes possible to ensure that the benefits of increased productivity are shared more equitably and contribute to improved nutritional outcomes for all members of society.

2.8 ENVIRONMENTAL CONSEQUENCES

The Green Revolution undoubtedly brought about significant benefits in terms of increased crop yields and avoided land conversion, as highlighted by Stevenson et al. (2013) and Burney et al. (2010). By adopting newly bred high-yielding rice and wheat cultivars, millions of hectares of land were spared from conversion to agriculture, and significant amounts of carbon emissions were avoided. However, alongside these benefits, the Green Revolution also had significant negative environmental consequences, as pointed out by its critics. The widespread use of synthetic fertilizers and pesticides, while enhancing crop productivity in the short term, contributed to various environmental issues, including soil degradation, water pollution, and loss of biodiversity.

Soil degradation, resulting from intensive agricultural practices such as excessive tillage and chemical inputs, can lead to reduced soil fertility, erosion, and loss of soil biodiversity. This, in turn, undermines the long-term productivity and resilience of agricultural systems. Water pollution is another concern associated with the use of synthetic fertilizers and pesticides. Runoff from agricultural fields can carry excess nutrients and chemicals into water bodies, leading to eutrophication, algal blooms, and contamination of drinking water sources, with detrimental effects on aquatic ecosystems and human health. Loss of biodiversity is also a significant consequence of intensive agricultural practices, as monoculture cropping systems and chemical inputs can disrupt natural ecosystems and reduce habitat diversity, threatening the survival of plant and animal species. These environmental issues highlight the importance of considering the long-term sustainability of agricultural systems. While improving crop yields through plant breeding is important for reducing greenhouse gas emissions and increasing food security, it is equally crucial to address the environmental impacts of agricultural intensification strategies.

Moving forward, sustainable agricultural practices that prioritize soil health, water conservation, and biodiversity conservation are essential for ensuring the resilience and sustainability of food production systems in the face of climate change and growing environmental challenges. This requires a shift toward agro-ecological approaches that promote biodiversity, reduce chemical inputs, and enhance ecosystem services, thereby supporting both environmental sustainability and food security goals.

2.9 MOVING TOWARD SUSTAINABLE AND NUTRIENT-DENSE AGRICULTURE

Recognizing the shortcomings of the Green Revolution, there is a growing awareness of the need for a more sustainable and nutrition-sensitive approach to agriculture. The sustainable intensification of agriculture, which emphasizes the integration of genetic, ecological, and socioeconomic principles into farming practices, offers a promising alternative. Diversifying crops, promoting resource-use-efficient farming, and incorporating traditional knowledge can enhance both the resilience of agricultural systems and the nutritional content of diets.

Presently, over 2 billion people worldwide suffer from micronutrient malnutrition, resulting in a range of detrimental outcomes such as poor health, diminished worker productivity, elevated mortality and morbidity rates, increased occurrences of chronic diseases (such as coronary heart disease, cancer, stroke, and diabetes), and lasting cognitive impairment in infants born to mothers deficient in micronutrients (Welch and Graham, 2000). These deficiencies not only impede national development efforts and contribute to sustained high population growth rates but also perpetuate a vicious cycle of poverty for a substantial number of underprivileged individuals globally. A crucial factor in this scenario is the global food system's failure to generate a well-balanced nutrient output that meets the nutritional needs of every individual, particularly resource-poor women, infants, and children in developing countries. Agriculture bears a share of responsibility in this inadequacy, as it has not explicitly prioritized achieving an adequate and balanced nutrient output in its production systems. Moreover, certain agricultural policies may have unintentionally led to a decline in nutrition and dietary diversity for the impoverished over the past four decades. Astonishingly, the nutrition and health sectors have yet to fully harness agriculture as a primary tool in their efforts to address global issues of poor nutrition and ill health.

In light of these challenges, a new paradigm for agriculture and nutrition is imperative. There is a pressing need to explore ways in which agriculture can actively contribute to sustainable solutions for failures within food systems. This entails adopting holistic food-based system approaches that tightly link agricultural production to the enhancement of human health, livelihoods, and overall well-being. Such a transformative approach has the potential to garner widespread support for agricultural research globally, as it not only addresses production concerns but also aligns with consumer issues, making it politically feasible and substantiating its potential impact on a global scale.

Plant breeding offers a considerable promise and represents a cost-effective and sustainable alternative when compared to supplementation and food fortification programs, for addressing global micronutrient deficiency. Breeding programs elsewhere have made significant strides in recent years. They have focused on screening and developing plant genotypes that boast augmented quantities and enhanced utilization (biological availability) of micronutrients in the edible parts of staple food crops, including cereal grains (Cakmak et al., 2002). Remarkable genetic variations have been identified in various food crops, specifically showcasing higher levels of essential micronutrients such as iron (Fe), zinc (Zn), and vitamin A, along with reduced levels of antinutrients like phytic acid and tannins. This variation has been leveraged to increase micronutrient levels and mitigate antinutritional substances. Both cross-breeding and genomic-based selection approaches, as well as genetic engineering techniques, are being employed to achieve these objectives. The manipulation of genes influencing nutritional quality traits in food crops has yielded significant progress, leading to the development of transgenic plants with elevated concentrations of crucial micronutrients, including iron, vitamin A, and sulfur-containing amino acids, and lower levels of antinutrients such as phytate. Looking ahead, a pivotal focus should be directed toward the development of plant genotypes that exhibit enhanced levels of multiple micronutrients. To expedite progress and enhance global awareness, it is crucial for the agriculture, health, and nutrition sectors at both national and international levels to emphasize the social, economic, and health consequences associated with micronutrient deficiency. Simultaneously, substantial funding should be provided to support the sustainable development of agriculture-based approaches aimed at preventing micronutrient deficiency. By fostering collaboration and prioritizing interdisciplinary efforts, we can pave the way for innovative solutions that address the multifaceted challenges posed by global micronutrient deficiency.

2.10 CONCLUSION

While the Green Revolution achieved its primary objective of increasing food production, it also brought to light the unintended consequences of overlooking nutritional quality. The mass production of rice and wheat, coupled with monoculture practices, contributed to widespread malnutrition in the Global South. Moving forward, it is essential to prioritize sustainable and nutrition-sensitive agricultural practices that address the diverse nutritional needs of communities while ensuring the long-term health of both people and the planet. Malnutrition emerges as a critical issue rooted in the inadequacies of food systems, which currently fall short in providing everyone with healthy, safe, affordable, and sustainable diets (Branca et al., 2020). The consequences of this shortfall extend beyond immediate health concerns; they encompass profound economic, social, and environmental costs. Failure to address these deficiencies will impede the growth and development of individuals and societies, exerting a lasting impact for decades. In this regard, as indicated by DeFries et al. (2015) alternative metrics of nutritional yield must replace standard yield metrics (weight per unit land) that measure the quantity of production because the former are adequate to assess progress toward sustainable and healthy diets.

Recognizing the intrinsic link between good nutrition and sustainable development, the Sustainable Development Goals (SDGs) articulate a compelling vision. These global priorities set forth the ambitious targets of eradicating malnutrition in all its forms and ensuring universal access to nutritious food by the year 2030. In doing so, the SDGs not only underscore the importance of nutrition in promoting human health but also uphold the fundamental human rights to adequate food, nutrition, and health. These objectives align with the burgeoning global movement to end hunger and malnutrition, further emphasizing the urgent need for collective action. Looking ahead, the United Nations has initiated the Decade of Action on Nutrition (2016–2025), a concerted effort aimed at expediting measures to attain the outlined goals. This initiative seeks to galvanize global commitment and accelerate progress in addressing malnutrition. By fostering collaboration among nations, organizations, and individuals, the UN Decade of Action on Nutrition (https://www.un.org/nutrition/) strives to create tangible advancements in nutrition, health, and food security. It stands as a testament to the collective's resolve to combat malnutrition and aligns with broader international efforts to build a more sustainable and equitable future.

REFERENCES

Branca, F., Demaio, A., Udomkesmalee, E., Baker, P., Aguayo, V.M., Barquera, S., et al. (2020). A new nutrition manifesto for a new nutrition reality. *Lancet* 395, 8–10. https://doi.org/10.1016/S0140-6736(19)32690-X

Burney, J.A., Davis, S.J., and Lobell, D.B. (2010). Greenhouse gas mitigation by agricultural intensification. *Proc. Natl. Acad. Sci. U. S. A.* 107, 12052–12057. https://doi.org/10.1073/pnas.0914216107

Cagauan, A. (1995). The impact of pesticides on rice field vertebrates with emphasis on fish. In: Pingali, P., and Roger, P. (eds.) *Impact of Pesticides on Farmer Health and the Rice Environment.* Kluwer, Boston, pp. 203–224.

Cakmak, I., Graham, R.D., and Welsch, R.M. (2002). Agricultural and molecular genetic approaches to improving nutrition and preventing micronutrient malnutrition globally. In: *Encyclopedia of Life Support Systems*, Vol. 1. UNESCO, pp. 1075–1099.

Davis, K.F., Chhatre, A., Rao, N.D., Singh, D., Ghosh-Jerath, S., Mridul, A., et al. (2019). Assessing the sustainability of post-green revolution cereals in India. *Proc. Natl. Acad. Sci. U. S. A.* 116, 25034–25041. https://doi.org/10.1073/pnas.1910935116.

Debnath, S., Dey, A., Khanam, R., Saha, S., Sarkar, D., Saha, J.K., et al. (2023). Historical shifting in grain mineral density of landmark rice and wheat cultivars released over the past 50 years in India. *Sci. Rep.* 13, 21164. https://doi.org/10.1038/s41598-023-48488-5

DeFries, R., Fanzo, J., Remans, R., Palm, C., Wood, S., and Anderman, T.L. (2015). Metrics for land-scarce agriculture. *Science* 349, 238–240. https://doi.org/10.1126/science.aaa5766

Eliazer Nelson, A.R.L., Ravichandran, K., and Antony, U. (2019). The impact of the Green Revolution on indigenous crops of India. *J. Ethnics Food* 6, 8. https://doi.org/10.1186/s42779-019-0011-9

Fischer, K. (2016). Why new crop technology is not scale-neutral—A critique of the expectations for a crop-based African Green Revolution. *Res. Policy* 45, 1185–1194. https://doi.org/10.1016/j.respol.2016.03.007

Gómez, M.I., Barrett, C.B., Raney, T., Pinstrup-Andersen, P., Meerman, J., Croppenstedt, A., et al. (2013). Post-green revolution food systems and the triple burden of malnutrition. *Food Policy* 42, 129–138. https://doi.org/10.1016/j.foodpol.2013.06.009

Hao, B., Ma, J., Si, L., Jiang, L., Wang, X., Yao, C., et al. (2022). Did wheat breeding simultaneously alter grain concentrations of macro- and micro-nutrient over the past 80 years of cultivar releasing in China? *Front. Plant Sci.* 13, 872781. https://doi.org/10.3389/fpls.2022.872781

Hazell, P.B.R. (2009). The Asian Green Revolution. *IFPRI Discussion Paper 00911.* International Food Policy Research Institute, Washington, DC. https://core.ac.uk/download/pdf/6257689.pdf

Headey, D.D., and Hoddinott, J. (2015). Agriculture, nutrition, and the Green Revolution in Bangladesh. *IFPRI Discussion Paper 01423.* International Food Policy Research Institute, Washington, DC. https://ssrn.com/abstract=2583850

Ortiz, R. (2011). Re-visiting the Green Revolution: Seeking innovations for a changing world. *Chron. Hortic.* 51, 6–11. http://doi.org/10.1016/j.scienta.2011.09.020

Pingali, P. (2019). The Green Revolution and crop biodiversity. In: Dasgupta, P., Raven, P., and McIvor, A. (eds.) *Biological Extinction: New Perspectives.* Cambridge University Press, Cambridge, pp. 175–192.

Pingali, P. (2023). Are the lessons from the Green Revolution relevant for agricultural growth and food security in the twenty-first century? In: Estudillo, J.P., Kijima, Y., and Sonobe, T. (eds.) *Agricultural Development in Asia and Africa. Emerging-Economy State and International Policy Studies.* Springer, Singapore, pp. 21–32. https://doi.org/10.1007/978-981-19-5542-6_2

Pingali, P.L. (2012). Green Revolution: Impacts, limits, and the path ahead. *Proc. Natl. Acad. Sci. U. S. A.* 109, 12302–12308. https://doi.org/10.1073/pnas.0912953109

Stevenson, J.R., Villoria, N., Byerlee, D., Kelley, T., and Maredia, M. (2013). Green Revolution research saved an estimated 18 to 27 million hectares from being brought into agricultural production. *Proc. Natl. Acad. Sci. U. S. A.* 110, 8363–8368. https://doi.org/10.1073/pnas.1208065110

Welch, R.M., and Graham, R.D. (2000). A new paradigm for world agriculture: Meeting human needs productive, sustainable, nutritious. *Field Crops Res.* 60, 1–10. https://doi.org/10.1016/S0378-4290(98)00129-4

3 Genotype (G), Environment (E), G × E Interaction (GEI), and G × Management (M) Interaction (GMI) Effects Influencing Nutritional, Sensory, and Textural Traits

Sangam L. Dwivedi and Autar K. Mattoo

3.1 INTRODUCTION

Seed quality (excluding physical characteristics) refers to the variation in protein and oil contents, protein and oil quality, carbohydrate and fiber, minerals (macro- and micronutrients), and vitamins. Together, these characteristics determine the quality of food and feed crops. The physical characteristics of the grains include their size, shape, color, and weight. The color, flavor, texture, taste, smell, and nutritive value, besides physical appearance (i.e., size, shape, firmness, absence of defects), are the main quality attributes of fruits and vegetables.

Plant genetic resources (both cultivated and crop wild relatives, or CWR) are the basic raw materials to support crop improvement programs. Crop genetic enhancers may draw functional allelic diversity from these gene pools to develop next-generation food and feed crops combining climate and nutritional resilience for betterment of human and animal health. The greater emphasis on yield and uniformity in breeding programs and adoption of monocropping rather than multicropping systems in the past had led to the decline in crop and nutritional diversity of food crops (Connors et al., 2021; Nicholson et al., 2021). The evidence to date suggests substantial genetic variation for nutritional traits (oil and protein contents and quality, carbohydrates, fibers, macro- and micronutrients, vitamins, and bioactive compounds) among cultigens and CWR of diverse crops (grain, fruit, and vegetable), with landraces and

 DOI: 10.1201/9781003467335-3

heirloom cultivars in many cases had greater amounts of nutrients than modern cultivars (Mattoo et al., 2022; Constantino et al., 2022; Dwivedi et al., 2019, 2022, 2023, and references therein; Medina-Lozano and Díaz, 2021), which may open doors for more research on exploring and testing variability and bioavailability of nutrients in food crops.

3.2 GEI AND GMI INTERACTIONS IMPACTING NUTRITIONAL QUALITY

The nutrients *per se* are the function of genotype (G), environment (E), GEI, and GMI, which potentially impact the nutritional quality of food crops. Understanding the effect of these variables is important for identifying and exploiting heritable genetic variation in plant breeding. Environmental effects are much larger than those of either G or GEI effects (Khan et al., 2022; Sellami et al., 2021). Variation in soil (texture, pH, and organic matter) and environmental parameters (temperature, solar radiation, and precipitation) potentially influence nutrient concentration that must be taken into consideration while explaining the variation for seed nutrients in germplasm or when assessing the nutritional quality of staple food crops grown in diverse agro-ecological conditions (dos santos Araújo et al., 2022; Gangashetty et al., 2023; Gelete et al., 2021; Gore et al., 2021; Osei et al., 2019). Variation in E and GMI have potential to impact both yield and quality (Assefa et al., 2018, 2019; Mahmood et al., 2022). The level of abiotic stresses and degree of tolerance or intolerance to stresses may also influence the nutritional quality of germplasm. Therefore, multi-site testing across diverse agro-ecologies is recommended to identify nutrient-dense germplasm for use in crop breeding.

3.3 ABIOTIC STRESS AND ELEVATED CARBON DIOXIDE (eCO_2) EFFECT ON SEED OIL, PROTEIN, AND CARBOHYDRATE CONTENTS

A review of early literature revealed that abiotic stresses caused variable influence on proximate composition, including minerals and bioactive compounds in grain crops. However, these effects may be variable depending on crop growth stage, the duration and intensity of stresses, and soil N. Drought- and/or heat-stressed crops produce have higher grain protein but of lower protein quality. Likewise, the product from similar stresses have reduced grain oil but improved oil quality. Drought and heat also impact macro- and micronutrients, but to a variable extent. Crop germplasm responded differently, giving various grain-quality attributes, under drought, heat, and eCO_2, which may provide researchers opportunities to identify germplasm with the least differences in grain quality under stress for direct cultivation or use in crop improvement programs (Dwivedi et al., 2013).

Subsequent reports involving large germplasms and multilocation assessment indicate that each environmental variable and/or stress factor alone or in combination have the potential to impact the nutritional quality of food crops. Drought and heat stress invariably reduce grain weight, while eCO_2 increases grain weight in staple food crops (Sehgal et al., 2018; Wang et al., 2022). Drought and heat stress reduce grain oil, protein, and starch contents in maize (Yousaf et al., 2022); moderate stress after anthesis enhances grain quality, while severe stress reduces wheat quality (Ozturk et al., 2021); heat rather than drought adversely affects canola oil quality (Elferjani and Soolanayakanahally, 2018); high night temperature reduces sugar content but has no impact on oil and protein in soybean (Yang et al., 2023). Drought and salinity stress impact C and N metabolism during grain filling, modifying the content and composition of the durum wheat (de Santis et al., 2021). eCO_2 caused variable effects in soybean seed chemistry: protein decreased while oil increased (Li et al., 2018); sucrose, glucose and fructose decreased, while raffinose and stachyose increased (Bellaloui et al., 2017). Increased soybean yield in response to eCO_2 correlated highly ($r = 0.95$) with an increase in symbiotically fixed N. Cultivars with a greater N_2-fixing efficiency during the late reproductive phase may provide a better adaptability to eCO_2 (Li et al., 2022).

3.4 ABIOTIC STRESS, eCO_2 EFFECT, AND SOIL TYPE'S ON SEED MINERAL CONTENT

Drought and heat stress reduced grain minerals in cereals and food legumes. The eCO_2 also had an adverse effect on minerals, but it increased seed phytate content, limiting minerals' bioavailability. Drought and N fertilizer modified the response to eCO_2 on minerals in diverse crops (Al-Hadeethi et al., 2019). Variable effects of eCO_2 (600 μmol mol^{-1}), drought, and/or heat stress on mineral contents are evident in many crops. eCO_2 significantly decreased Fe content, heat stress at ambient CO_2 (aCO_2; 400 μmol mol^{-1}) had opposite effect, while eCO_2 with heat stress (+2.7° C during the day and +3.4°C at night) restored Fe and Zn to levels obtained under aCO_2 and temperature in soybeans (Köhler et al., 2019). Structural (such as cellulose or fiber) and nonstructural (such as starch and sugars) are the two types of carbohydrates in plants. eCO_2 reduced grain protein and N concentrations, along with an increase in total nonstructural carbohydrates (TNC), shifted plant ionome toward lower mineral content, and increased TNC-mineral and C-minerals ratios in C_3 plants. TNC is readily digestible and absorbed in the human gut. eCO_2 had a greater effect on TNC, increased by 10–45% (Loladze, 2014).

The eCO2 significantly reduces grain processing and nutritional quality of rice. Reduced amylose content leads to lower gelatinization temperature and higher viscosity, which makes the cooked rice softer and more palatable. However, the increased chalkiness under eCO_2 reduces recovery and marketability of milled rice and increases percentage of broken rice grains with inferior nutritional quality (Hu et al., 2022; Kumar et al., 2023). The degree of chalkiness in rice adversely affects grain quality and milling yield.

eCO_2 and canopy warming had no influence on soil nutrient status in winter wheat. eCO_2 decreased K concentration in root and shoot but had no impact on N or P concentration. Canopy warming increased shoot N, P, and K, while it decreased root N, P, and K concentrations, suggesting that canopy warming rather than eCO_2 played a key role in nutrient translocation from root to shoot (Wang et al., 2023). Variation in climatic parameters (temperature, CO_2, O_3, and drought) alone or in combinations caused variable influence on nutritional quality of wheat grains. The wheat plants exposed to elevated ozone (eO_3) substantially increased N, Fe, Mg, Mn, P, and Zn, reduced K slightly, but had no influence on C. O_3, in the presence of eCO_2 and heat stress, enhanced Fe, Mn, P, and Zn to a variable degree. Plants exposed to chronic O_3 exposure and drought stress resulted in higher concentrations of all the minerals, except K (Galani et al., 2022a). It is evident that the produce from eCO_2 environments in principle will have more minerals, largely due to higher grain yield.

eCO_2 exerts significant influence on nutrient requirements in plants. C:N:P ratio is important for managing nutrient balance to maximize their efficiency in farming systems. Soybean cultivars responded differently to C:N:P ratio under eCO_2 environment: C:N ratio increased while C:P and N:P ratio decreased. Germplasm that did not alter C:N and C:P ratios in response to eCO_2 are the potential resource to develop soybean cultivars adapted to eCO_2 environments (Li et al., 2019). eCO_2 decreased N, P, and Zn contents irrespective of soil (chromosol, vertosol, calcarosol), crop (wheat, pea, canola), and year (FACE, free air CO_2 enrichment for seven years). P and Mg concentration of canola decreased significantly in chromosol but not in vertosol nor calcarosol soils. K, Fe, Mn, and Cu were unaffected by eCO_2 irrespective of crop and soil types. Non-leguminous crops were more affected by eCO_2, with a relatively greater reduction in fertile than infertile soils (Jin et al., 2019).

3.5 ABIOTIC STRESS AND eCO_2 EFFECT ON SEED BIOACTIVE COMPOUNDS

Bioactive compounds are plant secondary metabolites (or phytochemicals). They are nonessential nutrients but are critical to support plant growth, physiology, and stress tolerance and provide health benefits to humans (Dwivedi et al., 2022; Samitya et al., 2021). Colored grains, fruits, and vegetables are rich sources of plant-derived bioactive compounds (nonessential nutrients), such as polyphenols, dietary fiber, carotenoids, and vitamins. Drought and heat stress increased carotenoids and anthocyanins in cereal species (Beleggia et al., 2021). Warmer climates produced greater amounts of bioactive compounds, while cooler climates lowered phytic acid in wheat (Pu et al., 2019). Drought stress reduces polyphenols in the common bean (Ovando-Martínez et al., 2014). Manifold increase in grain anthocyanins in waterlogged conditions was noted than in the aerobic (well-drained) soils (Jaksomsak et al., 2022). Arid and semi-arid climate with higher temperature reduces flavonol and anthocyanin content in grape (Gashu et al., 2023).

3.6 ABIOTIC STRESS, eCO_2, AND CULTIVATION PRACTICES' EFFECTS ON BIOACTIVE COMPOUNDS IN FRUITS AND VEGETABLES

Abiotic stress: Cruciferous vegetables are rich sources of bioactive compounds that provide benefits to human health. The sulfur-containing glycosides, glucosinolates (GLSs) and organosulfur, are potential anticancer agents (Miekus et al., 2020). Water-deficit stress increased GLSs in *Brassica* leafy vegetables (Ammar et al.,

2023). The heat stress had variable response to bioactive compounds and mineral contents in broccoli (*Brassica oleracea* var. *italica*). Total phenols; soluble sugars; carotenoids; quercetin; sinapic, ferulic, *p*-coumaric, and gallic acid; and As, Co, Cr, Hg, K, Na, Ni, Pb, Se, and Sn increased, while total flavonoids, flavonols, phenolic acids, hydroxycinnamic acids, glucosinolates, and Ca, Cd, Cu, Mg, Mn, and P were reduced (Gmižić et al., 2023). *Amaranthus* species are better adapted to drought and heat stress. Investigating the effect of drought and heat stress, singly or in combination, on the accumulation of phenol and flavonoid compounds and the antioxidant capacity resulted in a strong positive but species- and compound-specific effect of drought and heat stress on accumulation of bioactive compounds. Heat stress at 40°C under well-watered conditions and combined drought and heat stress at 30°C and 35°C induced the accumulation of caffeic acid and rutin, suggesting that stressed produce of amaranth species, if consumed, is beneficial for human health (Netshimbupfe et al., 2022). Heat-stressed purple-skinned and purple-fleshed potatoes have reduced anthocyanins (Liu et al., 2019), while purple- or red-fleshed potatoes from cooler climates produce more anthocyanins (Hamouz et al., 2018).

Suboptimal environmental conditions adversely impact plant growth and development, reducing yield and nutritional quality. Abiotic stresses, however, induce several physiological, biochemical, and molecular responses and produce a wide range of bioactive compounds enabling crop adaptation. Products from suboptimal environments are beneficial to human health. Controlled abiotic stresses can therefore be used as tool to improve nutraceutical value of edible parts of the crop produce for betterment of human health (Ortega-Hernánde et al., 2021; Sharma et al., 2022; Toscano et al., 2019). Mechanically damaging the plant tissue or use of non-thermal technologies (high hydrostatic pressure, ultrasound, and pulsed electric fields) to induce a wound-like response may be used to increase health-promoting compounds in whole fruits and vegetables (Jacobo-Velázque and Benavides, 2022).

3.6.1 eCO_2

A detailed analysis of data involving 57 articles consisting of 1,015 observations found that eCO_2 increased the concentrations of fructose, glucose, soluble sugar, antioxidant capacity, phenols, flavonoids, ascorbic acid, and Ca in the edible part of vegetables, while protein, nitrate, Mg, Fe, and Zn decreased. The titratable acidity, chlorophyll, carotenoids, lycopene, anthocyanins, P, K, S, Cu, and Mn were not affected by eCO_2 (Dong et al., 2018). Further, eCO_2 (900 ppm) in tomatoes relative to aCO_2 (400 ppm) decreased proteins, carotenoids, lycopene, and minerals, while carbohydrates remained unaffected. The antioxidant and anti-inflammatory properties were not affected, contrasting a previous report (Zhang et al., 2014) about the effect of eCO_2 on nutrition and health properties of different plant species, probably due to differences in cultivation practices and genetic backgrounds within species (Boufeldja et al., 2023).

3.6.2 Ultraviolet Irradiation Exposure

Short-term exposure of ultraviolet (UV-B) irradiation can be used to enhance bioactive compounds in plants with no adverse impact on product appearance (Lee et al.,

2019; Son et al., 2020; Yoshida et al., 2022). *Amaranthus tricolor* is a nutritious baby-leaf vegetable that contains many bioactive compounds. Baby amaranth exposed to different irradiation intensities, irradiation durations, and cumulative energies, with a 24 h recovery period after irradiation, enhanced phenolics, flavonoids, anthocyanins, ascorbic acid, and antioxidant capacity of leaves without adverse effect on product appearance. The highest increase in bioactive compounds was noticed in leaves irradiated with UV-B at 6 W m^2 for 7 h with cumulative energy of 150 kJ m^2. Thus, short UV-B irradiation and a 24 h post-irradiation recovery period could be used as a benchmark to enhance bioactive compounds in baby leaf amaranth without causing abnormalities in leaves (Wittayathanarattana et al., 2022).

3.6.3 Conventional versus Organic Farming

Conventional (synthetic fertilizers, herbicides, pesticides) versus organic farming (environment-friendly cultivation practices, including the use of farmyard manures) may have the potential to alter the proximate composition of food crops. For example, organically produced purple grain rice had greater Fe, gamma-aminobutyric acid, total phenols, and anthocyanin but lower sugar content than that produced by conventional practice (Kunnam et al., 2023). However, appropriate management of N and Zn fertilizer may increase productivity and the grain N, Zn, and anthocyanin content in purple rice (Fongfon et al., 2021). The conventionally grown bell pepper fruits were superior in phenolic acids, while organically grown bell pepper fruits contain significantly more flavonoids and carotenoids in comparison to conventional ones (Hallmann et al., 2019). Contrarily, the organically grown bell pepper fruits contained significantly more dry matter, vitamin C, total carotenoids, phenolic acids, and flavonoids compared to conventionally grown fruits (Hallmann and Rembiałkowska, 2012). Although vegetables produced under the two production systems revealed no superiority in terms of dry matter, crude protein, and crude fiber contents, the organically grown vegetables were superior in minerals (Ca, Mg, Fe, Zn, Cu, Mn), bioactive compounds, and antioxidant capacity (Czech et al., 2022).

Significant effects of the production system on the total carotenoids and β-carotene content were detected in carrots, with higher concentrations in conventionally produced than organically grown roots (Średnicka-Tober et al., 2022). Tomatoes produced under reduced application of chemical fertilizer (75% of the conventional application) supplemented with *Trichoderma*-enriched bio-organic fertilizer resulted in fruit yields equivalent to that obtained using 100% of the chemical fertilizer, but with substantial improvement in soluble sugar, vitamin C, and nitrate accumulations compared to those of the control. *Trichoderma* bio-organic fertilizer with the appropriate rates of chemical fertilizers could be used to maximize fruit yield, quality, and fertilizer savings (Ye et al., 2020). The combined application of reduced chemical fertilizer and bioorganic fertilizer improved soil fertility, microbial community structure, yield, and quality in lettuce (Jin et al., 2022). The use of rhizobacterial strain (*Acinetobacter calcoaceticus* UTMR2), together with reduced doses of K fertilizer, had beneficial effect on plant growth, nutritional quality, bioactive compounds, and antioxidant capacity of tomatillo (*Physalis ixocarpa* Brot.) fruits (Ramírez-Cariño et al., 2023).

A comparison of plant-based food produced organically (ORG) and nonorganically (NORG) revealed that NORG fruits have higher phenolic compounds, whereas ORG fruits had more antioxidant activity. NORG legumes (*Ceratonia siliqua* cv. Banya de cabra) relative to ORG legumes was superior in phenolic compounds and antioxidant activity. ORG nuts, though, had more flavan-3-ols and flavonols than their NORG counterparts; the former were less antioxidant than the latter. Higher phenolics and anthocyanins in ORG vegetables reflect their higher antioxidant capacity than NORG vegetables (Cruz-Carrión et al., 2023). Organically produced rice grains had higher nutrients and bioactive compounds than those produced by conventional farming (Kunnam et al., 2023). The combined use of organic and inorganic fertilizers improved yield and quality (vitamin C, lycopene, total soluble solids, total soluble sugar, and sugar-acid ratio) by altering soil physicochemical properties and nitrification in cherry tomato (Wu et al., 2022).

Fertilization affected the glucosinolate content in *Brassica* species (De Pascale et al., 2007; Groenbaek and Kristensen, 2014; Gugała et al., 2020). The impact of natural occurring soil variation on glucosinolates (GLSs) content is unknown in *Brassica*. Up to twofold difference in GLS content between soils from different natural habitats were reported in *B. stricta*. The relative amounts of different compounds were less responsive. The physico-chemical soil properties rather than microbial communities contribute more to variation in total glucosinolate contents (Wagner and Mitchell-Olds, 2023). Amending soils with sledge sewage, chicken manure, and horse manure increased vitamin C content and total phenols compared to the control treatment in *Brassica* vegetables (Antonious, 2023). Application of biological fertilization (chicken manure and biological fertilizer with microorganisms), relative to chemical fertilization (NPK), substantially improved micronutrients, antioxidant activity, lycopene, and β-carotene contents of tomatoes (Rusu et al., 2023).

Vermicompost has great potential as an organic fertilizer to enhance soil health, productivity and nutritional quality fruits and vegetables. Vermicompost at 60–70% field capacity of water regime increased tomato yield and vitamin C content, while its effect on soil health varied with soil water regime (Yang et al., 2015). Although chemical fertilizers under salt stress did not increase tomato yield, cow manure and vermicompost substantially increased tomato yield and improved quality (vitamin C content, organic acid, soluble solids, and nitrate) by enhancing the photosynthetic capacity and promoting carbohydrate transport to fruits (Wu et al., 2023). The chicken manure and vermicompost, relative to conventional fertilizer with urea and no fertilizer, substantially increased fruit yield and improved fruit quality (vitamin C, soluble sugar, increased the sugar/acid ratio, and decreased nitrate concentration in fresh fruit) as well soil health (Wang et al., 2017). Though organic fertilization with reduced N did not influence fruit yield, it positively affected phenolic compounds in tomato cultivars less sensitive to drought stress (González-Coria et al., 2022).

Reducing agrochemicals, particularly N, may improve sustainable agricultural production with low cost but with high economic and environmental returns. An experiment involving controlled release of urea (CRU) via nano-fertilizer to the tomato plants at 100, 50, and 25% of the recommended dose of conventional urea improved fruit yield, fruit characteristics, and fruit marketability. The fruit firmness and acidity levels at 25% and 50% N levels were much higher than both at 100% N and the control. The vitamin C values of all CRU treatments were lower than the control. Nitrogen use efficiency of the CRU treatments were significantly higher (47–88%) than conventional fertilizer (33%). Clearly, CRU treatments at an N level of 25% of the recommended dose showed better plant growth, yield, and fruit quality of tomatoes than the conventional fertilizer (Helal et al., 2023).

Field management practices (optimized fertilization, soil tillage, farming systems, irrigation, etc.), in addition to selection of appropriate cultivars and fertilizers (including organic), substantially influence nutritional quality of harvested produce. However, such practices may be specific to the environment and farming systems, indicating complex interactions between crop genetics, agronomic practices, and site-specific pedoclimatic conditions (Aćin et al., 2023).

3.6.4 Root Microbiomes

Numerous reports suggest the beneficial effects of using microorganisms as biostimulants to increase yield and quality. Tomato plants inoculated with plant-growth-promoting rhizobacteria (PGPR) produced less biomass but larger fruits, increased antioxidant activity, less sugars, and more lycopene, while inoculation of cherry plants with PGPR had no effect on biomass productivity, sugar, or lycopene but had reduced antioxidant activity as compared to controls. Microorganisms isolated from agricultural soils may be used to manipulate the level of bioactive molecules in plants, but the effects seem to be species or variety specific (de la Osa et al., 2021; Gashash et al., 2022; Katsenios et al., 2021). Rhizosphere microbiomes are also reported to influence the taste and flavor in mustard (seed glucosinolates in *Brassica juncea*) and tea (Walsh et al., 2024; Xin et al., 2024).

3.6.5 Altitude, Temperature, and Daylight Effect on Bioactive Compounds

Environmental conditions, such as altitude, temperature, and daylight, may influence the accumulation of bioactive compounds among colored grain cereals. Delayed sowing (winter [optimal] vs. spring [delayed]) favored more accumulation of carotenoids and anthocyanins, but with strong genotype-dependent response noted. Substantial environmental influence was noticed for minerals. The pigmented genotypes accumulated more minerals than commercial cultivars, while the seed morphometric traits were less influenced by this agronomic practice. Drought and heat stress during grain filling period reduced grain yield, but it favored accumulation of anthocyanins, carotenoids, and minerals. Thus, optimization of sowing time could enhance the nutritional value of food crops, as evidenced in case of wheat grains, by increasing the concentration of micronutrients and bioactive compounds (Beleggia et al., 2021). Geospatial variation of soil micronutrients and landscape factors may also cause spatially dependent variation in nutrient deficiency across landscape (Botoman et al., 2022).

Soybeans are adapted to a wide range of latitudes (from 53°N to 35°S) across the world. Temperature decreases at higher altitude and small changes in altitude may affect environmental conditions (Alvares et al., 2013). A study involving

28 soybean cultivars grown in high-altitude (846–963 m asl) and low altitude (336–480 m asl) environments in Brazil increased the protein and oil contents, respectively, by ~6%. A high-altitude environment with low temperature promotes an increase in protein content, while a low-altitude environment with high temperature increases oil content. Thus, altitude caused positive and negative effects, respectively, on protein and oil contents in soybean (Capelin et al., 2022). Likewise, altitude (500 m asl vs. 1200 m asl) also caused variations in nutritional quality (oil, protein, total phenolics, antioxidants, and fatty acid profiles) of walnuts (Gundesli et al., 2023). Altitude gradients changed physicochemical and rheological properties of quinoa cultivars (Garcia-Parra et al., 2022). Variation in elevation levels caused significant differences in grain protein and mineral contents in finger millet landraces grown at different latitudes. A 3% increase in protein content was noted when altitude increased from 365 to 1856 m asl. Ca and Fe content increased by 0.47% and 0.33%, respectively, at 1856 m compared to 365 m altitude. Soil organic carbon, N, P, and K also increased with increasing altitudes (Luitel et al., 2020).

Tomatoes are adapted to a wide range of climatic conditions that may impact the yield and nutritional quality of their fruits. Production zones are defined based on variation in soil, landform, and climate characteristics and their influence on crop yields. Variation in production zones had a significant impact on nutritional and nutraceutical quality of tomato byproducts. Although the levels of phenolic compounds, vitamin C, and carotenoids in the peels were influenced by the cultivation zones, the nutraceutical properties of tomato byproducts were not affected by the production zones (Kaboré et al., 2022).

Western Hemisphere regions, which account for 85% of the global soybean production, had consistent variation in spatial patterns for total protein and amino acid composition, up to ~5% difference in protein among regions. Although there exists a negative correlation between protein content and essential amino acids, their (amino acids) amount per ton of grain increased with protein content, which may guide market and industry in benchmarking soybean protein quality across the Western Hemisphere (Cerrudo et al., 2023).

Genetically diverse melon cultivars, evaluated across United States, showed differences in sugar and carotenoid concentrations as a function of production location and cultigen. Such knowledge may help to improve the sensory and functional qualities of melons and identify suitable cultivars adapted to specific environments to maximize fruit quality (Singh et al., 2022).

3.6.6 Conservation Agriculture

Regenerative agriculture refers to nature-based agronomic practices that improve soil health and crop resilience and nutritional quality of crops produce, protect the environment, and sustain livelihoods. Organic farming, deficit irrigation, and alternate wetting and drying on a range of crops improved the nutritional quality of edible portions of several crops, though crop and context specific (Galani et al., 2022b; Manzeke-Kangara et al., 2023). Organic farming components included in this experiment were use of composts and manures, cover crops, crop rotations, crop residues and biochar's, reduced tillage, intercropping, use of arbuscular mycorrhizal fungi, and plant-growth-promoting bacteria.

3.6.7 Crop Rotation and Cropping System Diversification

Crop management practices (monocropping vs. crop rotation) and soil biota (loss of diversity and population density caused by ploughing) composition improved soybean yield and quality (Marro et al., 2020). The rice-wheat cropping system (RWCS) of the Indo-Gangetic Plains region is the predominant production system in South Asian countries, which caused widespread malnutrition as well as non-sustainable production system in the region. Large-scale adoption and diversification of the RWCS with pulses and vegetables could diversify the diets consumed, increase micronutrient intake, and make RWCS systems more viable and sustainable (Kataki, 2008). Legumes reduce the emission of greenhouse gases, allow the sequestration of carbon in soils, fixed atmospheric N to optimize application of N fertilizer, release in the soil high-quality organic matter, and facilitate soil nutrients' circulation and water retention (Stagnari et al., 2017). Optimizing crop productivity, sustaining biodiversity, profitability, and ecosystem viability are critical to minimize nutrient leaching and runoff pollution, causing eutrophication. Cropping system diversification may ameliorate eutrophication, optimize weed management, and enhance food and nutritional diversity. N and P runoff losses and soil erosion losses substantially reduced by adopting more diversified (corn–soybean–oat/clover 3 or corn–soybean–oat/alfalfa–alfalfa) crop rotation than 2-year corn–soybean system. The cropping system, however, had no influence on $NO3^-$ N leaching losses. The 3- and 4-year systems maintained or increased crop yields and net returns relative to the 2-year conventional system (Hunt et al., 2019).

3.7 CONCLUSION

Crop germplasms, including landraces and heirloom cultivars, are the source of novelty for nutritional quality. However, quality traits are highly influenced by location (environment) and by genotype by environment and genotype by management interactions, necessitating the need for multi-site testing to identify nutritionally enhanced germplasm for cultivation or use in breeding programs. Identifying geographical regions with similar agro-ecology should be a priority for large-scale production of nutritious crops. Products from suboptimal environments are best suited to formulate nutraceutical products for the betterment of human health.

REFERENCES

Aćin, V., Mirosavljević, M., Živančev, D., Jocković, B., Brbaklić, L., and Jaćimovićk, G. (2023). Field management practices to produce nutritional and healthier main crops. In: Rakszegi, M., Papageorgiou, M., and Rocha, J.M. (eds.) *Developing Sustainable and Health Promoting Cereals and Pseudocereals Conventional and Molecular Breeding.* Elsevier Inc, pp. 137–173.

Al-Hadeethi, I., Li, Y., Odhafa, A.K.H., Al-Hadeethi, H., Seneweera, S., and Lam, S.K. (2019). Assessment of grain quality in terms of functional group response to elevated

[CO_2], water, and nitrogen using a meta-analysis: Grain protein, zinc, and iron under future climate. *Ecol. Evol.* 9, 7425–7437. https://doi.org/10.1002/ece3.5210

Alvares, C.A., Stape, J.L., Sentelhas, P.C., et al. (2013). Köppen's climate classification map for Brazil. *Meteorol. Z.* 22, 711–728. https://doi.org/10.1127/0941-2948/2013/0507

Ammar, H.B., Arena, D., Treccarichi, S., Di Bella, M.C., Marghali, S., Ficcadenti, N., et al. (2023). The effect of water stress on the glucosinolate content and profile: A comparative study on roots and leaves of *Brassica oleracea* L. crops. *Agronomy* 13, 579. https://doi.org/10.3390/agronomy13020579

Antonious, G.F. (2023). The impact of organic, inorganic fertilizers, and biochar on phytochemicals content of three *Brassicaceae* vegetables. *Appl. Sci.* 13, 8801. https://doi.org/10.3390/app13158801

Assefa, Y., Bajjalieh, N., Archontoulis, S., Casteel, S., Davidson, D., Kovács, P., et al. (2018). Spatial characterization of soybean yield and quality (amino acids, oil, and protein) for United States. *Sci. Rep.* 8, 14653. https://doi.org/10.1038/s41598-018-32895-0

Assefa, Y., Purcell, L.C., Salmeron, M., Naeve, S., Casteel, S.N., Kovács, P., et al. (2019). Assessing variation in US soybean seed composition (protein and oil). *Front. Plant Sci.* 10, 298. https://doi.org/10.3389/fpls.2019.00298

Beleggia, R., Ficco, D.B.M., Nigro, F.M., Giovanniello, V., Colecchia, S.A., Pecorella, I., et al. (2021). Effect of sowing date on bioactive compounds and grain morphology of three pigmented cereal species. *Agronomy* 11, 591. https://doi.org/10.3390/agronomy11030591

Bellaloui, N., Mengistu, A., Abbas, H.K., and Kassem, M.A. (2017). Effects of drought and elevated atmospheric carbon dioxide on seed nutrition and 15N and 13C natural abundance isotopes in soybean under controlled environments. In: Kasai, M. (ed.) *Soybean—The Basis of Yield, Biomass and Productivity.* Intech Open. https://doi.org/10.5772/67511, ISBN978-953-51-3118-2

Botoman, L., Chagumaira, C., Mossa, A.W., Amede, T., Ander, E.L., Bailey, E.H., et al. (2022). Soil and landscape factors influence geospatial variation in maize grain zinc concentration in Malawi. *Sci. Rep.* 12, 7986. https://doi.org/10.1038/s41598-022-12014-w

Boufeldja, L., Boudard, F., Portet, K., Guzman, C., Morel, S., Berger, N., et al. (2023). The impact of elevated atmospheric carbon dioxide exposure on magic tomatoes' nutrition-health properties. *Int. J. Mol. Sci.* 24, 12815. https://doi.org/10.3390/ijms241612815

Capelin, M.A., Madella, L.A., Panho, M.C., Meira, D., Fernandes, R.A.T., Colonelli, L.L., et al. (2022). Impact of altitude on grain yield, oil, and protein content of soybean. *Aust. J. Crop Sci.* 16, 273–279. https://doi.org/10.21475/ajcs.22.16.02.3439

Cerrudo, A., Miller-Garvin, J., and Naeve, S.L. (2023). Western Hemisphere quality and production capacity of soybean protein. *Front. Sustain. Food Syst.* 7, 1223921. https://doi.org/10.3389/fsufs.2023.1223921

Connors, K., Jaacks, L.M., Prabhakaran, P., Veluguri, D., Ramanjaneyulu, G.V., and Roy, A. (2021). Impact of crop diversity on dietary diversity among farmers in India during the COVID-19 pandemic. *Front. Sustain. Food Syst.* 5, 695347. https://doi.org/10.3389/fsufs.2021.695347

Constantino, L.V., Shimizu, G.D., Macera, R., Fukuji, A.S.S., Zeffa, D.M., Koltun, A., et al. (2022). Genetic diversity and selection of heirloom tomato accessions based on the physical and biochemical fruit-related traits. *Bragantia* 81, e1422. https://doi.org/10.1590/1678-4499.20210193

Cruz-Carrión, A., de Azua, M.J.R., Muguerza, B., Mulero, M., Bravo, F.I., Arola-Arnal, A., et al. (2023). Organic vs. non-organic plant-based foods—A comparative study on phenolic content and cntioxidant Capacity. *Plants* 12, 183. https://doi.org/10.3390/plants12010183

Czech, A., Szmigielski, M., and Sembratowicz, I. (2022). Nutritional value and antioxidant capacity of organic and conventional vegetables of the genus *Allium*. *Sci. Rep.* 12, 18713. https://doi.org/10.1038/s41598-022-23497-y

de la Osa, C., Rodríguez-Carvajal, M.A., Gandullo, J., Aranda, C., Ollero, F.J., López-Baena, F.J., et al. (2021). Plant growth-promoting rhizobacteria modulate the concentration of bioactive compounds in tomato fruits. *Separations* 8, 223. https://doi.org/10.3390/separations8110223

De Pascale, S., Maggio, A., Pernice, R., Fogliano, V., and Barbieri, G. (2007). Sulphur fertilization may improve the nutritional value of *Brassica rapa* L. subsp. *sylvestris*. *Eur. J. Agron.* 26, 418–424. https://doi.org/10.1016/j.eja.2006.12.009

De Santis, M.A., Soccio, M., Laus, M.N., and Flagella, Z. (2021). Influence of drought and salt stress on durum wheat grain quality and composition: A review. *Plants* 10, 2599. https://doi.org/10.3390/plants10122599

Dong, J., Gruda, N., Lam, S.K., Li, X., and Duan, Z. (2018). Effects of elevated CO2 on nutritional quality of vegetables: A review. *Front. Plant Sci.* 9, 924. https://doi.org/10.3389/fpls.2018.00924

Dos Santos Araújo, M., de Aragão, W.F.L., dos Santos, S.P., Freitas, T.K.T., da Costa Saraiva, V., Damasceno-Silva, K.J., et al. (2022). Evaluation of adaptability and stability for iron, zinc and protein content in cowpea genotypes using GGE biplot approach. *Heliyon* 8, e11832. https://doi.org/10.1016/j.heliyon.2022.e11832

Dwivedi, S.L., Garcia-Oliveira, A.L., Govindraj, M., and Ortiz, R. (2023). Biofortification to avoid malnutrition in humans in a changing climate: Enhancing micronutrient bioavailability in seed, tuber, and storage roots. *Front. Plant Sci.* 14, 1119148. https://doi.org/10.3389/fpls.2023.1119148

Dwivedi, S.L., Goldman, I., and Ortiz, R. (2019). Pursuing the potential of heirloom cultivars to improve adaptation, nutritional, and culinary features of food crops. *Agronomy* 9, 441. https://doi.org/10.3390/agronomy9080441

Dwivedi, S.L., Mattoo, A.K., Garg, M., Dutt, S., Singh, B., and Ortiz, R. (2022). Developing germplasm and promoting consumption of anthocyanin-rich grains for health benefits. *Front. Sustain. Food Syst.* 6, 867897. https://doi.org/10.3389/fsufs.2022.867897

Dwivedi, S.L., Sahrawat, K.L., Upadhyaya, H.D., and Ortiz, R. (2013). Food, nutrition and agrobiodiversity under global climate change. *Adv. Agron.* 120, 1–128. http://doi.org/10.1016/B978-0-12-407686-0.00001-4

Elferjani, R., and Soolanayakanahally, R. (2018). Canola responses to drought, heat, and combined stress: Shared and specific effects on carbon assimilation, seed yield, and oil composition. *Front. Plant Sci.* 9, 1224. https://doi.org/10.3389/fpls.2018.01224

Fongfon, S., Prom-U-Thai, C., Pusadee, T., and Jamjod, S. (2021). Responses of purple rice genotypes to nitrogen and zinc fertilizer application on grain yield, nitrogen, zinc, and anthocyanin concentration. *Plants (Basel)*. 10, 1717. https://doi.org/10.3390/plants10081717

Galani, Y.J.H., Hansen, E.M.Ø., Droustas, I., Holmes, M., Challinor, A.J., Mikkelesen, T.N., et al. (2022a). Effects of combined abiotic stresses on nutrient content of European wheat and implications for nutritional security under climate change. *Sci. Rep.* 12, 5700. https://doi.org/10.1038/s41598-022-09538-6

Galani, Y.J.H., Ligowe, I.S., Kieffer, M., Kamalongo, D., Kambwiri, A.M., Kuwali, P., et al. (2022b). Conservation agriculture affects grain and nutrient yields of maize (*Zea Mays* L.) and can impact food and nutrition security in Sub-Saharan Africa. *Front. Nutr.* 8, 804663. https://doi.org/10.3389/fnut.2021.804663

Gangashetty, P.I., Yadav, C.B., Riyazaddin, M., Vermula, A., Asungre, P.A., Angarawai, I., et al. (2023). Genotype-by-environment interactions for starch, mineral, and agronomic traits in pearl millet hybrids evaluated across five locations in West Africa. *Front. Plant Sci.* 14, 1171773. https://doi.org/10.3389/fpls.2023.1171773

Garcia-Parra, M.A., Roa-Acosta, D.F., Bravo-Gomez, J.E., Hernández-Criado, J.L., and Villada-Castillo, H.S. (2022). Effects of altitudinal gradient on physicochemical and rheological potential of quinoa cultivars. *Front. Sustain. Food Syst.* 6, 862238. https://doi.org/10.3389/fsufs.2022.862238

Gashash, E.A., Osman, N.A., Alsahli, A.A., Hewait, H.M., Ashmawi, A.E., Alshallash, K.S., et al. (2022). Effects of plant-growth-promoting rhizobacteria (PGPR) and cyanobacteria on botanical characteristics of tomato (*Solanum lycopersicon* L.) plants. *Plants (Basel).* 11, 2732. https://doi.org/10.3390/plants11202732

Gashu, K., Verma, P.K., Acuña, T., Agam, A., Bustan, A., and Fait, A. (2023). Temperature differences between sites lead to altered phenylpropanoid metabolism in a varietal dependent manner. *Front. Plant Sci.* 14, 1239852. https://doi.org/10.3389/fpls.2023.1239852

Gelete, S.H., Mekbib, F., Fenta, B.A., and Teamir, M. (2021). Genotype-by-environment interaction on canning and cooking quality of advanced large-seeded common bean genotypes. *Heliyon* 7, e06936. https://doi.org/10.1016/j.heliyon.2021.e06936

Gmižić, D., Pinterić, M., Lazarus, M., and Šola, I. (2023). High growing temperature changes nutritional value of broccoli (*Brassica oleracea* L. convar. *botrytis* (L.) Alef. var. *cymosa* Duch.) seedlings. *Foods* 12, 582. https://doi.org/10.3390/foods12030582

González-Coria, J., Lozano-Castellón, J., Jaime-Rodríguez, C., Olmo-Cunillera, A., Laveriano-Santos, E.P., Pérez, M., et al. (2022). The effects of differentiated organic fertilization on tomato production and phenolic content in traditional and high-yielding varieties. *Antioxidants* 11, 2127. https://doi.org/10.3390/antiox11112127

Gore, P.G., Das, A., Bhardwaj, R., Tripathi, K., Pratap, A., Dikshit, H.K., et al. (2021). Understanding G × E interaction for nutritional and antinutritional factors in a diverse panel of *Vigna stipulacea* (Lam.) Kuntz germplasm tested over the locations. *Front. Plant Sci.* 12, 766645. https://doi.org/10.3389/fpls.2021.766645

Groenbaek, M., and Kristensen, H.L. (2014). Split dose fertilization with urea increases glucosinolate contents in white cabbage (*Brassica oleracea* L. var. *capitata*) under experimental pot conditions. *Sci. Hortic.* 168, 64–72. https://doi.org/10.1016/j.scienta.2014.01.016

Gugała, M., Sikorska, A., and Zarzecka, K. (2020). The effect of fertilization with sulphur, boron, and amino acids on the content of glucosinolate in winter rape seeds. *Agronomy* 10, 519. https://doi.org/10.3390/agronomy10040519

Gundesli, M.A., Ugur, R., and Yaman, M. (2023). The effects of altitude on fruit characteristics, nutrient chemicals, and biochemical properties of walnut fruits (*Juglans regia* L.). *Horticulturae* 9, 1086. https://doi.org/10.3390/horticulturae9101086

Hallmann, E., Marszalek, K., Lipowski, J., Jasińska, U., Kazimierczak, R., Średnicka-Tober, D., et al. (2019). Polyphenols and carotenoids in pickled bell pepper from organic and conventional production. *Food Chem.* 278, 254–260. https://doi.org/10.1016/j.foodchem.2018.11.052

Hallmann, E., and Rembiałkowska, E. (2012). Characterisation of antioxidant compounds in sweet bell pepper (*Capsicum annuum* L.) under organic and conventional growing systems. *J. Agric. Food Chem.* 92, 2409–2415. https://doi.org/10.1002/jsfa.5624

Hamouz, K., Lachman, J., David, B., and Josef, P. (2018). Effect of growing conditions and storage on the total anthocyanin content in potatoes with coloured flesh. *Plant Soil Environ.* 64, 435–440. https://doi.org/10.17221/384/2018-PSE

Helal, M.I.D., El-Mogy, M.M., Khater, H.A., Fathy, M.A., Ibrahim, F.E., Li, Y.C., et al. (2023). A controlled-release nanofertilizer improves tomato growth and minimizes nitrogen consumption. *Plants (Basel)* 12, 1978. https://doi.org/10.3390/plants12101978

Hu, S., Tong, K., Chen, W., Wang, Y., Wang, Y., and Yang, L. (2022). Response of rice grain quality to elevated atmospheric CO_2 concentration: A meta-analysis of 20-year FACE studies. *Field Crops Res.* 284, 108562. https://doi.org/10.1016/j.fcr.2022.108562

Hunt, N.D., Hill, J.D., and Liebman, M. (2019). Cropping system diversity effects nutrient discharge, soil erosion, and agronomic performance. *Environ. Sci. Technol.* 53, 1344–1352. https://doi.org/10.1021/acs.est.8b02193

Jacobo-Velázque, D.A., and Benavides, J. (2022). Non-thermal technologies as tools to increase the content of health-promoting compounds in whole fruits and vegetables while retaining quality attributes. *Foods* 10, 2904. https://doi.org/10.3390/foods10122904

Jaksomsak, P., Konseang, S., Dell, B., Rouached, H., and Prom-U-Thai, C. (2022). Grain and leaf anthocyanin concentration varies among purple rice varieties and growing condition in aerated and flooded soil. *Molecules* 27, 8355. https://doi.org/10.3390/molecules27238355

Jin, J., Armstrong, R., and Tang, C. (2019). Impact of elevated CO_2 on grain nutrient concentration varies with crops and soils—A long-term FACE study. *Sci. Total Environ.* 651(Pt 2), 2641–2647. https://doi.org/10.1016/j.scitotenv.2018.10.170

Jin, N., Jin, L., Wang, S., Li, J., Liu, F., Liu, Z., et al. (2022). Reduced chemical fertilizer combined with bioorganic fertilizer affects the soil microbial community and yield and quality of lettuce. *Front. Microbiol.* 13, 863325. https://doi.org/10.3389/fmicb.2022.863325

Kaboré, K., Konaté, K., Bazié, D., Dakuyo, R., Sanou, A., Sama, H., et al. (2022). Effects of growing zones on nutritional and bioactive compounds of by-products of two tomato cultivars. *J. Agric. Food Res.* 10, 100414. https://doi.org/10.1016/j.jafr.2022.100414

Kataki, P.K. (2008). Shifts in cropping system and its effect on human nutrition: Case study from India. *J. Crop Prod.* 6, 119–144. https://doi.org/10.1300/J144v06n01_08

Katsenios, N., Andreou, V., Sparangis, P., Djordjevic, N., Giannaoglou, M., Chanioti, S., et al. (2021). Evaluation of plant growth promoting bacteria strains on growth, yield and quality of industrial tomato. *Microorganisms* 9, 2099. https://doi.org/10.3390/microorganisms9102099

Khan, M.M.H., Rafi, M.Y., Ramlee, S.I., Juso, M., and Mamun, M.A. (2022). Hereditary analysis and genotype × environment interaction effects on growth and yield components of Bambara groundnut (*Vigna subterranea (L.) Verdc.*) over multi-environments. *Sci. Rep.* 12, 15658. https://doi.org/10.1038/s41598-022-19003-z

Köhler, I.H., Huber, S.C., Bernacchi, C.J., and Baxter, I.R. (2019). Increased temperatures may safeguard the nutritional quality of crops under future elevated CO_2 concentrations. *Plant J.* 97, 872–886. https://doi.org/10.1111/tpj.14166

Kumar, G., Basak, N., Priyadarsani, S., Bagchi, T.B., Kumar, A., Pradhan, S.K., et al. (2023). Alteration in the physico-chemical traits and nutritional quality of rice under anticipated rise in atmospheric CO_2 concentration: A review. *J. Food Compost. Anal.* 121, 105332. https://doi.org/10.1016/j.jfca.2023.105332

Kunnam, J., Teamkao, P., and Aninbon, C. (2023). Chemical composition and bioactive compounds of "riceberry" rice produced under organic and conventional practices. *J. Agric. Sci.* 161, 13–21. https://doi.org/10.1017/S0021859622000648

Lee, J.H., Oh, M.M., and Son, K.H. (2019). Short-term ultraviolet (UV)-A light-emitting diode (LED) radiation improves biomass and bioactive compounds of kale. *Front. Plant Sci.* 10, 1042. https://doi.org/10.3389/fpls.2019.01042

Li, Y., Yu, Z., Jin, J., Zhang, Q., Wang, G., Liu, C., et al. (2018). Impact of elevated CO_2 on seed quality of soybean at the fresh edible and mature stages. *Front. Plant Sci.* 9, 1413. https://doi.org/10.3389/fpls.2018.01413

Li, Y., Yu, Z., Liu, X., Mathesius, U., Wang, G., Tang, C., et al. (2022). Elevated CO_2 increases nitrogen fixation at the reproductive phase contributing to various yield responses of soybean cultivars. *Front. Plant Sci.* 8, 1546. https://doi.org/10.3389/fpls.2017.01546

Li, Y., Yu, Z., Yang, S., Wang, G., Liu, X., Wang, C., et al. (2019). Impact of elevated CO_2 on C:N:P ratio among soybean cultivars. *Sci. Total Environ.* 694, 133784. https://doi.org/10.1016/j.scitotenv.2019.133784

Liu, Y., Lin-Wang, K., Espley, R.V., Wang, L., Li, Y., Liu, Z., et al. (2019). StMYB44 negatively regulates anthocyanin biosynthesis at high temperatures in tuber flesh of potato. *J. Exp. Bot.* 70, 3809–3824. https://doi.org/10.1093/jxb/erz194

Loladze, I. (2014). Hidden shift of the ionome of plants exposed to elevated CO_2 depletes minerals at the base of human nutrition. *eLife* 3, e02245. https://doi.org/10.7554/eLife.02245

Luitel, D.R., Siwakoti, M., and Jha, P.K. (2020). Nutrients in finger millet and soil at different elevation gradients in Central Nepal. *CABI Agric. Biosci.* 1, 20. https://doi.org/10.1186/s43170-020-00018-3

Mahmood, T., Ahmad, T., and Trethowan, R. (2022). Genotype x environment x management (GEM) reciprocity and crop productivity. *Front. Agron.* 4, 800365. https://doi.org/10.3389/fagro.2022.800365

Manzeke-Kangara, M.G., Joy, E.J.M., Lark, R.M., Redfern, S., Eilander, A., and Broadley, M.R. (2023). Do agronomic approaches aligned to regenerative agriculture improve the micronutrient concentrations of edible portions of crops? A scoping review of evidence. *Front. Nutr.* 10, 1078667. https://doi.org/10.3389/fnut.2023.1078667

Marro, N., Cofré, N., Grilli, G., Alvarez, C., Labuckas, D., Maestri, D., et al. (2020). Soybean yield, protein content and oil quality in response to interaction of arbuscular mycorrhizal fungi and native microbial populations from mono- and rotation-cropped soils. *Appl. Soil Ecol.* 152, 103575. https://doi.org/10.1016/j.apsoil.2020.103575

Mattoo, A.K., Dwivedi, S.L., Dutt, S., Singh, B., Garg, M., and Ortiz, R. (2022). Anthocyanin-rich vegetables for human consumption—Focus on potato, sweetpotato and tomato. *J. Mol. Sci.* 23, 2634. https://doi.org/10.3390/ijms23052634

Medina-Lozano, I., and Diaz, A. (2021). Nutritional value and phytochemical content of crop landraces and traditional varieties. In: *Landraces—Traditional Variety and Natural Breed.* https://doi.org/10.5772/intechopen.95514

Miękus, N., Marszałek, K., Podlacha, M., Iqbal, A., Puchalski, C., and Świergiel, A.H. (2020). Health benefits of plant-derived sulfur compounds, glucosinolates, and organosulfur compounds. *Molecules* 25, 3804. https://doi.org/10.3390/molecules25173804

Netshimbupfe, M.H., Berner, J., Van Der Kooy, F., Oladimaji, O., and Gouws, C. (2022). Influence of drought and heat stress on mineral content, antioxidant activity and bioactive compound accumulation in four African *Amaranthus* Species. *Plants (Basel)* 12, 953. https://doi.org/10.3390/plants12040953

Nicholson, C.C., Emery, B.F., and Niles, M.T. (2021). Global relationships between crop diversity and nutritional stability. *Nat. Commun.* 12, 5310. https://doi.org/10.1038/s41467-021-25615-2

Ortega-Hernández, E., Antunes-Ricardo, M., and Jacobo-Velázquez, D.A. (2021). Improving the health-benefits of kales (*Brassica oleracea* L. var. *acephala* DC) through the application of controlled abiotic stresses: A review. *Plants (Basel)* 10, 2629. https://doi.org/10.3390/plants10122629

Osei, M.K., Annor, B., Adjebeng-Danquah, J., Danquah, A., Danquah, E., Blay, E., et al. (2019). Genotype × environment interaction: A prerequisite for tomato variety development. In: Nyaku, S.T., and Danquah, A. (eds.) *Recent Advances in Tomato Breeding and Production.* IntechOpen. http://doi.org/10.5772/intechopen.76011

Ovando-Martínez, M., Guzmán-Maldonado, S.H., Simsek, S., Bello-Pérez, L.A., and Osorio-Díaz, P. (2014). Effect of water regimes on dietary fiber, polyphenols and antioxidant capacity of black and pinto beans. *Agric. Sci.* 5, 342–352. https://doi.org/10.4236/as.2014.54036

Ozturk, A., Erdem, E., Aydin, M., and Karaoglu, M.M. (2021). The effects of drought after anthesis on the grain quality of bread wheat depend on drought severity and drought resistance of the variety. *Cereal Res. Commun.* 50, 105–116. https://doi.org/10.1007/s42976-021-00155-z

Pu, Z., Liu, Q., Li, Z., Chen, S., Liu, Y., Qi, P., et al. (2019). Planting locations with higher temperature produce more bioactive compounds and antioxidant capacities of wheat. *Agronomy* 9, 538. https://doi.org/10.3390/agronomy9090538

Ramírez-Cariño, H.F., Ocha-Velasco, C.E., Gurrero-Analco, J., Moniribot-Villaneueva, J.L., Calderón-García, C., González-Terreros, E., et al. (2023). Combined effect of the potassium dose and plant biofertilization by *Acinetobacter calcoaceticus* on the growth, mineral content, nutritional quality, antioxidant activity, and metabolomic features of tomatillo fruits (*Physalis ixocarpa* Brot.). *Plants (Basel)* 12, 466. https://doi.org/10.3390/plants12030466

Rusu, O.-R., Mangalagiu, I., Amăriucăi-Mantu, D., Teliban, G.-C., Cojocaru, A., Burducea, M., et al. (2023). Interaction effects of cultivars and nutrition on quality and yield of tomato. *Horticulturae* 9, 541. https://doi.org/10.3390/horticulturae9050541

Samitya, M., Aluko, R.E., Dhewa, T., and Moreno-Rojas, J.M. (2021). Potential health benefits of plant food-derived bioactive components: An overview. *Foods* 10, 839. https://doi.org/10.3390/foods10040839

Sehgal, A., Sita, K., Siddique, K.H.M., Kumar, R., Bhogireddy, S., Varshney, R.K., et al. (2018). Drought or/and heat-stress effects on seed filling in food crops: Impacts on functional biochemistry, seed yields, and nutritional quality. *Front. Plant Sci.* 9, 1705. https://doi.org/10.3389/fpls.2018.01705

Sellami, M.H., Lavini, A., and Pulvento, C. (2021). Phenotypic and quality traits of chickpea genotypes under rainfed conditions in South Italy. *Agronomy* 11, 962. https://doi.org/10.3390/agronomy11050962

Sharma, D., Shree, B., Kumar, S., Kumar, V., Sharma, S., and Sharma, S. (2022). Stress induced production of plant secondary metabolites in vegetables: Functional approach for designing next generation super foods. *Plant Physiol. Biochem.* 192, 252–272. https://doi.org/10.1016/j.plaphy.2022.09.034

Singh, J., Metrani, R., Jayaprakasha, G.K., Crosby, K.M., Jifon, J.L., Ravishankar, S., et al. (2022). Profiling carotenoid and sugar contents in unique *Cucumis melo* L. cultigens harvested from different climatic regions of the United States. *J. Food Compost. Anal.* 106, 104306. https://doi.org/10.1016/j.jfca.2021.104306

Son, K.-H., Furuyama, S., Hayashi, K., Ishigami, Y., Hikosaka, S., and Goto, E. (2020). Enhancement of accumulation of bioactive compounds in red leaf lettuce by manipulation of UV light before harvest. *Acta Hortic.* 1271, 79–84. https://doi.org/10.17660/ActaHortic.2020.1271.11

Średnicka-Tober, D., Kopczyńska, K., Góralska-Walczak, R., Hallmann, E., Barański, M., Marszalek, K., et al. (2022). Are organic certified carrots richer in health-promoting phenolics and carotenoids than the conventionally grown ones? *Molecules* 27, 4184. https://doi.org/10.3390/molecules27134184

Stagnari, F., Maggio, A., Galieni, A., and Pisante, M. (2017). Multiple benefits of legumes for agriculture sustainability: An overview. *Chem. Biol. Technol. Agric.* 4, 2. https://doi.org/10.1186/s40538-016-0085-1

Toscano, S., Trivellini, A., Cocetta, G., Bulgari, R., Francini, A., Romano, D., et al. (2019). Effect of preharvest abiotic stresses on the accumulation of bioactive compounds in horticultural produce. *Front. Plant Sci.* 10, 1212. https://doi.org/10.3389/fpls.2019.01212

Wagner, M.R., and Mitchell-Olds, T. (2023). Soil variation among natural habitats alters glucosinolate content in a wild perennial mustard. *J. Exp. Bot.* 74, 1723–1740. https://doi.org/10.1093/jxb/erac520

Walsh, C., Vandenburgh, C., Grant, L., Katz, E., Kliebenstein, D.J., and Fierer, N. (2024). Microbial terroir: Associations between soil microbiomes and the flavor chemistry of mustard (*Brassica juncea*). *New Phytol.* 243, 1951–1965. https://doi.org/10.1111/nph.19708

Wang, J., Li, L., Lam, S.K., Shi, X., and Pan, G. (2023). Changes in plant nutrient status following combined elevated [CO_2] and canopy warming in winter wheat. *Front. Plant Sci.* 14, 1132414. https://doi.org/10.3389/fpls.2023.1132414

Wang, X., Li, X., Zhong, Y., Blennow, A., Liang, K., and Liu, F. (2022). Effects of elevated CO_2 on grain yield and quality in five wheat cultivars. *J. Agron. Crop. Sci.* 208, 733–745.

Wang, X.-X., Zhao, F., Zhang, G., Zhang, Y., and Yang, L. (2017). Vermicompost improves tomato yield and quality and the biochemical properties of soils with different tomato planting history in a greenhouse study. *Front. Plant Sci.* 8, 1978. https://doi.org/10.3389/fpls.2017.01978

Wittayathanarattana, T., Wanichananan, P., Supaibulwatana, K., and Goto, E. (2022). Enhancement of bioactive compounds in baby leaf *Amaranthus tricolor* L. using short-term application of UV-B irradiation. *Plant Physiol. Biochem.* 182, 202–215. https://doi.org/10.1016/j.plaphy.2022.04.003

Wu, D., Chen, C., Liu, Y., Zhang, G., and Yang, L. (2023). Vermicompost improves tomato yield and quality by promoting carbohydrate transport to fruit under salt stress. *Horticulturae* 9, 1015. https://doi.org/10.3390/horticulturae9091015

Wu, W., Lin, Z., Zhu, X., Li, G., Zhang, W., Chen, L., et al. (2022). Improved tomato yield and quality by altering soil physicochemical properties and nitrification processes in the combined use of organic-inorganic fertilizers. *Eur. J. Soil Biol.* 109, 103384. https://doi.org/10.1016/j.ejsobi.2022.103384

Xin, W., Zhang, J., Yu, Y., Tian, Y., Li, H., Chen, X. et al. (2024). Root microbiota of tea plants regulate nitrogen homeostasis and theanine synthesis to influence tea quality. *Curr. Biol.* 34, 868–880.e6. https://doi.org/10.1016/j.cub.2024.01.044

Yang, L., Song, W., Xu, C., Sapey, E., Jiang, D., and Wu, C. (2023). Effects of high night temperature on soybean yield and compositions. *Front. Plant Sci.* 14, 1065604. https://doi.org/10.3389/fpls.2023.1065604

Yang, L., Zhao, F., Chang, Q., Li, T., and Li, F. (2015). Effects of vermicomposts on tomato yield and quality and soil fertility in greenhouse under different soil water regimes. *Agric. Water Manag.* 160, 98–105. https://doi.org/10.1016/j.agwat.2015.07.002

Ye, L., Zhao, X., Bao, E., Li, J., Zou, Z., and Cao, K. (2020). Bio-organic fertilizer with reduced rates of chemical fertilization improves soil fertility and enhances tomato yield and quality. *Sci. Rep.* 10, 177. https://doi.org/10.1038/s41598-019-56954-2

Yoshida, H., Shimada, K., Hikosaka, S., and Goto, E. (2022). Effect of UV-B irradiation on bioactive compounds of red perilla (*Perilla frutescens* (L.) Britton) cultivated in a plant factory with artificial light. *Horticulturae* 8, 725. https://doi.org/10.3390/horticulturae8080725

Yousaf, M.I., Riaz, M.W., Jiang, Y., Yasir, M., Aslam, M.Z., Hussain, S., et al. (2022). Concurrent effects of drought and heat stresses on physio-chemical attributes, antioxidant status and kernel quality traits in Maize (*Zea mays L.*) Hybrids. *Front. Plant Sci.* 13, 898823. https://doi.org/10.3389/fpls.2022.898823

Zhang, Z., Liu, L., Zhang, M., Zhang, Y., and Wang, Q. (2014). Effect of carbon dioxide enrichment on health-promoting compounds and organoleptic properties of tomato fruits grown in greenhouse. *Food Chem.* 153, 157–163. https://doi.org/10.1016/j.foodchem.2013.12.052

4 Combining Lexicons (Breeders' Sensory Test by Trained Human Panel), Volatile Compounds, and Physicochemical Attributes for Sensory Evaluations

Autar K. Mattoo and Sangam L. Dwivedi

4.1 INTRODUCTION

In recent years, there has been a growing demand for good quality fruit and vegetables. The consumers are willing to pay extra price for fruit and vegetables with good flavor attributes. Fruit and vegetables with pleasant aromas are preferred, while those with unpleasant aromas are rejected. Sugars, acids, and volatile organic compounds (VOCs) contribute to differences in flavor characteristics. Terpenoids, phenylpropanoids/benzenoids, and fatty acid and amino acid derivatives are characteristic VOCs (Mostafa et al., 2022). Many genes and TFs regulate the synthesis of VOCs (Imran et al., 2022; Li et al., 2023). Genotype, soil and environmental factors, abiotic stress, fruit ripening and harvest time, and postharvest processing, and storage conditions all influence flavor characteristics (Mostafa et al., 2022).

Sensory analysis is a scientific discipline that measures and explores the stimuli perceived by the senses against different variants which can be validated through instrumental methods. Trained human panelists with specific sense of differentiating flavors are assigned to define sensory attributes. The panel usually consists of 8–12 members. They are trained for varying durations (~2 h per session) and sessions (at least 10 sessions) to discriminate sensory attributes using descriptive analysis. The panel uses descriptive methods, mainly with intensity scale (9- or 5-point scale) to describe sensory attributes and validate the training of their sensory panel. The test samples are replicated, preferably in triplicate. The descriptive sensory attributes should be supported by hedonic (a 9-point scale describing consumer acceptance and satisfaction) test, typical for consumer perception.

DOI: 10.1201/9781003467335-4

Instrumental sensory devices (e.g., e-noses, e-tongues, e-eyes) and immersive (e.g., virtual sensory booth utilizing virtual reality; Zulkarnain et al., 2024a, 2024b) techniques are gaining headway as sensory profiling techniques (Marques et al., 2022; Rodrigues et al., 2024). The data obtained from discriminative and descriptive techniques are then channeled through various statistical techniques (multivariate analysis of variance, cluster analysis, correspondence analysis, multidimensional scaling analysis, hierarchical cluster analysis, partial least squares regression, multiple linear regression, generalized Procrustes analysis, categorical principal components analysis, and structural equation modeling) to assess how the consumers perceive a product (Marques et al., 2022). Several consumer-based approaches—e.g., check-all-that-apply (CATA), rate-all-that-apply (RATA), flash profile (FP), free listing (FL), and conventional descriptive analysis (DA)—have emerged for assessing sensory profiles. The performance of the consumer methods is similar to that of DA. It effectively profiled differences in consumer perception of sensory attributes, while RATA exhibited the highest similarity to the DA (regressor vector [RV] values = 0.96), facilitating quicker and less tedious processes compared to FP or FL (Kim et al., 2023).

A lexicon refers to a set of standardized vocabularies developed by highly trained panelists to describe sensory attributes of food products. It is a very vital tool to assess sensory attributes of food crops as well. To date, several lexicons have been developed and involve sensory perception of a variety of food products, including fruits and vegetables. It is an effective communication and a guidance tool for new product development processes, quality control, product improvement, and monitoring changes during product shelf-life and for developing new germplasm with specific flavor attributes (Suwonsichon, 2019).

Cost and complexity in the determination of volatiles limit breeders' ability to initiate large-scale genetic improvement of flavor traits. Trained sensory panels are often used to assess taste and flavor at later stage of breeding programs. It is quite possible that favorable alleles associated with improved flavor may have been lost during handling of early generation breeding populations. Breeders' sensory test based on sweetness, sourness, and aroma, conducted by a team of 2–3 trained persons and scored on 1 to 9 scale with increments of 0.5 (1 = not detectable and 9 = highest intensity), significantly correlates with the physico-chemical and sensory traits, meaning that it can be used as a tool for improving the flavor of tomato and possibly other crops as well (Hagenguth et al., 2022).

4.2 COMBINING MACHINE LEARNING TECHNIQUES WITH IMAGE-BASED ANALYSIS TO ASSESS FOR SENSORY ATTRIBUTES

The use of trained human panel for assessing sensory attributes is costly, time-consuming, low-throughput, and subject to human error, bias, and fatigues. Therefore, it remains as a biggest challenge, particularly when applying it in early stage of breeding populations. Cutting-edge technology, such as artificial-intelligence-based machine learning is becoming popular in the fields of agriculture and food and nutrition science and for developing customized dietary needs (Esmaeily et al., 2024).

Machine learning models accurately predicted flesh color and mealiness based on images of boiled sweet potato roots. It captures images of boiled sweet potato cross-sections using the DigiEye imaging system, pre-processes data for background elimination, and features extraction to predict the flesh color and mealiness. The trained linear regression and random forest regression models, respectively, attained R^2 values of 0.92 and 0.87 for flesh color. For mealiness, the random forest regressor and gradient boosting model attained R^2 values of 0.85 and 0.80, respectively. The performance of these models matched the desirable R^2 threshold of 0.80 for acceptable comparability to human sensory panels. The International Potato Center (https://cipotato.org/) routinely uses machine learning models to automate and increase throughput for analyzing flesh-color and mealiness for assessing sensory attributes in sweet potato breeding programs (Nakatumba-Nabende et al., 2023). Likewise, ResNet-R&H machine learning model, relative to other machine learning models, efficiently differentiated fresh edamame (vegetable soybean) from frozen soybeans (Bu et al., 2024).

4.3 DEVELOPING CROP-SPECIFIC LEXICON

Crop-specific common lexicon may facilitate access to sensory attributes, consumer acceptance, and liking or disliking of the food products.

4.3.1 Grain Crops

Certain class of rice (basmati and jasmine types) grains are highly favored and command a premium price because of unique flavor characteristics. A rice flavor lexicon consists of 24 attributes and expanded by sensory panelists to characterize the flavor of cooked rice differing in forms, types, and specialties. Nineteen of the 24 descriptive attributes were aromatic, while five attributes relate to tastes and oral feeling. Eighteen aromatic descriptors were common in most rice accessions, whereas some unique to a specific-rice type (Limpawattana and Shewfelt, 2010). A sensory lexicon with 27 attributes, validated by trained panel, has been suggested to characterize sorghum flavor as grain and grain-based products (file:///C:/Users/sanga/Downloads/47_ChambersIV(1).pdf). A trained panel of 9 human subjects was used to develop quinoa lexicon based on differences in aroma, taste/flavor, texture, and color. This lexicon differentiated the most (Commercial Red) and least (QQ74) preferred cooked quinoa genotypes based on aroma (caramel, nutty, buttery, grassy, earthy, woody), taste (sweet, bitter, grain-like, nutty, earthy, toasty), and firmness (cohesive, pasty, adhesive, crunchy, chewy, astringent, moist). Higher intensities of grassy aroma and of firm and crunchy texture drive overall consumer acceptance of cooked quinoa (Wu et al., 2017).

Vegetable soybeans, popularly named as edamame in the USA, is becoming popular in developed economies but is also very familiar food in Southeast and East Asian countries. Thirteen aromatic attributes, three feeling factors and the basic tastes, further validated by trained panelists, defines the edamame lexicon (Krinsky et al., 2006). The freshness of edamame is an important quality parameter. Deep-learning-based image recognition technology is a fast, efficient, and low-cost method for analysis of the freshness of food. Of the four machine learning models tested, the deep learning model ResNet18 (named as ResNet-R&H), relative to other models, has higher classification accuracy and computational efficiency to differentiate fresh from frozen edamame pods/seeds. It is based on the residual networks (ResNet) structure and incorporates the fusion data of RGB (red, green, and blue) image recognition technology and hyperspectral images. The ResNet-R&H achieves accuracy of 97.6%, demonstrating a significant improvement compared to the distinct utilization of hyperspectral and RGB data (Bu et al., 2024). Peas and pea-derived products are used globally. Peas are a great source of protein and dietary fiber. Pea protein is not allergy-inducing in humans. The pea lexicon based on sensory and instrumental analyses of texture and of volatile compounds describes 205 descriptive sensory attributes. They are grouped into five categories: smell/odor (27%), flavor (51%), taste (10%), texture (8%), and visual (4%), which could be used for sensory evaluation of peas and other grain legumes (Lara and Tsiami, 2024).

4.3.2 Fruits and Vegetables

Bell pepper (*Capsicum annuum*), a popular vegetable worldwide, contains unique sensory profiles and is rich in antioxidant properties. Forty-six descriptive attributes (aroma, 14; flavor, 19; texture, 9; chemesthetic, 4) describe fresh, pickled, and roasted bell pepper products. Smoky, burnt, and cooked tomato flavors are typical of roasted red bell peppers, while the floral attributes are characteristic of fresh red bell peppers, and vinegar, pungency, and spicy attributes characterized pickled peppers (Uppili et al., 2023).

Goji berry (*Lycium barbarum*) pulp, relative to fresh goji berries, is increasingly popular among customers due to its superior nutritional value. The sensory lexicon for goji berry pulp contains 83 aromatic compounds and 17 aroma-active compounds as the key ingredients. The key aroma compounds are 3-ethylphenol, methyl caprylate, 2-hydroxy-4-methyl ethyl valerate, benzeneacetic acid, ethyl ester, hexanal, *(E,Z)*-2,6-nonadienal, acetylpyrazine, butyric acid, 2-ethylhexanoic acid, 2-methyl-1-propanol, 1-pentanol, phenylethyl alcohol, and 2-nonanone (Wang et al., 2024).

Edible mushrooms, a nutritious and versatile ingredient in many food products, are low in calories but rich in minerals and vitamins, and are considered as healthy food (Valverde et al., 2015). A highly trained, descriptive sensory panel classified 27 flavor attributes, including musty, fishy, shellfish, woody, nutty, brown, green, cardboard, burnt/ashy, potato, umami, protein (vegetable), yeasty, bitter, salty, sweet aromatics, sour, and astringent. Dried mushrooms were either as bitter, burnt, musty/dusty, astringent, old leather, and fresh mushroom, whereas fresh mushroom were classified as umami, sweet, earthy/potato, earthy/damp, yeasty, and fermented (Chun et al., 2020). Velvet foot (*Flammulina velutipes*) is another species of mushroom that is widely cultivated worldwide for its rich nutritional value. A lexicon for *F. velutipes* detected 74 volatile compounds that categorized aromas as fruity, alcoholic, milky, floral, sulfurous, and musty (Wang et al., 2022).

Twenty-two sensory attributes differentiated cantaloupe melons (*Cucumis melo* var. *cantalupensis*). CaldeoDF, Caldeo, Globstar, RML0499, and Caribbean Gold are characterized for higher firmness, crunchiness, and denser textures, while

Caldeo, CaldeoDF, Globstar, Magellan, and Sol Real for higher overall intensity, sweetness and oversweet flavor, longer-lasting aftertaste, and salivating mouthfeel. Impac melon is known for higher watermelon flavor, bitterness, and sourness. Sweet, oversweet flavor, and intense orange and richness of color drives the overall liking score (Ayres et al., 2019).

The peach (*Prunus persica*) lexicon based on trained descriptive panel and diverse cultivars monitored throughout growing season identified 29 attributes. It includes 18 aroma and flavor attributes, describing mature, underripe, or overripe peaches, appearance (based on redness of fruits), three feeling factors, and seven texture-related attributes. Texture and peach identity were the primary differentiating factors. Fruity, sweet, tart, citrus, and sour flavors, along with texture and feeling factors, differentiated peach cultivars. Thus, it provides a platform for researchers and producers to understand the desirable sensory traits in peaches (Belisle et al., 2017).

A highly trained descriptive panel uses a consensus based descriptive analysis to develop lexicon for potatoes by assessing diverse cultivars. The lexicon includes 5 appearance, 17 aroma, 21 flavor, 13 texture, 5 mouthfeel, and 3 aftertaste attributes. The most common among all samples were cooked potato, cardboard, umami, starchy, salty, metallic, bitter, and astringent attributes, while specific attributes, such as vegetable complex, beany, nutty, toasted, sweet potato, eggy, and cauliflower, were unique to certain cultivars (Sharma et al., 2020).

Raspberry (*Rubus idaeus*) and blueberry (*Vaccinium corymbosum*) are widely produced and/or consumed worldwide. Multilingual sensory lexicon, if available, facilitates comparative assessment of sensory traits through panels operating in different countries using their native language. It ensures consistency of sensory evaluations across countries and/or with international panelists. A multilingual (English, Finnish, French, German, Italian, Spanish, Turkish) lexicon for berries rate only those flavor that are perceived and reduced sensory fatigue. Such a discerning efficacy of the lexicon across countries may provide new opportunities to enrich berries descriptor lists (Lippi et al., 2023). The strawberry (*Fragaria* × *ananassa*) lexicon by a trained panel describes a set of 25 descriptors (2 appearance, 8 aroma, 10 flavor, and 5 texture/mouthfeel attributes) associated with strawberry sensory traits (Oliver et al., 2018).

A lexicon involving sweet potatoes (*Ipomea batatas*) with varying flesh color (orange, purple, yellow) was developed to assess the impact of flesh colors on consumer acceptance. The lexicon differentiated sweet potatoes. Appearance positively impacted liking scores of products. Flavor and texture are the driving attributes for liking. The consumer-based overall liking clusters were three. Clusters with smooth texture, brown sugar and dried apricot flavor, and sweet taste were liked, while those with bitter or umami taste, astringent mouthfeel, vanilla aroma, and residual fibers were disliked. Thus, the development of sweet potato lexicon facilitated the characterization of sensory attributes and assisted in unfolding key consumer liking attributes (Leksrisompong et al., 2012). More recently, sensory-guided selection criteria for breeding consumer-preferred sweet potatoes developed in Uganda comprises 27 sensory attributes to characterize and discriminate accessions for sensory profiles. Significant correlation was detected between sensory firmness by hand and mouth with TPA peak positive force ($r = 0.695$ and $r = 0.648$, respectively) and positive area (r = 0.748, r = 0.715, respectively), with few accessions liked the most. An average peak positive force of 3700 gf is proposed as a minimum texture value for screening sweet potatoes, which correspond to at least 46% of consumers perceiving sweet potatoes as just-about-right in firmness and a minimum overall liking of 6 on average. Thus, combining descriptive sensory analysis with instrumental texture analysis facilitates efficient screening of sweet potatoes in breeding programs (Nakitto et al., 2022).

A lexicon for fresh and processed tomatoes defines 33 sensory attributes (aroma, 5; texture, 10; flavor, 18, including 6 taste and mouthfeel) for fresh and processed tomatoes, represented a wide range of sensory characteristics (Hongsoongnern and Chambers IV, 2008).

4.3.3 Fresh Leafy Vegetables

A highly trained descriptive sensory panel described 32 sensory attributes that described the flavor of the fresh leafy vegetables. It contains five 'green' attributes, mouthfeel characteristics (pungent, bite, tooth-etch, heat/burn), tastes (bitter, umami), some unique flavors related to specific vegetables (cabbage, celery, lettuce, spinach, parsley, beet, radish leaves), and several other attributes such as citrus, piney, woody, water-like, musty/earthy, floral, sulfur, metallic, soapy, petroleum-like, and overall sweet (Talavera-Bianchi, 2010). A lexicon representing arugula, Belgium endive, bok choy, chard, collards, escarole, kale, lettuce, mustard, radicchio, spinach, and watercress leafy vegetables have been recently reported (https://leafylexicon.com/products/the-illustrated-guide-to-leafy-greens-the-common-varieties) to assess flavor profiles of leafy vegetables.

4.4 SIGNIFICANT MILESTONES ACHIEVED IN BREEDING FOR SENSORY TRAITS

Plant breeders have been successful in enhancing the productivity of staple food crops, with most of the genetic improvement driven by industry demand. In doing so, they ignored sensory traits, concentrating mainly on productivity traits, uniformity, stress tolerance, and shipping quality. It ensured farm profits and affordable food to consumers. In other words, sensory qualities have been sacrificed in exchange of increasing the productivity of food crops. Self-sufficiency in food sectors have encouraged plant breeders turn to improving edible produce quality, including sensory traits and consumers preferences. The emphasis is to develop next-generation food crops combining productivity, stress tolerance, and improve nutritional quality, including flavor, color, aroma, and texture (Folta and Klee, 2016).

- Tomato introgression lines (ILs) carrying *Lycopersicon pennellii* allele show up to 60-fold variation in 2-phenylethanol and phenylacetaldehyde. ILs containing low levels of these compounds contribute to desirable aroma, while those with higher concentration (>0.005 ppm), unpleasant (nauseating) aroma (Tadmor et al., 2002).
- Modern tomato cultivars are deficient in good organoleptic properties. Andean landraces are excellent breeding targets for improving sensory traits of modern tomatoes (Cortina et al., 2018).

- QTL clusters for flavor-related traits on chr 6, containing five sensory and nine volatile QTLs, are ideal candidates for discovery of SNPs associated with flavor and sensory traits (Hagenguth et al., 2024).
- Breeders' sensory test (BST), conducted by a small team with a high number of small sample sizes, indicated significant correlations with total soluble solids and titratable acidity, with higher genetic + environmental variation (EV) than EV itself. These data suggest BST, as a promising tool to select for flavor in early stages of breeding. Discerning pathways for volatile biosynthesis genes unlocked rare volatile compounds associated with unique flavor, while genomic selection for increased sugar content may improve flavor intensity (Porter et al., 2023).
- Apples, blueberries, wine grapes, tomatoes, peanuts, and edamame (vegetable soybean) are significantly improved for sensory attributes through plant-breeding programs in the United States (Carneiro et al., 2020).
- Significant decline has been seen in the roast peanut (Virginia market class) flavor and a modest improvement in other critical flavor attributes over years. New germplasm (Chimera, NC Bunch) may provide an opportunity to breed for Virginia-type peanut cultivars with improved roast peanut flavor (Fritz et al., 2022).
- A single gene, *MdTDT, MdSOT2*, or *Ma1*, is sufficient to cause a dramatic decrease in the accumulation of citrate, sorbitol, or malate, respectively, while pyramiding of multiple genes with minor effects and little hitchhiking effect on fruit taste, improve fruit (large) size (Liao et al., 2021). Combining fruit chemicals and consumer sensory panel information to train machine learning models efficiently predicted fruit flavor and may allow plant breeders to integrate flavor attributes in early generation breeding programs to design the next-generation of fruits and vegetables with exceptional flavor profiles (Colantonio et al., 2022).
- *GmBADH2* is a functionally characterized gene and marker associated with aroma in soybean (Qian et al., 2022).

4.5 CONCLUSION

To date, significant progress has been achieved toward developing crop-specific lexicons in grain, fruit, and vegetable crops, including green leafy vegetables. Various AI-based machine learning tools and statistical models and software are being tried to facilitate the use of lexicons in early breeding generations and to minimize the risk of loss of rare alleles associated with unique flavor. Also, significant breeding milestones, both genetic and genomic resources, are documented to facilitate the use of these resources in improving the flavor profiles of edible-grade food products via crop breeding.

REFERENCES

Ayres, E.M.M., Lee, S.M., Boyden, L., and Guinard, J.-X. (2019). Sensory properties and consumer acceptance of Cantaloupe melon cultivars. *J. Food Sci.* 84, 2278–2288. https://doi.org/10.1111/1750-3841.14724

Belisle, C., Adhikari, K., Chavez, D., and Phan, U.T.X. (2017). Development of a lexicon for flavor and texture of fresh peach cultivars. *J. Sens. Stud.* 32, e12276. https://doi.org/10.1111/joss.12276

Bu, Y., Hu, J., Chen, C., Bai, S., Chen, Z., Hu, T., et al. (2024). ResNet incorporating the fusion data of RGB & hyperspectral images improves classification accuracy of vegetable soybean freshness. *Sci. Rep.* 14, 2568. https://doi.org/10.1038/s41598-024-51668-6

Carneiro, R.C.V., Duncan, S.E., O'Keefe, S.F., Yin, Y., Neill, C.L., and Zhang, B. (2020). Sensory and consumer studies in plant breeding: A guidance for Edamame development in the U.S. *Front. Sustain. Food Syst.* 4, 124. https://doi.org/10.3389/fsufs.2020.00124

Chun, S.-S., Chambers, E., and Han, I. (2020). Development of a sensory flavor lexicon for mushrooms and subsequent characterization of fresh and dried mushrooms. *Foods* 9, 980. https://doi.org/10.3390/foods9080980

Colantonio, V., Ferrão, L.F.V., Tiemana, D.M., Bliznyuk, N., Sims, C., Klee, H.J., et al. (2022). Metabolomic selection for enhanced fruit flavor. *Proc. Natl. Acad. Sci. U. S. A.* 119, e2115865119. https://doi.org/10.1073/pnas.2115865119

Cortina, P.R., Santiago, A.N., Sance, M.M., Peralta, I.E., Carrari, F., and Asia, R. (2018). Neuronal network analyses reveal novel associations between volatile organic compounds and sensory properties of tomato fruits. *Metabolomics* 14, 57. https://doi.org/10.1007/s11306-018-1355-7

Esmaeily, R., Razavi, M.A., and Razavi, S.H. (2024). A step forward in food science, technology and industry using artificial intelligence. *Trends Food Sci. Technol.* 143, 104286. https://doi.org/10.1016/j.tifs.2023.104286

Folta, K.M., and Klee, H.J. (2016). Sensory sacrifices when we mass-produce mass produce. *Hortic. Res.* 3, 16032. https://doi.org/10.1038/hortres.2016.32

Fritz, K.R., Dean, L.L., Hendrix, K.W., Andres, R.J., Newman, C.S., Oakley, A.T., et al. (2022). Flavor quality and composition of accession resources in the North Carolina State University peanut breeding program. *Crop Sci.* 62, 1880–1890. https://doi.org/10.1002/csc2.20774

Hagenguth, J., Kanski, L., Kahle, H., Becker, H.C., and Horneburg, B. (2024). Flavour improvement in early generations of fresh market tomatoes (*Solanum lycopersicum*): I. Identification of QTL for sensory attributes, physicochemical measurements and volatile compounds. *Plant Breed.* 143, 334–349. https://doi.org/10.1111/pbr.13168

Hagenguth, J., Kanski, L., Kahle, H., Naumann, M., Pawelzik, E., Becker, H.C., et al. (2022). Breeders' sensory test: A new tool for early selection in breeding for tomato (*Solanum lycopersicum*) flavour. *Plant Breed.* 141, 96–107. https://doi.org/10.1111/pbr.12994

Hongsoongnern, P., and Chambers IV, E. (2008). A lexicon for texture and flavor characteristics of fresh and processed tomatoes. *J. Sens. Stud.* 23, 583–599. https://doi.org/10.1111/j.1745-459X.2008.00174.x

Imran, M., Liu, Y., Shafiq, S., Abbas, F., Ilahi, S., Rahman, N., et al. (2022). Transcriptional cascades in the regulation of 2-AP biosynthesis under Zn supply in fragrant rice. *Physiol. Plant.* 174, e13721. https://doi.org/10.1111/ppl.13721

Kim, D.-E., Kwak, H.-S., Lim, M., and Lee, Y. (2023). Comparison of check-all-that-apply (CATA), rate-all-that-apply (RATA), flash profile, free listing, and conventional descriptive analysis (DA) for the sensory profiling of sweet pumpkin porridge. *Foods* 12, 3556. https://doi.org/10.3390/foods12193556

Krinsky, B.F., Drake, M.A., Civille, G.V., Dean, L.L., Hendrix, K.W., and Sanders, T.H. (2006). The development of a lexicon for frozen vegetable soybeans (Edamame). *J. Sens. Stud.* 21, 644–653. https://doi.org/10.1111/j.1745-459X.2006.00088.x

Lara, S.W., and Tsiami, A. (2024). A lexicon of descriptive sensory terms for peas (*Pisum sativum* L.): A systematic review. *Foods* 13, 2290. https://doi.org/10.3390/foods13142290

Leksrisompong, P.P., Whitson, M.E., Truong, V.D., and Drake, M.A. (2012). Sensory attributes and consume acceptance of sweet potato cultivars with varying flesh colors. *J. Sens. Stud.* 27, 59–69. https://doi.org/10.1111/j.1745-459X.2011.00367.x

Li, Y., He, L., Song, Y., Zhang, P., Chen, D., Guan, L., et al. (2023). Comprehensive study of volatile compounds and transcriptome data providing genes for grape aroma. *BMC Plant Biol.* 23, 171. https://doi.org/10.1186/s12870-023-04191-1

Liao, L., Zhang, W., Zhang, B., Fang, T., Wang, X.-F., Cai, Y., et al. (2021). Unraveling a genetic roadmap for improved taste in the domesticated apple. *Mol. Plant.* 14, 1454–1471. https://doi.org/10.1016/j.molp.2021.05.018

Limpawattana, M., and Shewfelt, R.L. (2010). Flavor lexicon for sensory descriptive profiling of different rice types. *J. Food Sci.* 75, S199–S205. https://doi.org/10.1111/j.1750-3841.2010.01577.x

Lippi, N., Senger, E., Karhu, S., Mezetti, B., Cianciabella, M., Denoyes, B., et al. (2023). Development and validation of a multilingual lexicon as a key tool for the sensory analyses and consumer tests of blueberry and raspberry fruit. *Agriculture* 13, 314. https://doi.org/10.3390/agriculture13020314

Marques, C., Correia, E., Dinis, L.-T., and Vilela, A. (2022). An overview of sensory characterization techniques: From classical descriptive analysis to the emergence of novel profiling methods. *Foods* 11, 255. https://doi.org/10.3390/foods11030255

Mostafa, S., Wang, Y., Zeng, W., and Jin, B. (2022). Floral scents and fruit aromas: Functions, compositions, biosynthesis, and regulation. *Front. Plant Sci.* 13, 860157. https://doi.org/10.3389/fpls.2022.860157

Nakatumba-Nabende, J., Babirye, C., Tusubira, J.F., Mutegeki, H., Nabiryo, A.L., Murinadnyi, S., et al. (2023). Using machine learning for image-based analysis of sweetpotato root sensory attributes. *Smart Agric. Technol.* https://doi.org/10.1016/j.atech.2023.100291

Nakitto, M., Johanningsmeier, S.D., Moyo, M., Bugaud, C., de Kock, H., Dahdouh, L., et al. (2022). Sensory guided selection criteria for breeding consumer-preferred sweet potatoes in Uganda. *Food Qual. Prefer.* 101, 104628. https://doi.org/10.1016/j.foodqual.2022.104628

Oliver, P., Cicerale, S., Pang, E., and Keast, R. (2018). Developing a strawberry lexicon to describe cultivars at two maturation stages. *J. Sens. Stud.* 33, e12312. https://doi.org/10.1111/joss.12312

Porter, M., Fan, Z., Lee, S., and Whitaker, V.M. (2023). Strawberry breeding for improved flavor. *Crop Sci.* 1949–1963. https://doi.org/10.1002/csc2.21012

Qian, L., Jin, H., Yang, Q., Zhu, L., Yu, X., Fu, X., et al. (2022). A sequence variation in *GmBADH2* enhances soybean aroma and is a functional marker for improving soybean flavor. *Int. J. Mol. Sci.* 23, 4116. https://doi.org/10.3390/ijms23084116

Rodrigues, S.S.Q., Dias, L.G., and Teixeira, A. (2024). Emerging methods for the evaluation of sensory quality of food: Technology at service. *Curr. Food Sci. Technol. Rep.* 2, 77–90. https://doi.org/10.1007/s43555-024-00019-7

Sharma, C., Chambers IV, E., Jayanty, S.S., Sathuvalli, V., Kalyan, R., Holm, D.G., et al. (2020). Development of a lexicon to describe the sensory characteristics of a wide variety of potato cultivars. *J. Sens. Stud.* 35, e12577. https://doi.org/10.1111/joss.12577

Suwonsichon, S. (2019). The importance of sensory lexicons for research and development of food products. *Foods* 8, 27. https://doi.org/10.3390/foods8010027

Tadmor, Y., Fridman, E., Gur, A., Larkov, O., Lastochkin, E., Ravid, U., et al. (2002). Identification of malodorous, a wild species allele affecting tomato aroma that was selected against during domestication. *J. Agric. Food Chem.* 50, 2005–2009. https://doi.org/10.1021/jf011237x

Talavera-Bianchi, M., Chambers IV, E., and Chambers, D.H. (2010). Lexicon to describe flavor of fresh leafy vegetables. *J. Sens. Stud.* 25, 163–183. https://doi.org/10.1111/j.1745-459X.2009.00249.x

Uppili, B., LaFountain, L.J., and Johanningsmeier, S.D. (2023). A sensory lexicon to characterize the quality of fresh and preserved peppers. *J. Food Sci.* 88, 5216–5235. https://doi.org/10.1111/1750-3841.16814

Valverde, M.E., Hernández-Pérez, T., and Paredes-López, O. (2015). Edible mushrooms: Improving human health and promoting quality life. *Int. J. Microbiol.* 2015, 376387. https://doi.org/10.1155/2015/376387

Wang, R., Zhang, Y., Lu, H., Liu, J., Song, C., Xu, Z., et al. (2022). Comparative aroma profile analysis and development of a sensory aroma lexicon of seven different varieties of *Flammulina velutipes*. *Front. Nutr.* 9, 827825. https://doi.org/10.3389/fnut.2022.827825

Wang, S., Su, Q., Zhu, Y., Liu, J., Zhang, X., Zhang, Y., et al. (2024). Sensory-guided establishment of sensory lexicon and investigation of key flavor components for goji berry pulp. *Plants* 13, 173. https://doi.org/10.3390/plants13020173

Wu, G., Ross, C.F., Morris, C.F., and Murphy, K.M. (2017). Lexicon development, consumer acceptance, and drivers of liking of quinoa varieties. *J. Food Sci.* 82, 993–1005. https://doi.org/10.1111/1750-3841.13677

Zulkarnain, A.H.B., Kókai, Z., and Gere, A. (2024a). Immersive sensory evaluation: Practical use of virtual reality sensory booth. *MethodsX* 12, 102631. https://doi.org/10.1016/j.mex.2024.102631

Zulkarnain, A.H.B., Kókai, Z., and Gere, A. (2024b). Assessment of a virtual sensory laboratory for consumer sensory evaluations. *Heliyon* 10, e25498. https://doi.org/10.1016/j.heliyon.2024.e25498

5 Cataloguing and Factoring Chefs' and Consumers' Preferences as a Guide for Developing New Cultivars with Unique Nutritional and Health-Promoting Characteristics

Rodomiro Ortiz and Sangam L. Dwivedi

5.1 INTRODUCTION

The world is currently undergoing a transformative shift in dietary habits and food preferences, marked by a heightened emphasis on health and nutrition (Reyes et al., 2021). In response to this pervasive trend, the agrifood industry is increasingly investing in pioneering innovative approaches to grow new cultivars and eat healthy diets deriving for their use. These endeavors aim not only to meet the evolving nutritional needs of consumers but also to align with their culinary preferences (Stetkiewicz et al., 2023). This chapter endeavors to underscore the paramount importance of meticulously cataloguing and integrating the preferences of chefs and consumers alike into the developmental process of novel cultivars, endowed with distinct nutritional profiles and health-enhancing attributes. It delves into the dynamic convergence of culinary creativity, agricultural science, and consumer demands, envisioning a future characterized by sustainable, health-conscious food choices.

Reyes et al. (2021) identified 13 emerging subjects that are influencing global nutrition initiatives, thereby providing a comprehensive framework for addressing the multifaceted challenges in food systems. These subjects encompass a diverse array of strategies and interventions, ranging from prioritizing agricultural production

DOI: 10.1201/9781003467335-5

across a spectrum of nutrient-rich foods to fostering links between smallholders and small and medium-sized enterprises (SMEs) in agrifood value chains.

Among the key focal points highlighted by Reyes et al. (2021) is the imperative to safeguard nutrient-rich wild foods and species-rich ecosystems, both on land and in oceans. This underscores the critical role of biodiversity conservation in ensuring the availability of diverse and nutritious food sources. Additionally, the reengineering of safety net or social protection programs to enhance nutritional outcomes reflects a concerted effort to address food insecurity and malnutrition through targeted interventions. Furthermore, efforts to reduce food waste and loss, coupled with initiatives aimed at improving food quality and safety, are integral components of the broader agenda to promote sustainable and healthy food systems. Toughening regulations for the advertising and marketing of food products, alongside initiatives to increase transparency in food labeling, are pivotal in empowering consumers to make informed dietary choices. Moreover, incentivizing healthy eating behaviors through subsidies and promotional campaigns, while simultaneously discouraging the consumption of unhealthy products through taxation, represents a multifaceted approach to shaping consumer preferences and behaviors. The development of consumer demand for healthy foods through nutrition education and civic engagement serves to reinforce the importance of fostering a culture of health and wellness within communities. Additionally, initiatives aimed at boosting the acceptability of healthy diets, promoting traditional foods and processing methods that offer nutritional benefits, and investing in metrics research to inform policymaking further underscore the holistic nature of efforts to advance global nutrition initiatives. By addressing these emerging subjects comprehensively, stakeholders can collaboratively work toward building more resilient, equitable, and sustainable agrifood systems that prioritize the health and well-being of present and future generations.

Over the years, the cultivation and consumption of crops have undergone profound transformations, driven by shifting dietary paradigms and an escalating awareness of health and wellness. Amid the quest for dietary alternatives that are both nutritious and appealing, the collaborative synergy among chefs, farmers, and consumers emerges as a pivotal force. As elucidated by Benson (2019), chefs hold a profound appreciation for the versatility of potatoes within the culinary realm, often regarding them as a 'blank palette for creativity'. The potato, with its myriad culinary possibilities, serves as a cornerstone in crafting diets that are not only wholesome but also exquisitely flavorful. Thus, it is within the intricate interplay of these stakeholders that the foundation for innovative cultivar development is laid.

The ensuing sections will meticulously explore the symbiotic relationship between chefs, farmers, and consumers, elucidating how their collective preferences profoundly influence the trajectory of cultivar innovation. By cataloguing and attentively incorporating their insights and preferences, the agricultural landscape can evolve in tandem with the ever-changing culinary and nutritional landscape, fostering a harmonious equilibrium between gastronomic delight and nutritional well-being.

5.2 THE ROLE OF CHEFS IN CULTIVAR DEVELOPMENT

The influence of cooking shows on dining decisions, as highlighted by Gajdzik et al. (2023), emphasizes the significant impact chefs wield over consumers' food choices, especially when taste and flavor outweigh health concerns in decision-making. Indeed, chefs play a crucial role in conserving plant genetic resources by bridging the gap between consumers and agriculture, thereby introducing novel plant-derived foods into diets (Moreau and Speight, 2019). Within their culinary domains, chefs possess a unique ability to forge connections between seeds, farms, and plates through innovative means. Leveraging their culinary expertise and influence, chefs can effectively advocate for the utilization of diverse and sustainable ingredients, thereby contributing to biodiversity preservation and the promotion of healthier eating habits. This symbiotic relationship has the potential to cultivate fresh, sustainable, and health-conscious consumption patterns that harness the biodiversity inherent within agrifood systems worldwide. Furthermore, chefs demonstrate a keen focus on prioritizing food quality, taste, and preferences, often exhibiting a strong preference for locally sourced foods (Curtis and Cowee, 2009). This steadfast commitment to quality and locality not only enriches the gastronomic experience for consumers but also fosters sustainability by bolstering regional agriculture and safeguarding heirloom food cultivars.

As noted by Brouwer et al. (2015), the selection of germplasm tailored to local production and end-users aids in constructing resilient agrifood systems. By selecting germplasm adapted to local conditions and preferences, agrifood systems become more resilient to environmental challenges and market fluctuations. This approach promotes the preservation of traditional agricultural practices and cultivars while also ensuring food security and promoting biodiversity conservation.

5.2.1 Gastronomic Modernization

Chefs are unequivocally at the vanguard of culinary innovation, continuously pushing the envelope when it comes to exploring novel flavors, textures, and presentation techniques. With a keen eye on consumer trends, they tirelessly endeavor to capture the essence of 'big flavors in small bites', anticipating and responding to the evolving palate preferences (Benson, 2019). Indeed, chefs wield considerable influence in shaping the culinary landscape, dictating what consumers seek out in their everyday dining experiences.

Their unparalleled expertise in harmonizing diverse ingredients to craft dishes that are not only aesthetically pleasing but also sensorially gratifying renders them invaluable partners in the endeavor to develop new cultivars. In this capacity, chefs serve as conduits between the realms of gastronomy and agriculture, offering invaluable insights into the characteristics that can elevate the culinary and nutritional profiles of crops. Through their discerning palate and deep understanding of flavor profiles, aromas, and textures, chefs can effectively articulate their preferences, thereby guiding plant breeders in refining cultivars to align with the exacting standards of the culinary world. Moreover, chefs play a pivotal role in driving market demand, wielding their influence to champion the adoption of novel cultivars among both their peers and consumers alike. By showcasing the culinary potential of newly developed cultivars, chefs stimulate interest and curiosity, thereby catalyzing a ripple effect that permeates throughout the food industry. Consequently, farmers are incentivized to grow these innovative crops, knowing that there exists a receptive market eager to embrace new gastronomic experiences. Thus, the symbiotic relationship between chefs, farmers, and consumers emerges as a potent catalyst for fostering a culture of innovation and experimentation within the realm of agriculture, ultimately propelling us toward a more vibrant and diverse food landscape.

5.2.2 Collaborative Partnerships

Such cooperation between chefs and plant breeders is becoming more common, leading to a dynamic exchange of knowledge. Chefs provide insights into emerging food trends, flavor profiles, and cooking techniques, while plant breeders offer their expertise in cultivar development. This synergy can result in the creation of cultivars that not only thrive in different agricultural environments but also align with the creative vision of chefs. The pursuit of a better quality of life by consumers has led to an increased demand for vegetables with higher nutritional and sensory qualities. However, consumer dissatisfaction with the flavor quality of many modern cultivars, such as tomatoes for fresh markets, has sparked breeders' interest in improving sensory quality (Sinesio et al., 2021). Additionally, there has been a growing demand for traditional heirloom cultivars, which are often associated with better aroma, flavor, and taste. To address these concerns and meet the preferences of growers, consumers, and chefs, who all prioritize sustainable agrifood systems, an interdisciplinary assessment is crucial. This assessment should encompass agronomic, physicochemical, and sensory characteristics to develop new cultivars that align with the desired qualities. By integrating diverse perspectives and expertise, breeders can develop cultivars that not only excel in terms of yield and disease resistance but also offer superior sensory attributes, such as flavor and aroma. This approach ensures that the resulting cultivars meet the expectations of all stakeholders along the agrifood chain, ultimately contributing to the promotion of sustainable and nutritious agrifood systems (Constantino et al., 2022).

5.3 CONSUMER PREFERENCES AND HEALTH CONSIDERATIONS

Consumers' attitudes and preferences toward various products are influenced by several factors, including age, gender, and education level (Wendin et al., 2020). These demographic characteristics play significant roles in shaping consumer behavior and decision-making processes. Different age groups often have distinct preferences and priorities when it comes to product selection. For example, younger consumers might prioritize trends, convenience, and novelty, while older consumers may prioritize quality, reliability, and familiarity. Gender can also impact consumer preferences, with research indicating that males and females may have different tastes, interests, and buying habits. Marketers often tailor their strategies to appeal to specific gender demographics based on these differences. Likewise, education level may influence consumers' knowledge, awareness, and attitudes toward products. Individuals with higher levels of education may be more discerning and informed consumers, seeking products that align with their values, health considerations, or environmental concerns. Understanding how these factors intersect can help plant-breeding enterprises to tailor their product offerings to better meet the diverse needs and preferences of their target audience.

Country-based product profiles for developing new root, tuber and banana (RTB) cultivars were established through a comprehensive market analysis, considering gender roles and preferences, as well as prioritizing food quality traits (Arnaud et al., 2023). The analysis incorporated a variety of data types, including biochemical, instrumental textural, and sensory data, alongside survey responses. This multifaceted approach provided valuable insights for guiding RTB breeding efforts.

By integrating diverse perspectives and expertise, breeders can breed cultivars that not only excel in terms of edible yield and host plant resistance to pathogens and pests or tolerance to abiotic factors but also exhibit superior nutritional profiles and sensory attributes. This holistic approach ensures that the resulting cultivars meet the expectations of all stakeholders involved in the agrifood chain, from producers to end consumers. Furthermore, by emphasizing sustainability, breeders can contribute to the long-term viability of agricultural practices while addressing the evolving needs and preferences of consumers in pursuit of healthier and more satisfying food options. In this regard, Kelley et al. (2001) utilized surveys to evaluate the perceptions of consumer and professional chefs regarding edible-flower species, considering their taste, flavor, and appearance. Their study revealed only minor rating differences between the two groups, thus indicating the feasibility of employing a homogeneous marketing approach.

5.3.1 Changing Consumer Trends

Consumers are increasingly conscious of their dietary choices, seeking products that align with their health and wellness goals. Preferences for organic, locally sourced, and sustainably produced foods are on the rise. Understanding these trends is crucial for developing cultivars that resonate with the values of today's consumers.

5.3.2 Nutritional Requirements

Cataloguing nutritional preferences is paramount in cultivar development. Consumers are more informed about the nutritional content of their food, and cultivars with enhanced nutritional profiles, such as higher levels of antioxidants, vitamins, and minerals, are likely to gain popularity. Incorporating these preferences into the breeding process ensures that new cultivars not only taste good but also contribute to overall health and well-being.

5.4 INTEGRATING CHEFS' AND CONSUMERS' PREFERENCES IN CULTIVAR DEVELOPMENT

Plant-breeding programs often prioritize traits such as edible yield and host plant resistance to pathogens and pests or tolerance to abiotic factors, thus neglecting consumer demand for quality traits. However, this approach can result in low adoption levels of released cultivars. To address this gap, Forsythe et al. (2021) propose an interdisciplinary five-step approach to identify demand for quality traits among diverse users along the agrifood value chain, particularly focusing on RTB crops. Their methodology includes an evidence review, consultations with key informants and rural communities, processing diagnosis with experienced processors, and consumer testing in both urban and rural areas. Following these steps, the quality characteristics undergo prioritization by the user groups to develop a product profile that guides selection. Additionally, sensory methods, as advocated by Bechoff et al. (2023), should be employed to evaluate both quality traits and their acceptability by RTB user groups. These methods include a combination of qualitative approaches such as surveys and rankings, as well as quantitative sensory tests conducted with a trained panel. It is essential to note that some sensory traits, such as texture, can be assessed using laboratory equipment, while others like aroma or mealiness require evaluation by a sensory panel due to the difficulty in predicting them, as high-throughput methods for studying them are lacking.

Chefs are contributing valuable perspectives to vegetable breeding initiatives in recent years. Recognizing that the quality of their dishes is closely linked to the ingredients they use, chefs have advocated for cultivars with improved food quality, making their culinary endeavors more fulfilling and successful (Dwivedi et al., 2019). In some cases, consumers and chefs have directly influenced cultivar development. For instance, Beans (2017) highlights examples such as a new tomato cultivar that emerged from a cross between a modern breeding line and an heirloom cultivar, identified through a taste panel involving 100 consumers. Similarly, a squash cultivar was developed through extensive dialogue between a plant breeder and a chef, who trialed the product in restaurants during its development stages. These collaborative breeding efforts represent a departure from traditional plant-breeding practices, expanding the scope to encompass input from end-users, such as consumers and chefs. Such partnerships are likely to yield cultivars that are not only more appealing to consumers but also possess enhanced culinary quality. Ultimately, this approach fosters innovation in vegetable breeding and contributes to the development of cultivars that meet the evolving needs and preferences of both growers and consumers.

5.4.1 Genetic Modification and Traditional Breeding Techniques

The development of new cultivars involves a combination of genetic modification and traditional breeding techniques. Plant breeders use advanced methods to manage the genetic makeup of crops, enhancing desirable traits. However, it is equally essential to preserve the authenticity of flavors and textures through crossbreeding, ensuring that cultivars maintain their natural appeal.

5.4.2 Sensory Analysis and Consumer Trials

Sensory analysis, including taste tests, aroma evaluations, and texture assessments, plays a crucial role in validating cultivar development. Engaging consumers in trials provides valuable feedback on the sensory attributes of new cultivars. This iterative process allows for adjustments based on real-world preferences, ensuring that the final products meet both culinary and consumer expectations.

5.5 CHALLENGES AND PROSPECTS

Projects like the Seed to Kitchen Collaborative (SKC) exemplify the benefits of including various stakeholders from the agrifood system, including chefs, farmers, and the general public ('eaters'), in the priority setting for target traits in breeding endeavors (Hill, 2020). SKC utilizes participatory plant-breeding and participatory varietal selection, incorporating both local perspectives and knowledge of plant-breeding techniques. The primary objective is to develop flavorful vegetable cultivars that are well-suited to low-input systems. Partnerships with growers are integral to the success of SKC, as they serve as primary consumers of the information produced by the project. Additionally, other relevant end-users, such as eaters and chefs, provide valuable feedback on the aroma, taste, flavor, texture, and overall sensory experience of newly bred germplasm and potential cultivars. This collaborative approach ensures that the resulting cultivars meet the needs and preferences of all stakeholders involved, thereby leading to the development of more desirable and marketable vegetable varieties.

A genuinely participatory plant-breeding initiative should involve collaboration among researchers, farmers, and end-users at every stage of the breeding process, aiming to address the needs of the agrifood system (Colley et al., 2021). This inclusive approach is likely to enhance both crop diversity and farmers' access to newly developed germplasm. It fosters a holistic understanding of the diverse requirements and preferences within agricultural communities, leading to the development of cultivars that are better suited to local conditions and consumer demands. Furthermore, an interdisciplinary mindset is essential to prevent overlooking the influence of social, political, and agricultural factors on the success of plant-breeding efforts aimed at delivering healthy and nutritious cultivars. By incorporating perspectives from various disciplines, such as agronomy, sociology, and political science, breeders can develop a more nuanced understanding of the broader context in which their work operates, thus increasing the likelihood of producing cultivars that contribute positively to sustainable agrifood systems and public health.

Plant-breeding initiatives aimed at resource-poor farmers and end-users can significantly benefit from gender and social targeting strategies, as highlighted by Orr et al. (2018). Often, the lack of information about the target demographic poses challenges for plant-breeding enterprises in prioritizing traits for genetic improvement, leading to a disconnect between the intended users and the outputs of their breeding efforts, such as newly bred germplasm and cultivars. Segmenting the market by defining end-user groups with homogeneous preferences can address this issue effectively. By understanding the specific needs and preferences of different user segments, plant breeders can tailor their breeding programs to better meet the demands of their target audience. This targeted approach facilitates the positioning of newly bred germplasm and potential cultivars in the market, ensuring that they align closely with the preferences of their intended users. Ultimately, this strategy enhances the adoption and success of new cultivars among resource-poor farmers and end-users.

5.5.1 Balancing Flavor and Nutrition

One of the challenges in developing new cultivars is striking the right balance between flavor and nutrition. While enhancing nutritional content is crucial, cultivars must not compromise on taste and culinary versatility. Achieving this delicate balance requires a nuanced understanding of the interactions between different compounds in crops.

5.5.2 Globalization and Cultural Considerations

As the world becomes more interconnected, cultivar development must consider global culinary trends and cultural preferences. What appeals to chefs and consumers in one region may differ significantly from another. Adapting cultivars to diverse cultural contexts ensures their widespread acceptance and adoption.

5.5.3 Sustainable Agriculture

Sustainability is a key consideration in cultivar development. The ecological impact of cultivation practices and the resilience of crops to changing environmental conditions are essential factors. Integrating sustainable agriculture practices into cultivar development contributes to long-term food security and environmental health.

5.6 CONCLUSION

Cataloguing and factoring of chefs' and consumers' preferences are integral to the development of new cultivars with unique nutritional and health-promoting characteristics. The collaboration between culinary experts, plant breeders, and consumers creates a synergistic approach that ensures cultivars are not only scientifically advanced but also resonate with the diverse tastes and health-conscious choices of the modern world. As we navigate the future of food, this cooperative model holds the key to cultivating a sustainable and flavorful food landscape.

REFERENCES

Arnaud, E., Menda, N., Tran, T., Asiimwe, A., Kanaabi, M., Meghar, K., et al. (2023). Connecting data for consumer preferences, food quality, and breeding in support of market-oriented breeding of root, tuber, and banana crops. *J. Sci. Food Agric.* 104, 4514–4526. https://doi.org/10.1002/jsfa.12710

Beans, C. (2017). Vegetable breeders turn to chefs for flavor boost. *Proc. Natl. Acad. Sci. U. S. A.* 114, 10506–10508. https://doi.org/10.1073/pnas.1714536114

Bechoff, A., Adinsi, L., Ngoh Newilah, G., Nakitto, M., Deuscher, Z., Ssali, R., et al. (2023). Combined use of sensory methods for the selection of root, tuber and banana varieties acceptable to end-users. *J. Sci. Food Agric.*104, 4700–4708. https://doi.org/10.1002/jsfa.12723

Benson, L. (2019). A chef's perspective on potatoes and diet. *Am. J. Potato Res.* 96, 98–99. https://doi.org/10.1007/s12230-019-09718-7

Brouwer, B.O., Murphy, K.M., and Jones, S.S. (2015). Plant breeding for local food systems: A contextual review of end-use selection for small grains and dry beans in Western Washington. *Renew. Agric. Food Syst.* 31, 172–184. https://doi.org/10.1017/S1742170515000198

Colley, M.R., Dawson, J.C., McCluskey, C., Myers, J.R., Tracy, W.F., and Lammerts van Bueren, E.T. (2021). Exploring the emergence of participatory plant breeding in countries of the Global North—A review. *J. Agric. Sci. (Cambridge)* 159, 320–338. https://doi.org/10.1017/S0021859621000782

Constantino, L.V., de Araujo, S.R., Fukuji, A.S.S., Nogueira, A.F., Barros de Lima Filho, R., and Zeffa, D.M. (2022). Post-harvest characterization and sensory analysis of Roma tomato cultivars under organic cultivation: A strategy using consumers and chefs. *Int. J. Gastron. Food Sci.* 29, 100564. https://doi.org/10.1016/j.ijgfs.2022.100564

Curtis, K.R., and Cowee, M.W. (2009). Direct marketing local food to chefs: Chef preferences and perceived obstacles. *J. Food Distrib. Res.* 40, 1–11. https://doi.org/10.22004/ag.econ.99784

Dwivedi, S., Goldman, I., and Ortiz, R. (2019). Pursuing the potential of heirloom cultivars to improve adaptation, nutritional, and culinary features of food crops. *Agronomy* 9, 441. https://doi.org/10.3390/agronomy9080441

Forsythe, L., Tufan, H., Bouniol, A., Kleih, U., and Fliedel, G. (2021). An interdisciplinary and participatory methodology to improve user acceptability of root, tuber and banana varieties. *Int. J. Food Sci. Technol.* 56, 1115–1123. https://doi.org/10.1111/ijfs.14680

Gajdzik, B., Jaciow, M., Wolniak, R., and Wolny, R. (2023). Gastronomic curiosity and consumer behavior: The impact of television culinary programs on choices of food services. *Foods* 13, 115. https://doi.org/10.3390/foods13010115

Hill, K.M. (2020). *Stakeholder Roles in Participatory Plant Breeding: Perspectives from the Seed to Kitchen Collaborative*. MSc Thesis. University of Wisconsin—Madison, Madison, WI.

Kelley, K.M., Behe, B.K., Biernbaum, J.A., and Poff, K.L. (2001). Consumer and professional chef perceptions of three edible-flower species. *HortScience* 3, 162–166. https://doi.org/10.21273/HORTSCI.36.1.162

Moreau, T., and Speight, D. (2019). Cooking up diverse diets: Advancing biodiversity in food and agriculture through collaborations with chefs. *Crop Sci.* 59, 2381–2386. https://doi.org//10.2135/cropsci2019.06.0355

Orr, A., Cox, C.M., Ru, Y., and Ashby, J. (2018). Gender and social targeting in plant breeding. *GBI Working Paper. No. 1*. CGIAR Gender and Breeding Initiative, Lima Perú. www.rtb.cgiar.org/gender-breeding-initiative

Reyes, L.I., Constantinides, S.V., Bhandari, S., Frongillo, E.A., Schreinemachers, P., Wertheim-Heck, S., et al. (2021). Actions in global nutrition initiatives to promote sustainable healthy diets. *Glob. Food Sec.* 21, 100585. https://doi.org/10.1016/j.gfs.2021.100585

Sinesio, F., Cammareri, M., Cottet, V., Fontanet, L., Jost, M., Moneta, E., et al. (2021). Sensory traits and consumer's perceived quality of traditional and modern fresh market tomato varieties: A study in three European countries. *Foods* 10, 2521. https://doi.org/10.3390/foods10112521

Stetkiewicz, S., Menary, J., Nair, A., Rufino, M., Fischer, A.R.H., Cornelissen, M., et al. (2023). Food system actor perspectives on future-proofing European food systems through plant breeding. *Sci. Rep.* 13, 5444. https://doi.org/10.1038/s41598-023-32207-1

Wendin, K., Mustafa, A., Ortman, T., and Gerhardt, K. (2020). Consumer awareness, attitudes and preferences towards heritage cereals. *Foods* 9, 742. https://doi.org/10.3390/foods9060742

6 Crop Wild Relatives, Landraces, and Heirloom Germplasm

The Source of Novelty for Taste and Flavor

Rodomiro Ortiz and Sangam L. Dwivedi

6.1 INTRODUCTION

Crop wild relatives, landraces, and heirloom germplasm represent invaluable components of the agricultural heritage, each contributing uniquely to the genetic diversity of crops. This chapter explains the significance of these genetic resources as sources of novelty, specifically in enhancing taste and flavor in crops. Their intricate interplay of genetic diversity offers a rich tapestry of flavors that have sustained cultures and intrigued palates for generations.

Crop wild relatives, landraces, and heirloom germplasm contribute to maintaining and expanding the genetic diversity of crops. This diversity is crucial for adapting to changing environmental conditions and related stress, pests, and pathogens. Landraces and heirloom germplasm, in particular, have adapted over time to local climates and soil conditions (Medina-Lozano and Díaz, 2021). This adaptation ensures the resilience of crops and supports sustainable agriculture. Heirloom germplasm carries cultural significance, representing the agricultural practices, culinary traditions, and historical context of a community or region. Preserving heirlooms is essential for maintaining cultural diversity in agriculture.

The use of locally adapted landraces and heirloom cultivars promotes sustainable agriculture by reducing the dependence on external inputs. Crop wild relatives are often used in breeding programs to introduce beneficial traits such as disease resistance or drought tolerance into cultivars (Flint-García et al., 2023) Landraces and heirlooms also offer valuable genetic material for breeding efforts (Dwivedi et al., 2019), thereby contributing to crop improvement. inputs and fostering agricultural systems that are in harmony with the local ecosystem. Hence, crop wild relatives, landraces, and heirloom germplasm collectively represent a reservoir of genetic diversity that is critical for the sustainability and adaptability of agriculture. Recognizing and preserving these genetic resources are essential for addressing current and future challenges in food security and agricultural sustainability.

 DOI: 10.1201/9781003467335-6

6.2 UNDERSTANDING CROP WILD RELATIVES

Crop wild relatives (CWR) are the wild progenitors of cultivated crops, harboring a reservoir of genetic diversity forged by evolution (Dwivedi et al., 2008). They share common ancestry with domesticated plants but have not been subject to intensive human selection. These plants thrive in natural ecosystems, adapting to diverse environmental conditions. CWR are reservoirs of genetic diversity, harboring traits that can be valuable for crop improvement (Haddad et al., 2021). They play a crucial role in providing genetic resources for developing crops with enhanced resilience to pests, diseases, and environmental stresses. As climate change and evolving agricultural challenges pose threats to food security, the genetic diversity found in CWR becomes increasingly important for developing more robust and adaptable cultivars. Likewise, the genetic makeup of CWR often includes unique traits that can influence taste and flavor in their domesticated counterparts. For instance, the wild relatives of tomatoes (*Solanum lycopersicum*) exhibit traits such as increased sugar content and unique aromatic compounds, contributing to the development of tastier bred cultivars (Bai and Lindhout, 2007).

CWR have evolved in natural environments and harbor genetic diversity that can contribute unique flavors and traits to their cultivated counterparts. While the taste and flavor of crops are complex traits influenced by various factors, including environmental conditions, the following are given some details of key CWR and their potential genetic contributions to taste and flavor.

Wild tomatoes: They are known for their small size and intense flavor. For example, *Solanum pimpinellifolium* contributes to the genetic diversity of cultivated tomatoes (*Solanum lycopersicum*) by providing genes associated with enhanced sugar content, unique aromatic compounds, and a more robust, tangy flavor (Li et al., 2023). These traits contribute to the development of flavorful tomato cultivars.

Wild strawberry: Fragaria vesca is a close relative of cultivated strawberry (*Fragaria × ananassa*). Wild strawberries are often appreciated for their intense sweetness and aromatic profile. Through crossbreeding with the cultigen pool, the genetic diversity of wild strawberry germplasm contributes to improving the flavor, sweetness, and aroma of strawberry cultivars (Negri et al., 2015).

Wild blueberry: Vaccinium angustifolium has a distinct flavor profile characterized by a sweet-tart taste. The wild relatives of cultivated blueberry germplasm contribute genetic diversity that can influence the balance of sugars and acids, contributing to a more complex and desirable flavor in blueberry cultivars (Cabezas et al., 2021).

Wild chili pepper: Capsicum annuum var. *glabriusculum* is known for their heat and unique flavor compounds. The genetic diversity of wild chili peppers contributes to the variation in capsaicin content, which influences the pungency and flavor intensity (Comparini et al., 2021). Incorporating traits from *Capsicum* CWR can lead to the development of chili cultivars with distinctive flavors.

Wild apple: Malus sieversii, the wild apple native to Central Asia, is considered the primary ancestor of cultivated apples. Wild apples contribute genetic diversity related to fruit texture, acidity, and aromatic compounds (Ma et al., 2015). Traits from wild apples influence the taste, texture, and flavor complexity in modern apple cultivars.

Wild rice: Zizania spp. is a close relative of cultivated rice. Wild rice germplasm often has a nutty and robust flavor (Yu et al., 2020). The genetic diversity of wild rice can contribute to the development of rice cultivars with enhanced aromatic compounds, nuttiness, and a unique flavor profile.

6.3 GENETIC TRAITS OF LANDRACES, HERITAGE CULTIVARS, AND CROP WILD RELATIVES

It is essential to note that the genetic contributions from CWR are complex and involve multiple genes influencing various traits. Breeding programs often use these CWR to introduce desirable characteristics, including taste and flavor, into cultivated crops. The successful integration of these traits requires careful selection and breeding efforts to ensure the development of crops with improved sensory quality.

6.3.1 Exploring Landraces

Landraces are traditional, locally adapted cultivars that have evolved through generations of farmers' selection. These cultivars have adapted to specific agro-climatic conditions and often carry unique genetic traits, thus resulting also in a rich array of flavors. For example, the ancient bere barley, which was replaced by modern high-yielding cultivars, has been getting recent attention due to its ability to withstand poor soils (high pH, salinity) as well as producing high-value foods and drinks (Martin et al., 2023). Landraces are also significant for their cultural and historical importance, representing a living heritage of agricultural biodiversity. They contribute to food security by providing farmers with crops that are well-suited to local conditions. In this regard, Sathya (2013) provides an account on how ancient rice landraces were used due to their diversity in size and color, and how today they are sources for breeding nutritious cultivars for a changing climate.

As described here, landraces offer a source of genetic diversity that can be tapped into for breeding programs, ensuring the development of crops tailored to specific ecological niches. The adaptability of landraces to local environments often involves the expression of unique genetic traits influencing taste and flavor (Cebolla-Cornejo et al., 2013). The famous basmati rice (*Oryza sativa*) landrace, grown in the Indian subcontinent, is celebrated for its distinctive aroma and long grain, illustrating how landraces contribute to culinary excellence. Moreover, landrace chili peppers showcase a wide range of heat levels, flavors, and aromas. The genetic diversity within these landraces influences the production of capsaicin, the compound responsible for heat, as well as other flavor compounds. Different landraces contribute to the diverse and nuanced tastes of chili peppers, allowing for a variety of culinary uses and preferences. Likewise, wheat landraces contribute to the flavor and texture of bread (Frankin et al., 2023). The genetic diversity within these landraces influences factors, such as gluten content, protein composition, and starch characteristics, thus impacting the taste, texture, and aroma of bread. Traditional bread made from landraces of wheat often exhibit unique flavors that are cherished in local cuisines.

Aroma and volatile compounds: The genetic diversity within landraces contributes to the variability in the production of aromatic compounds and volatile substances. These compounds, such as terpenes and esters, are responsible for the unique aromas that characterize different landraces. For example, basmati rice is known for its distinctive aroma, long grains, and fluffy texture. Landraces of aromatic rice such as basmati exhibit variation in the expression of genes related to aroma. The specific combination of these genes results in the distinct fragrance that sets basmati rice apart.

Sugar content and sweetness: The genetic diversity within landraces influences the composition of sugars, such as sucrose, fructose, and glucose, in fruits and vegetables. Variations in sugar content contribute to differences in sweetness and overall flavor. It is known that landraces of apples can exhibit a wide range of sugar profiles (Ma et al., 2015). This genetic diversity influences the balance between sweetness and tartness, defining the unique flavor of each apple type.

Acidity and tartness: Genetic diversity within landraces also affects the production of organic acids, such as citric acid and malic acid. Variation in acidity contributes to the tartness or sourness of fruits and vegetables. Tomato landraces, for example, can vary in their acidity levels and sugar content, as well as the presence of unique flavor compounds due to genetic differences (Villena et al., 2023). Some landraces may have a tangier flavor, while others might be milder in acidity. Each heirloom variety offers therefore a distinctive taste experience, from sweet and fruity to tangy and complex, showcasing the impact of genetic diversity on taste enhancement

Flavor compounds and nutritional content: The genetic makeup of landraces influences the production of various flavor compounds, including phenolic compounds and flavonoids. These compounds not only contribute to taste but also play a role in the nutritional content of crops. For example, strawberry landraces exhibit genetic diversity that affects the levels of antioxidants and flavonoids. This diversity contributes to the unique taste and nutritional benefits of the different landraces.

Textural traits and mouthfeel: Genetic variation within landraces influences textural traits, such as crispness, juiciness, or tenderness. These traits contribute to the overall mouthfeel and eating experience. Potato landraces can have diverse textures when cooked because of their variation in starch content. Some may be mealy, while others are waxy, impacting the overall taste experience. The genetic diversity within landrace potatoes influences their taste and cooking characteristics, contributing to a broad spectrum of flavors and textures in culinary applications.

Adaptation to local climates and effects on flavor: Landraces are often adapted to specific local climates, and their genetic diversity reflects adaptations to these conditions. This adaptation can influence the timing of flowering, ripening, and the synthesis of flavor compounds. It is worth highlighting that wine grape landraces exhibit genetic diversity that contributes to their adaptation to specific terroirs. The interaction between the grapevine and the local environment influences the flavor profile (sugar levels, acidity, and synthesis of flavor compounds) of the resulting wine. Hence, wines produced from different landrace varieties showcase the nuanced tastes and unique regional characteristics that are highly prized by connoisseurs.

These examples demonstrate the role of landraces in taste enhancement across different crops, emphasizing how genetic diversity within landraces contributes to the unique and culturally significant flavors in agricultural produce.

6.3.2 Unveiling Heirloom Germplasm

Heirloom germplasm comprises traditional, open-pollinated cultivars that have been passed down through relatives or neighbors in the community. Heirloom germplasm is often treasured for their historical significance, their cultural ties, and, importantly, their exceptional taste, unique flavors, diverse colors, and various shapes, thus representing a connection to agricultural practices of the past. Heirlooms may lack the uniformity of modern high-yielding hybrids, but their diverse genetic makeup gives rise to a spectrum of flavors, tastes, colors, and shapes. For example, heirloom tomatoes, with their eccentric shapes and vibrant colors, offer a taste experience that transcends the standardized offerings found in supermarkets. The cultivation and preservation of heirlooms contribute to the conservation of genetic resources and cultural identity. The following are details of some genetic traits commonly associated with the unique taste and flavor in heirloom crops:

Aromatic compounds: Genes responsible for the synthesis of aromatic compounds, such as terpenes and esters, contribute to the unique fragrance and aroma of heirloom crops. For example, heirloom tomatoes often contain genes that lead to the production of aromatic compounds, resulting in a rich, flavorful scent (Rodríguez-Burruezo et al., 2005). Similarly, aromatic herbs like basil or mint heirlooms exhibit distinct fragrance due to specific gene combinations.

Sugar content and brix levels: Genetic variation in sugar composition contributes to sweetness and overall flavor of heirloom germplasm. Heirloom watermelons, like the Moon and Stars variety, are known for their high sugar content, thereby resulting in exceptionally sweet and flavorful fruit. Similarly, heirloom apples may exhibit a diverse range of sugar profiles, thus influencing their taste (Lončarić et al., 2023).

Acidic and tart flavors: The expression of genes related to the production of organic acids, such as citric acid and malic acid, contributes to the acidic and tart flavors in heirloom cultivars. Certain heirloom apples, such as the Ashmead's Kernel, are known for their distinct tartness. Heirloom tomatoes (e.g., the Green Zebra) may exhibit a balance of acidity and sweetness, thereby creating a unique flavor profile.

Phenolic compounds and flavonoids: Genes control in heirloom germplasm the synthesis of phenolic compounds and flavonoids that contribute to their color, taste, and antioxidant features. Strawberry and blueberry heirlooms often have higher levels of anthocyanins and other flavonoids, providing not only unique flavors but also vibrant colors and potential health benefits.

Texture and mouthfeel: Genetics influences the texture and mouthfeel of heirloom cultivars. Factors such as starch composition, water content, and cell structure contribute to their overall eating experience. Heirloom purple potatoes from Peru have a distinct texture that sets them apart from modern cultivars grown elsewhere (https://kanchayperutravel.com/peru-potatoes/; accessed on October 18, 2024). Similarly, heirloom beans, such as the Scarlet Runner, exhibit unique textures and mouthfeel when cooked (https://hort.extension.wisc.edu/articles/scarlet-runner-bean-phaseolus-coccineus/; accessed on October 18, 2024). These beans are prized for their tender texture and rich, savory flavor. The beans have a unique taste that sets them apart, making them a favorite for both culinary and ornamental purposes.

Unique enzymes and metabolites: They are influenced by specific genes, which can contribute to complex and distinctive flavors in heirloom cultivars. For example, a garlic heirloom cultivar such as the Russian Red or German White (https://specialtyproduce.com/produce/German_White_Garlic_12798.php; accessed on October 18, 2024), may contain specific enzymes that produce unique sulfur compounds, thus contributing to their pungent and robust flavors.

There are numerous examples that showcase the remarkable diversity of flavors found in heirloom cultivars, emphasizing the unique and distinctive taste profiles that endear them to home gardeners, chefs, and food enthusiasts alike. Notably, Brandywine tomatoes, particularly Brandywine Pink, are esteemed for their rich, sweet, and complex flavor (https://specialtyproduce.com/produce/Brandywine_Heirloom_Tomatoes_1800.php). Exhibiting a harmonious balance of sweetness and acidity, they offer a robust taste highly sought after by tomato aficionados. Similarly, Yellow Pear tomatoes, characterized by their small, pear-shaped fruits, present a sweet and mild flavor. These heirloom tomatoes are often relished fresh in salads or as snacks due to their delightful sweetness. Furthermore, Cherokee Purple tomatoes are renowned for their deep, complex flavor, featuring a rich, smoky taste with a hint of sweetness. This makes them a popular choice for slicing and enjoying fresh in salads or sandwiches. In the same vein, Black Krim tomatoes are prized for their complex, smoky flavor, and deep, dark color. Possessing a distinctive taste with a perfect balance of sweetness and acidity, these heirloom tomatoes contribute to the diverse flavors found in this category. Moreover, Moon and Stars watermelons are celebrated for their exceptional sweetness and unique appearance. The deep-red flesh is sweet and succulent, offering a flavor often described as classic and reminiscent of old-fashioned watermelons. Likewise, French Breakfast radishes are esteemed for their mild, slightly peppery flavor. With elongated roots that boast a crisp and tender texture, their delicate taste makes them versatile for use in salads or as a snack. Equally noteworthy are Jimmy Nardello's peppers, which are Italian frying heirlooms known for their sweet, fruity flavor. Enjoyed when sautéed or grilled, their unique sweetness has made them a favorite among culinary devotees. While Honeycrisp apples are not traditional heirlooms, their popularity has surged due to their crisp texture and exceptionally sweet, juicy flavor. The unique combination of sweetness, tartness, and a burst of juice distinguishes them in the apple world. Finally, Delicata squash, also known as sweet potato squash, boasts a creamy and sweet flavor. With smooth and tender flesh, and edible skin, it offers a taste that is often described as nutty and rich.

The interplay of genetics for the characteristics in these heirloom cultivars of various crops contributes to the diversity and complexity of taste and flavor that characterize them. Preserving and celebrating their inherent genetic diversity is key to maintaining culinary heritage and ensuring a rich array of flavors for future generations.

6.4 GENETIC MECHANISMS INFLUENCING TASTE AND FLAVOR

The enhancement of taste and flavor in crops, facilitated by CWR, landraces, and heirloom germplasm, is intricately linked to underlying genetic mechanisms. Research has identified specific genes and pathways associated with flavor compounds, sugars, and volatile organic compounds in crops. These unique alleles and gene combinations found in CWR, landraces, and heirlooms can lead to the expression of novel traits, creating distinct flavors that captivate the senses. The following are given the main features of key genes and pathways associated with each of these components.

Aroma: There are genes encoding enzymes that ate responsible for the synthesis of terpenes, thereby contributing to the aromatic profile of crops. Other genes are also involved in pathways that produce aromatic precursors, such as phenylalanine, which play a crucial role in aroma development.

Taste: Sweetness (*SUC*) genes control the synthesis and transport of sugars in apples and grapes, thus influencing the perception of sweetness in these crops. Furthermore, there are genes involved in amino acid metabolism that contribute to the production of compounds that enhance savory and umami tastes. For example, amino acid metabolism pathways in tomatoes contribute to the umami taste.

Cell wall: Genes involved in the synthesis of cell wall components, such as cellulose and pectin, influence the texture of fruits and vegetables. The *EXPANSIN* family of genes is involved in cell wall modification (Sampedro and Cosgrove, 2005), influencing the crispness of apples and the tenderness of tomatoes. Genes such as *GBSS* in regulate starch synthesis in potato tubers, which impacts on their texture.

Flavor: Genes in the phenylpropanoid pathway regulate the production of phenolic compounds and flavonoids, contributing to flavor and color, while ethylene synthesis genes influencing ripening and flavor development in fruits.

Texture: Genes involved in the synthesis of cell wall components, such as cellulose and pectin, influence the texture of fruits and vegetables, while, as noted, genes regulating starch synthesis impact the texture of starchy crops like potatoes.

Capsaicin synthesis: Capsaicinoid biosynthesis genes, such as *Pun1* and *Pun2*, have a role in the biosynthesis of capsaicinoids, thereby contributing to the pungency of chili peppers. The *Pun1* gene is responsible for the production of capsaicin, determining the level of spiciness.

Umami taste: Glutamate receptor genes, such as *mGluR4*, are involved in the perception of umami taste. Genes regulating the metabolism of amino acids, particularly glutamic acid, also influence umami taste. For example, tomatoes, especially those rich in glutamic acid, contribute to umami taste.

Understanding the intricacies of the above genes and their related pathways provides valuable insights to plant-breeding programs aiming specific sensory attributes. Managing these genes leads to the development of crops with improved aroma, taste, texture, and flavor profiles, thus catering to diverse consumer preferences.

6.5 TASTE AND FLAVOR ENHANCEMENT IN SPECIFIC CROPS

Examining specific crops provides a deeper insight into how CWR, landraces, and heirloom germplasm contribute to taste and flavor. The diverse flavors of cocoa, for instance, are influenced by the wild relatives of cacao (*Theobroma cacao*), which harbor genes associated with the synthesis of compounds contributing to chocolate flavor. In the case of chili peppers (*Capsicum* spp.), landraces exhibit variations in capsaicin content, influencing the pungency and flavor intensity. These examples underscore the practical implications of harnessing genetic diversity for culinary delight.

The diversity in aromatic compounds depend on the germplasm. CWR exhibit diverse aromas based on adaptation to natural habitats; landraces reflect regional adaptations, with diverse aromatic profiles; and in heirloom cultivars, specific aromatic traits are preserved due to their cultural significance. Similarly, the taste and flavor diversity are primarily shaped in CWR by natural selection for survival. In contrast, regional taste preferences have been emphasized by centuries of farmers' selection in landraces, whereas culinary heritage has guided the selection for specific taste attributes in heirloom germplasm. Ecological adaptations define the texture of CWR, but the diverse texture of landraces has been shaped by regional crop husbandry, and cultural and historical factors account for the specific textures of heirloom germplasm. Elucidating the genetic factors influencing aroma, taste, flavor, and texture in CWR, landraces, and heirlooms have implications for plant breeding. Incorporating target traits from each category contributes to the development of crops with enhanced sensory characteristics, adaptability, and cultural relevance. Additionally, preserving their genetic diversity within each of them is essential for sustainable agriculture as well as food and nutrition security.

6.6 CHALLENGES AND CONSERVATION OF GENETIC RESOURCES

Despite their significance, CWR, landraces, and heirloom germplasm face numerous challenges. Modern agricultural practices often prioritize high-yielding and uniform cultivars, thus leading to the neglect of traditional and locally adapted crops. The encroachment of urbanization, climate change, and habitat destruction further threaten the habitats of CWR. Conservation efforts, including the establishment of seed banks and the engagement of the local community, are crucial for safeguarding these genetic resources and the flavors they offer.

6.7 PROSPECTS AND IMPLICATIONS

The continued exploration and utilization of CWR, landraces, and heirloom germplasm holds promise for the future of agriculture and culinary diversity. Integrating these genetic resources into plant-breeding programs can lead to the development of crops with enhanced taste, nutritional value, and resilience. Biotechnological advancements, guided by ethical considerations, offer opportunities to unlock the full potential of these genetic treasures. Embracing sustainable agriculture practices that acknowledge the value of diverse genetic resources is essential for ensuring a flavorful and resilient food future.

6.8 CONCLUSIONS

The crop wild relatives, landraces, and heirloom germplasm emerge as pivotal sources of novelty for taste and flavor in the realm of agriculture. Their genetic diversity, shaped by natural selection and human cultivation, underpins the rich tapestry of flavors that define our culinary experiences. Recognizing and preserving these plant genetic resources is not just an act of conservation but an investment in the continuation of diverse and delectable flavors that have sustained society for centuries. As we navigate the challenges of modern agriculture, let us embrace the richness offered by these genetic treasures, ensuring that future generations can savor the unique tastes cultivated by the wisdom of nature and the stewardship of farmers.

REFERENCES

Bai, Y., and Lindhout, P. (2007). Domestication and breeding of tomatoes: What have we gained and what can we gain in the future? *Ann. Bot.* 100, 1085–1094. https://doi.org/10.1093/aob/mcm150

Cabezas, D., de Bem Oliviera, I., Acker, M., Lyrene, P., and Munoz, P.R. (2021). Evaluating wild germplasm introgression into autotetraploid blueberry. *Agronomy* 11, 614. https://doi.org/10.3390/agronomy11040614

Cebolla-Cornejo, J., Roselló, S., and Nuez, F. (2013). Phenotypic and genetic diversity of Spanish tomato landraces. *Sci. Hortic. (Amsterdam)* 162, 150–164. https://doi.org/10.1016/j.scienta.2013.07.044

Comparini, D., Taiti, C., Lanza, M., Vita, F., Pondolfi, C., Luti, S., et al. (2021). Comparison of wild and domesticated hot peppers fruit: Volatile emissions, pungency and protein profiles. *Adv. Hortic. Sci.* 35, 305–327. https://doi.org/10.36253/ahsc-10630

Dwivedi, S., Goldman, I., and Ortiz, R. (2019). Pursuing the potential of heirloom cultivars to improve adaptation, nutritional, and culinary features of food crops. *Agronomy* 9, 441. https://doi.org/10.3390/agronomy9080441

Dwivedi, S.L., Stalker, H.T., Blair, M.W., Bertioli, D., Upadhyaya, H.D., Nielen, S., et al. (2008). Enhancing crop gene pools with beneficial traits using wild relatives. *Plant Breed. Rev.* 30, 179–230.

Flint-Garcia, S., Feldmann, M.J., Dempewolf, H., Morrell, P.L., and Ross-Ibarra, J. (2023). Diamonds in the not-so-rough: Wild relative diversity hidden in crop genomes. *PLoS Biol.* 21, e3002235. https://doi.org/10.1371/journal.pbio.3002235

Frankin, S., Cna'ani, A., Bonfil, D.J., Tzin, V., Nashef, K., Degen, D., et al. (2023). New flavors from old wheats: Exploring the aroma profiles and sensory attributes of local Mediterranean wheat landraces. *Front. Nutr.* 10, 1059078. https://doi.org/10.3389/fnut.2023.1059078

Haddad, N.E., Sanchez-Garcia, M., Visioni, A., Jilal, A., Amil, R.E., Sall, A.T., et al. (2021). Crop wild relatives crosses: Multi-location assessment in durum wheat, barley, and lentil. *Agronomy* 11, 2283. https://doi.org/10.3390/agronomy11112283

Li, Y., Liang, Y., Ni, L., Huang, H., Wei, Y., Wang, M., et al. (2023). Generating novel tomato germplasm using the ancestral wild relative of *Solanum pimpinellifolium*. *Horticulturae* 9, 34. https://doi.org/10.3390/horticulturae9010034

Lončarić, A., Flanjak, I., Kovač, T., Tomac, I., Gotal Skoko, A.-M., Babojelić, M.S., et al. (2023). Unveiling apple diversity: The quality of juice produced from old vs. Commercial apple cultivars. *Plants (Basel)* 12, 3733. https://doi.org/10.3390/plants12213733

Ma, B., Chen, J., Zheng, H., Fang, T., Ogutu, C., Li, S., et al. (2015). Comparative assessment of sugar and malic acid composition in cultivated and wild apples. *Food Chem.* 172, 86–91. https://doi.org/10.1016/j.foodchem.2014.09.032

Martin, P., Russell, J., Wishart, J., Brown, L.K., Wallace, M., Iannetta, P.P.M., et al. (2023). Back to the future: Using ancient Bere barley landraces for a sustainable future. *Plants People Planet.* https://doi.org/10.1002/ppp3.10432

Medina-Lozano, I., and Díaz, A. (2021). Nutritional value and phytochemical content of crop landraces and traditional varieties. In: Elkelish, A. (ed.) *Landraces—Traditional Variety and Natural Breed.* IntechOpen, London. https://doi.org/10.5772/intechopen.95514

Negri, A.S., Allegra, D., Simoni, L., Rusconi, F., Tonelli, C., Espen, L., et al. (2015). Comparative analysis of fruit aroma patterns in the domesticated wild strawberries "Profumata di Tortona" (*F. moschata*) and "Regina delle Valli" (*F. vesca*). *Front. Plant Sci.* 6, 56. https://doi.org/10.3389/fpls.2015.00056

Rodríguez-Burruezo, S., Prohens, J., Roselló, J., and Nuez, F. (2005). Heirloom varieties as sources of variation for the improvement of fruit quality in greenhouse-grown tomatoes. *J. Hortic. Sci. Biotechnol.* 80, 453–460. https://doi.org/10.1080/14620316.2005.11511959

Sampedro, J., and Cosgrove, D.J. (2005). The expansin superfamily. *Genome Biol.* 6, 242. https://doi.org/10.1186/gb-2005-6-12-242.

Sathya, A. (2013). Are the Indian rice landraces a heritage of biodiversity to reminisce their past or to reinvent for future? *Asian Agri-History* 17, 221–232.

Villena, J., Moreno, C., Roselló, S., Beltrán, J., Cebolla-Cornejo, J., and Moreno, M.M. (2023). Breeding tomato flavor: Modelling consumer preferences of tomato landraces. *Sci. Hortic. (Amestardam)* 308, 111597. https://doi.org/10.1016/j.scienta.2022.111597

Yu, X., Chu, M., Chu, C., Du, Y., Shi, J., Liu, X., et al. (2020). Wild rice (*Zizania* spp.): A review of its nutritional constituents, phytochemicals, antioxidant activities, and health-promoting effects. *Food Chem.* 30, 127293. https://doi.org/10.1016/j.foodchem.2020.127293.

7 Allelic Variation Associated with Culinary, Sensory, and Textural Traits in Cereals and Food Legumes

Sangam L. Dwivedi and Autar K. Mattoo

7.1 INTRODUCTION

Culinary, sensory, and textural traits determine food choice and intake behavior. Culinary traits from breeding perspectives include variation in seed size, shape, color, uniformity, hydration capacity, swelling index, and cooking time. Aroma influences food choice and stimulate sensory-specific appetites. Texture moderates eating behavior (rate at which food is consumed) and the energy consumed to get sense of satiation and metabolic response to food. Sensory cues moderate food intake and improve dietary behavior, while taste influence liking and energy intake to fullness. Higher taste intensity and duration is linked to lower food intake. Fat has low impact on sensory intensity but contribute significantly to dietary energy. While it is easy to rank taste intensities but challenging to discriminate from fat. The fat discrimination declines further when combined with high-taste intensity. Intense taste and high fat content promote overeating (Forde and de Graaf, 2022). The culinary and sensory traits link scientific knowledge on human perception and appreciation mechanisms to culinary practices (Schifferstein et al., 2020).

Plant genetic resources (PGR), both from cultivated and crop wild relatives, provide natural variation for traits of interest to support crop improvement programs. Key to successful breeding and genetics of crops depend on continued supply of genetic diversity, including new or improved variability for target traits. Over 0.7 million legume germplasms, including crop wild relatives, are conserved in 276 gene banks distributed worldwide (https://www.google.com/search?client=firefox-b-d&q=current+status+of+food+legume+germplasm). Unlike yield and yield attributing and stress tolerance traits, the researchers in the past paid little attention (except for total oil and protein contents) in studying aroma, sensory, and textural traits in the management of plant genetic resources and their deployment in breeding and genetics of food crops. However, in recent times there has been increased emphasis in finding natural variation in gene pools and dissecting the genetic and molecular basis

DOI: 10.1201/9781003467335-7

of sensory and textural traits. The deployment of these resources and sensory knowledge may provide breeders opportunities to develop next-generation food crops, combining culinary, sensory, and nutritional (both essential and nonessential) traits, as discussed in this chapter.

7.2 NATURAL VARIABILITY FOR KERNEL HARDNESS, AROMA, SENSORY, AND TEXTURAL TRAITS IN THE CEREAL GENE POOL

Limited studies to date detected sufficient variability for grain hardness, texture, and fragrance among barley, foxtail millet, rice, sorghum, and wheat germplasm.

7.2.1 Kernel Hardness and Texture

Hardness refers whether endosperm of the grain is physically hard or soft. Kernel hardness is the main classification parameter for global wheat trade. A non-destructive imaging technique, such as near infrared using single-kernel characterization system (SKCS) and partial least squares regression and artificial neural network models, is recommended to measure variation in grain hardness, classified into three categories of soft, medium, and hard, based on hardness index, with moderate prediction (Erkinbaev et al., 2019). Variations in endosperm texture determine milling, baking, and end-use quality. The grains are classified as very hard, hard, or soft. Hardness also influences the world trade of wheat grains. Soft wheats are used to make cakes, cookies, and pastries, hard wheats for making breads, and very hard durum wheats for pasta. Tetraploid wheat (*Triticum turgidum*) exhibits a very hard texture, whereas bread wheat (*T. aestivum*) texture varies from hard to soft (Guzmán et al., 2022).

Wheat and its relatives contain endosperm-specific proteins, puroindolines, and sequence variation in puroindolines regulates majority of the wheat grain texture variation, while in barley, it is hordoindolines, the homologs of puroindolines, that control substantial allelic diversity associated with grain texture (Beecher et al., 2002). Grain hardness varies from very soft to very hard. A grain hardness score of 34 SKCS or less is characterized as very soft, 35 to 54 as soft, 55 to 74 as medium hard, 75 to 89 as hard, and 90 or more very hard. Sufficient genetic variability for grain hardness is reported in barley and wheat gene pools. Environment has significant influence on grain hardness, for example, the frequencies of kernel hardness of Chinese wheat germplasm increased along with the latitude across China, while those of soft kernel decreased along with the increase of latitude (Ma et al., 2017).

7.2.2 Sensory Traits

Sufficient genetic variation for fragrance is reported in sorghum and rice germplasm. The recessive *fgr* gene on chromosome 8, a non-functional *badh2* allele, caused fragrance in rice, while its functional *badh2* allele encoding putative betaine aldehyde dehydrogenase (BADH2) impart nonfragrant rice. Assessment of fragrant and nonfragrant rice germplasm for the *badh2* locus resulted into two classes of fragrant rice, one containing the known *badh2* allele (*badh2-E7*) and the other having a novel null *badh2* allele (*badh2-E2*). The functional markers developed herein could be easily deployed to distinguish nonfragrant from fragrant rice as well to differentiate two kinds of rice fragrance (Shi et al., 2008). Over 80% of the 516 rice fragrant accessions contain *badh2.7* allele. Re-sequencing in the *badh2* region resulted three alleles (*badh2.7, badh2.2*, and *badh2.4–5*), which originated in the *japonica* gene pool, and later spread into the *indica* gene pool via hybridization. Likewise, loss-of-function alleles of *badh2* are also detected in the *indica* gene pools, which were then transferred into *japonica* (Phitaktansakul et al., 2022).

Fragrant sorghum with a special aroma (pandan-like fragrance) was first reported in the landrace cultivar ambemohar and IS-19912 from the province of Maharashtra, India (Kottur, 1919). Ambemohar has coconut-juice-like fragrance in its inflorescence and grains (Monkhan et al., 2021). E228 (IC 568489), a traditional sorghum cultivar collected from Bundelkhand region of India, has fragrant grains. This cultivar gets its flavor from the same compound 2-acetyl-pyrroline (2AP), which gives basmati rice its unique aroma. Twenty-five volatile compounds, including 2AP, contributed to the aroma in E228 than nonfragrant M35–1. The 2AP content in the grain was 6.5 ± 0.4 pbb (Zanan et al., 2016).

Fragrant foxtail millet germplasms are reported to contain 62 volatile compounds, with 18 aldehydes, 6 alcohols, 9 ketones, 5 acids, 10 hydrocarbons, 10 benzene derivatives, and 4 other components as main flavor components (Li et al., 2021).

Bread aroma is the main characteristics perceived by consumers. Diverse wheat cultivars showed large differences in bread odor and flavor, with old and modern cultivars were comparable for these traits. A combination of metabolite and SNP genotyping profiles, instead of the SNP genotype profile only, better predicts bread aroma (Longin et al., 2020). A comparison of a representative subset of Israeli and Palestinian landraces of wheat for dough quality, rheology, aroma, and taste using both whole and refined flours and breads revealed that bread wheat loaves baked using landrace flour scored higher in quality parameters, whereas in durum wheat lines, opposite tend was evident. Loaves baked from landraces Diar Alla and Hittia Soada had different aroma than control loaves prepared from modern flours. A significant association between specific aroma classes and sensory attributes, scored by panelists consisting of 40 human subjects, suggests that landraces have the potential to improve flavor and aroma in wholemeal bread (Frankin et al., 2023).

TABLE 7.1
Genetic Variability for Culinary, Sensory, and Textural Traits in Selected Cereal Crops from 2002 to 2023

Germplasm with specific culinary, sensory, and textural traits identified	Reference
Barley (*Hordeum vulgare*)	
Allelic variation for grain hardness between cultivated and wild *Hordeum* species, with two distinct clusters containing 17 *Hinb* genes	Terasawa et al. (2012)

(*Continued*)

Table 7.1 ***(Continued)***
Genetic Variability for Culinary, Sensory, and Textural Traits in Selected Cereal Crops from 2002 to 2023

Germplasm with specific culinary, sensory, and textural traits identified	Reference
Japanese two- and six-rowed barley lines, respectively, contain *Hinb-1* and *Hinb-2* alleles of *Hina*; significantly higher grain hardness in two-rowed barley lines with the *Hinb-2b* than those with Hinb-2a	Takahashi et al. (2010)
Variable grain hardness grain hardness and dry matter digestibility in the Steptoe × Morex mapping population	Beecher et al. (2002)
Buckwheat (*Fagopyrum esculentum*)	
Significant variability for unique flavor associated with variation in flavonoids (rutin)	Zhang et al. (2023b)
Foxtail millet (*Setaria italica*)	
39 common volatile compounds	Kang et al. (2023)
17 kinds of differentially volatile compounds among colored (black, green, white, and yellow) grains	Jin et al. (2023a)
Significant genotypic variation in volatile compounds	Li et al. (2021)
Rice (*Oryza sativa*)	
Fifty-three aromatic rice accessions containing *badh2* gene for fragrance display different levels of aroma expression, accessions with strong fragrance	Ndikuryayo et al. (2023)
15 volatile compounds discriminated four haplotypes in *badh2* region of Korean rice collection	Phitaktansakul et al. (2022)
Sufficient genetic variation at *badh2* locus for fragrance among rice accessions	Shao et al. (2013)
Fragrant lines containing *badh2* allele (*badh2-E7*) or null *badh2* allele (*badh2-E2*) discovered	Shi et al. (2008)
Sorghum (*Sorghum bicolor*)	
Coconut-juice-like fragrance in ambemohar	Monkhan et al. (2021)
Fragrant sorghum accession E228 (IC 568489)	Zanan et al. (2016)
Wheat (*Triticum aestivum*)	
22 differential volatile compounds in colored grains (black, green, and yellow)	Jin et al. (2023b)
Substantial variation in kernel hardness (-2 to 44) among RILs involving soft durum wheat lines	Ibba et al. (2019)
Mediterranean landraces identified with distinct aroma and sensory attributes	Frankin et al. (2023)
Grain hardness among European spring wheat germplasm varied from hard, medium, and soft texture	Nucia et al. (2021)
Substantial variation in aroma between old and modern wheat cultivars	Longin et al. (2020)
Wide variation in kernel hardness (1.4–107.2) among diverse panel of Chinese germplasm	Ma et al. (2017)

7.3 NATURAL VARIABILITY FOR CULINARY, SENSORY, AND TEXTURAL TRAITS IN THE FOOD LEGUME GENE POOL

7.3.1 Culinary and Sensory Traits

Wide variations in grain color and cooking time have been reported in the common bean (Bassett et al., 2021a, 2021b; Ribeiro and Maziero, 2023; Sadohara et al., 2021, 2022; Wiesinger et al., 2016). Cotyledon cell wall thickness and composition have a significant role in genetic variability for cooking time of soaked beans, whereas seed coat layer thickness relates to the genetic variability for cooking time of unsoaked dry beans (Bassett et al., 2021a). The accessions with unique characteristics were identified; for example, white bean accessions have starchy and sweet flavors (Bassett et al., 2021b), fast-cooking yellow bean Cebo Cela provides more nutrients with each serving compared with a slow-cooking yellow bean Canario or significantly higher bioavailable iron in faster cooking compared to slower cooking beans in the same market class (Wiesinger et al., 2016), there is greater variability in seed coat than in cotyledon compounds (Rivera et al., 2018), and the Middle American gene pool is richer in Ca than the Andean gene pool (Rivera et al., 2016). Correlation analysis revealed that white seed coat positively correlated with brightness and negatively with mealiness; mealiness correlated negatively with seed coat roughness and rate of water absorption; Andean-origin beans had lower seed coat brightness and seed coat roughness and greater seed coat perceptibility, mealiness, flavor, and aroma than Mesoamerican-origin beans (Rivera et al., 2016). Variation in seed proximate composition, together with differences in culinary and sensory traits in the gene pool, offers an opportunity to develop crop cultivars combining desirable culinary, sensory, and nutritional traits. Near infrared (NIR) spectroscopy calibrations using variable selection and partial least squares regression may be used to predict cooking times of fresh beans (Wafula et al., 2020).

The acceptability of pulses is often limited due to presence of beany flavor that contribute to negative perceptions of pulses. A variety of volatile compounds, produced from the oxidation of unsaturated free fatty acids and the degradation of amino acids during seed development, storage, and milling, determines the flavor profile of the processed products. A better understanding of the volatile compounds involved in the off-flavors (or unpleasant flavors) and their origins may allow to develop efficient strategies to minimize their impact on overall perception for more acceptable healthy food design (Karolkowski et al., 2021). Red kidney beans are known for its unique flavor, characterized by smoky, sulfury, and earthy aroma. Methional with flavor dilution (FD) factor 21 contribute to characteristic odor of the cooked red kidney beans, while p-vinyl guaiacol (FD 13) toward smoky odor flavor. Diethyl sulfide (FD 10) contributes to the sulfury flavor and 2-ethyl-3-methyl pyrazine (FD 13) to the earthy flavor in cooked red kidney beans (Mishra et al., 2019).

In faba bean (*Vicia faba*), seven key aroma compounds (3-methylbutanoic acid, (*E*)-non-2-enal, hexanal, methional, 3-methylbutanal, sotolon, and 2-methylbutan-1-ol) were identified and dominated by broth-like, fatty, or cheesy/sweaty odors (Ritter et al., 2024). Substantial variation in cooking time (CT) and related traits (swelling capacity, capacity, swelling index, hydration index) were recorded among faba bean germplasm, with germplasm showing minimum CT and maximum swelling capacity.

Swelling capacity significantly positively correlated with swelling index and hydration capacity (Çilesiz et al., 2023).

A comparison of volatile compounds in black bean, mung bean, and soybean revealed five major volatile compounds (hexanal, 2-methyl-1-butanol, 1-hexanol, 1-octen-3-ol, and benzaldehyde), with highest total volatile compound concentrations in the extracts of black bean, mung bean, and soybean, respectively (Han et al., 2022). Mung bean cultivar Sona has characteristics aroma and popularly cultivated in the pockets of river basin at Bhutnir char village of Malda District in the West Bengal province, India. A comparison of volatile compounds between Sona mung and non-scented mung unfolded 26 and 20 volatile compounds, respectively. 2AP, octanal, 1-pentanol, decanal, phenylmethanol. and 2-nonen-1-ol contributed to the aroma of Sona mung bean (Attar et al., 2017).

Nine and 20 volatile compounds were reported in raw and roasted peas, respectively. Six aroma compounds, including 3-methylbutanoic acid and hexanal, significantly contributed to the aroma of raw peas, while 15 aroma compounds, including pyrazines and pyranones, significantly contributed to the characteristic aroma of roasted peas. Roasting increased the variety of key aroma compounds significantly and contributed a nutty flavor to peas (Bi et al., 2020). The common sensory attributes in pea were beany, green peas, grassy, green peppers, cucumber, and fatty. The key odorants were mainly aldehydes, ketones, and methoxypyrazines (Zhogoleva et al., 2023). Thus, by knowing the key aroma compounds, the specific strategies could be developed to optimize, reduce, or mask the volatile compounds emanating unpleasant flavor by crossbreeding or during the food preparations.

7.3.2 Seed Size, Color, Oil, and Oil Quality in Peanut

The regular consumption of peanuts improves human health. Oil content and oil quality are important seed quality traits. Oil quality is influenced by variation in fatty acid profile, particularly oleic (O) and linoleic (L) acid contents. A higher O/L ratio ensures longer shelf-life and marketability of the peanut-based products. The discovery of two closely related breeding lines with exceptionally higher O/L ratio of ~40 (Nordon et al., 1987) and its exploitation in crop breeding led to release of several high-oleate peanut cultivars across globe (Dwivedi et al., 2014; Nawade et al., 2018). In addition, there is a substantial market for in-shell consumption of peanut, for which, small-seeded red-colored peanuts with 3–4-seeded pods are preferred (Zhao et al., 2017). A substantial part of peanuts, in addition to oil, is also consumed as peanut butter and confectionery products, and seed size matters for this class of consumption (Dwivedi et al., 2014). A wide range of variation for seed color, 100-seed weight, and oil content and fatty acid profiles were reported in germplasm (Li et al., 2022; Pattee et al., 2002; Wang et al., 2022). Peanut flavor is one of the most important quality traits. Substantial variability in flavor profiles is reported among peanut gene pool, with some germplasm/cultivars possessing unique flavor, for example, Chinese peanuts Taihua 4 and Taihua 9922 (Ding et al., 2023); US peanuts Chimera and NC Bunch (Fritz et al., 2022); 86x45B-10-1-2-2-b2-B, UF714021, and X90053 (Pattee et al., 2002); or Malawian peanuts Nsinjiro and Baka (Gama and Adhikari, 2019), with significantly higher intensity of roasted flavor. Profiling of secondary metabolite compounds in raw peanuts has large influence on peanut flavor and nutritional quality and variation in metabolite compounds could be used as yardstick to differentiate market types, e.g., 52 of 365 metabolites differentiated Virginia and Runner market type peanuts (Klevorn et al., 2019). Interestingly, subspecies *fastigiata* and *vulgaris* score for flavor better than subspecies *hypogaea* types (Isleib et al., 2000).

7.3.3 Edamame (Vegetable) Soybean

Vegetable soybeans are a class of specialty soybeans. They are popularly known as edamame in Japan and mao dou in China. Physiologically mature green pods are harvested as fresh beans at the R6 to R7 growth stages when beans are filled 80–90% of the pod capacity, and whole pods or shelled beans are used as a fresh or frozen vegetable. Such beans are highly nutritious, with excellent taste and flavor (Agyenim-Boateng et al., 2023; Nair et al., 2023). Unlike grain soybean, it is perishable. The production and consumption of vegetable soybeans are mainly concentrated in East and Southeast Asia. Over the years, the interest in cultivation of vegetable soybean has constantly increased in other regions, particularly in Europe and the USA. The lack of germplasm or cultivars with suitable characteristics is a major constraint in vegetable soybean production and expansion. Both morphological characters (pod size, color, and ratio of 2–3 seed per pod) and chemical composition (sweetness, taste, flavor, texture, and nutrients) determine the nutritional quality of vegetable soybean (Agyenim-Boateng et al., 2023; Nair et al., 2023).

To date, there is sufficient genetic variability for sensory descriptors (seed size, green color, sweetness, aftertaste, starch, nuttiness, beaniness, and chewiness) among edamame beans, with some accessions most preferred by panelists (Ahomondji et al., 2023; Guo et al., 2022; Kumar et al., 2011; Rao et al., 2002; Wang et al., 2023). Appearance and color characteristics of edamame soybeans have significant influence in consumer acceptability, with dark-green edamame beans most preferred by the consumers. The green color intensity (color index) and ‘a^*’ color coordinate correlates to appearance liking scores, which suggests that the use of color data supports breeding selection criteria and possibly predicts consumer acceptability (Carneiro et al., 2022).

Aroma greatly influences the market value of vegetable soybeans. A comparison of aromantic and non-aromatic beans at two seed development stages revealed significant differences in aroma profiles. The differentially accumulated metabolites (DAM) grouped into nine classes (flavonoids, lipids, amino acids and derivatives, saccharides and alcohols, organic acids, nucleotides and derivatives, phenolic acids, alkaloids, and vitamins), contributing to phenotypic differences. Metabolites and genes related to the 2-acetyl-1-pyrroline (2AP) synthesis were significantly different in aromatic and non-aromatic soybeans, which may be exploited for aromatic vegetable soybean breeding (Zhao et al., 2022).

Variation in sucrose and alanine contents in vegetable soybeans significantly impacts the sweetness flavor of edamame-derived products. Analysis of sweet edamame beans reveals significant variability for sucrose and alanine contents across seasons, with few accessions combining high sucrose and Ala contents (Wang et al., 2023). Higher 2-AP contents in Xiangdou, ZHE1754, Zhexian 65018–33, and Qvxian No. 1 resulted in pleasant flavors in these edamame beans. The ecotype of cultivar, the direction of breeding selection, and their geographic origin greatly influence the volatile compounds of the edamame bean (Yuan et al., 2021).

TABLE 7.2
Genetic Variability for Culinary, Sensory, and Textural traits in Food Legume Crops from 2000 to 2024

Germplasm with specific culinary, sensory, and textural traits identified	Reference
Common bean (*Phaseolus vulgaris*)	
Significant ($P \leq 0.001$) differences (36–361 minutes) for cooking time among 427 breeding lines, cultivars, and landraces	Amongi et al. (2023)
Identified lines with high culinary quality and greater mineral contents	Ribeiro and Maziero (2023)
Extensive variation in sensory attributes across and within seed types, accessions with extreme flavor intensity across environments	Bassett et al. (2021b)
Substantial variation for light-colored seed coat at harvest and at storage	De Almeida et al. (2021)
Greater variability in seed coat than in cotyledon compounds	Rivera et al. (2018)
Sufficient genetic variation for darkened seed coat among RILs	Erfatpour et al. (2018)
Substantial variability for culinary and sensory traits of the cooked beans among landrace core collection	Rivera et al. (2016)
Faba bean (*Vicia faba*)	
Seven key aroma compounds with characteristics broth-like, fatty, and cheesy/sweaty odors	Ritter et al. (2024)
Large range variation in cooking time (51–288 min) among 399 germplasms, with landraces from Şanlıurfa and Bilecik having minimum cooking time	Çilesiz et al. (2023)
Mung bean (*Vigna radiata*)	
Sona with characteristic aroma containing 26 volatiles	Attar et al. (2017)
Pea (*Pisum sativum*)	
Over 100 volatile compounds, aldehydes the main molecule; 39 aroma-active compounds;	Zhogoleva et al. (2023)
9 and 21 aroma compounds associated with different flavors of raw and roasted peas	Bi et al. (2020)
Peanut (*Arachis hypogaea*)	
A total of 42 kinds of volatile flavor compounds detected among 10 Chinese germplasm, with Taihua 4 and Taihua 6 showing the most	Ding et al. (2023)
A wide range of variation for seed coat color, 100-seed weight, oil, and oil quality	Wang et al. (2022)
11 flavor attributes, with superior accessions Chimera and NC Bunch for improving flavor in Virginia types	Fritz et al. (2022)
Wide variations in the 100-seed weight, total oil content, and fatty acid profile	Li et al. (2022)
Normal-oleic peanut oil showed stronger characteristic flavor than high-oleic peanut oil	Hu et al. (2021)
Significant variation for sensory traits among cultivars	Gama and Adhikari (2019)
52 of 365 metabolites significantly different between Runner and Virginia market types, contributing to raw peanut flavor	Klevorn et al. (2019)
Large phenotypic variation for four seed-related traits	Zhao et al. (2017)

(*Continued*)

Table 7.2 ***(Continued)***
Genetic Variability for Culinary, Sensory, and Textural traits in Food Legume Crops from 2000 to 2024

Germplasm with specific culinary, sensory, and textural traits identified	Reference
Significant variation among interspecific derived lines for the roasted peanut, sweet, and bitter sensory attributes	Tallury et al. (2008)
Sufficient variability for sensory traits (sweet, bitterness) among US peanut germplasm and breeding lines, specific lines with superior flavor score identified	Pattee et al. (2001)
Subsp. *fastigiata* and subsp. *vulgaris* cultivars with better flavor than subsp. *hypogaea* types	Isleib et al. (2000)
Soybean (*Glycine max*)	
Substantial variability for sucrose and alanine contents among sweet edamame accessions	Wang et al. (2023)
Significant variation in sensory attributes among genotypes, with few having better sensory attributes	Ahomondji et al. (2023)
Significant genotypic differences in appearance and overall liking, with R14–16195 the most liked accession	Carneiro et al. (2022)
Substantial differences in nutritional, organoleptic, and aroma characteristics among vegetable soybean varieties	Guo et al. (2022)
Differences in 2AP accumulation between aromatic and non-aromatic vegetable soybeans	Zhao et al. (2022)
63 volatile compounds impacting aroma (15 esters, 15 aldehydes, 13 alcohols, 15 ketones, one acid, and four other compounds), with characteristics pleasant aroma in four accessions	Yuan et al. (2021)
Large variability among vegetable soybeans for sucrose, aspartic acid, glutamic acid, glycine, alanine, and for total isoflavone contents	Kumar et al. (2011)
Sufficient variation for yield and nutritional traits including sugars, with Japanese edamame better source for use in breeding	Rao et al. (2002)

7.4 GENES AND MARKERS ASSOCIATED WITH CULINARY, SENSORY, AND TEXTURAL TRAITS AND DISCOLORATION OF DOUGH OR DOUGH-BASED PRODUCTS IN CEREALS

7.4.1 Kernel Hardness and Texture

Hinb-1 and *Hinb-2* of *Hordoindoline* (*Hin*) genes regulate grain hardness in barley. In a study of variation in *Hin* genes involving two *H. vulgare* subspecies (cultivated barley and *H. vulgare spontaneum*) and 10 wild relatives, Terasawa et al. (2012) detected two *Hinb* genes in *H. bogdanii*, *H. brachyantherum*, and *H. roshevitzii* that they named them as *Hinb-A* and *Hinb-B*, classified into two clusters, cluster A containing 7 and cluster B 10 *Hinb* genes. Contrarily, *Hinb-1* and *Hinb-2* in *H. vulgare* were included in cluster B, suggesting sufficient allelic diversity for grain hardness in barley gene pool.

HinA/HinB/Gsp locus, mapped on the short arm of chromosome 7 (5H), explains 22% of phenotypic variation of grain hardness in barley (Beecher et al., 2002). *HINb* is associated with both major endosperm texture and grain hardness in barley. It is highly expressed at mid-stage endosperm development stage. Spatio-temporal dynamics of transcripts and protein accumulation reveals a continuous increase of *HIN* transcript and HIN protein abundance across developmental stages. HIN proteins are localized mainly at the vacuolar membrane in the aleurone, in the subaleurone, and at the periphery of starch granules in the endosperm. The transcript and protein levels of HINs at subaleurone levels correlates poorly. The expression and localization pattern of HINs, however, correlates during endosperm development. Thus, understanding spatio-temporal dynamics of HINs protein is essential to improve barely grains for high-end product quality (Shabrangy et al., 2018).

Two genes, *Pina-D1* and *Pinb-D1*, at the *hardness locus* (*Ha*), regulate grain hardness in wheat. Soft endosperms are characterized by the presence of 'a' allele in both the gene, while deletion in the *Pina-D1* gene or single mutation in *Pinb-D1* characterizes hard, medium, and soft grain endosperm. The distribution of puroindoline genes and grain hardness in European wheat resulted in three classes of grain hardness—hard, medium, and soft—and differential allelic combinations of puroindoline genes among these classes (Nucia et al., 2021). An earlier reported revealed six *Pina* and nine *Pinb* alleles in Chinese wheat germplasm. The most common alleles were *Pina-D1a* and *Pina-D1b* for *Pina* and *Pinb-D1b*, *Pinb-D1a*, and *Pinb-D1p* for *Pinb*. A null genotype, *Pina-D1x*/*Pinb-D1ah*, possesses hardest grains, which contributed to the release of more hard type varieties with different prevalence of *Pin-D1* combinations in diverse regions (Ma et al., 2017). Two major QTLs, mapped on chr 3AL and 6AS, and allele-specific markers targeting these regions were developed to select for kernel hardness in durum wheat (Ibba et al., 2019).

7.4.2 Sensory Traits

Grain aroma in rice is one of the highly valued quality traits. A recessive gene (*fgr*) on chromosome 8 confers fragrance. It encodes betaine aldehyde dehydrogenase (BAD) that shows significant polymorphisms between fragrant and nonfragrant accessions. Non-functional alleles of betaine aldehyde dehydrogenase (BADH2) are associated with fragrance. Loss of function of the BADH2 enzyme leads to accumulation of 2-acetyl-1-pyrroline, 2AP (Bradbury et al., 2005a), and a perfect marker, allele-specific amplification (ASA), differentiates fragrant (ASA_{257}) from nonfragrant (ASA_{355}) alleles (Bradbury et al., 2005b). Eight putatively nonfunctional alleles of the BADH2 were obtained from the distinct geographic and genetic origins. The fragrance trait is selected on multiple occasions. However, the alleles remained isolated within local rice gene pools. *badh2.1* is the predominant allele present in most fragrant rice cultivars, including basmati and jasmine types. Haplotype analysis suggests a single origin of the *badh2.1* allele within *japonica* group, subsequently introgressed into *indica*, suggesting a close evolutionary relationship between *indica* (basmati type) and *japonica* gene pools (Kovach et al., 2009). Up to 19 alleles of the *badh2* to date accounted for unique spectrum of fragrance among aromatic rice cultivars (Akwero et al., 2020), and functional markers linked with some of these alleles have been developed (Bradbury et al., 2005b; He and Park, 2015; Li et al., 2020; Shao et al., 2011; Shi et al., 2008).

A single dominant gene and a major QTL, *qAro6.1* mapped on chromosome 6 between markers SB3567 and SB3570, regulates coconut-juice-like fragrance in sorghum cultivar ambemohar, and *BADH1* a candidate gene for the coconut-juice-like fragrance in ambemohar (Monkhan et al., 2021). In buckwheat, a pseudo cereal, *Fh06G015130* is associated with unique flavor, caused by variation in flavonoids (Zhang et al., 2023a).

TABLE 7.3

Genes and Markers Associated with Aroma, Culinary, Sensory, and Textural Traits in Cereals from 2002 to 2023

SNPs associated with specific culinary, sensory, and textural traits	Reference
Barley (*Hordeum vulgare*)	
Hinb, highly expressed in the mid-stage developed endosperm and associated with both major endosperm texture and grain hardness	Shabrangy et al. (2018)
Japanese two- and six-rowed barley lines, respectively, contain *Hinb-1* and *Hinb-2* alleles, significantly higher grain hardness in two-rowed barley lines with the *Hinb-2b* than those with *Hinb-2a*	Takahashi et al. (2010)
Several genomic regions associated with grain hardness, with largest QTL mapped to the HinA/HinB/Gsp locus on the short arm of chromosome 7 (5H)	Beecher et al. (2002)
Buckwheat (*Fagopyrum esculentum*)	
Fh06G015130 regulates unique flavor-associated flavonoids	Zhang et al. (2023a)
Rice (*Oryza sativa*)	
A typical aromatic gene *badh2-E2* for fragrance in an early maturing fragrant rice QG10	Jiang et al. (2023)
A 7-bp deletion in the second exon *of badh2-E2* type of deletion caused fragrance in *japonica* rice, Changxianggeng 1813	Lu et al. (2022a)
Four novel haplotypes for fragrance, 316 eQTL and 13 pQTLs in the *BADH2* coding region	Phitaktansakul et al. (2022)
badh2-E7 allele among 20 aromatic accessions	Akwero et al. (2020)
A recessive allele of *BADH2* regulates fragrance in rice, *badh2* alleles associated with the accumulation and synthesis of 2-acetyl-1-pyrroline2AP	Okpala et al. (2019)
Novel haplotypes and candidate genes influencing textural properties	Misra et al. (2018)
A new *badh2* allele (*badh2-p*) with an 8 bp insertion in the promoter region of the *BADH2* gene	Bindusree et al. (2017)
A novel allele (*badh2-E12*) associated with fragrance	He and Park (2015)
Eight putatively nonfunctional alleles of the *BADH2* gene associated with fragrance with distinct geographic and genetic origins	Kovach et al. (2009)
Sorghum (*Sorghum bicolor*)	
A major QTL, *qAro6.1*, for the fragrance on chr. 6 between markers SB3567 and SB3570	Monkhan et al. (2021)
SbBADH2 gene associated with fragrance in E228	Zanan et al. (2016)

(Continued)

Table 7.3 ***(Continued)***
Genes and Markers Associated with Aroma, Culinary, Sensory, and Textural Traits in Cereals from 2002 to 2023

SNPs associated with specific culinary, sensory, and textural traits	Reference
A 1,444 bp deletion that causes non-functional premature stop codon in *SbBADH2* is associated with fragrance in IS19912	Yundaeng et al. (2013)
Wheat (*Triticum aestivum*)	
Two genes *Pina-D1* and *Pinb-D1* at the *Ha* locus control kernel hardness, with differential allelic combinations determine varying degree of hardness (hard, medium, soft)	Nucia et al. (2021)
Two major genomic regions on chromosomes 3AL and 6AS with additive effect, associated with kernel hardness in durum wheat	Ibba et al. (2019)
6 *Pina* and 9 *Pinb* alleles contribute to varying degree of kernel hardness	Ma et al. (2017)

7.5 QTLS, CANDIDATE/FUNCTIONALLY CHARACTERIZED GENES, AND MARKERS ASSOCIATED WITH AROMA, CULINARY, SENSORY, AND TEXTURAL TRAITS IN FOOD LEGUMES

7.5.1 Culinary and Sensory Traits

Seed hardness is scored as visual appeal of cooked beans. It inherits independent of cooking quality. Hyperspectral imaging technology provides a non-destructive, simple, fast, and economically viable platform to estimate the water uptake and cooking time (CT) of dry bean. This platform obtained a relatively good predictions of water uptake (R_{pred} = 0.789 ± 4.4%) and CT (R_{pred} = 0.886 ± 7.9 min) for soaked beans. The prediction models for CT of unsoaked bean are less robust and accurate (R_{pred} = 0.708 ± 10.6 min). A further validation involving 110 similar dry bean samples confirmed the suitability of the technique for predicting CT of soaked beans, with improved prediction (R_{pred} = 0.872 ± 3.7 min). A periodical update of these prediction models and testing other multivariate prediction methods, however, are needed to improve model robustness and generalization because of substantial variation for water absorption and CT in dry beans (Mendoza et al., 2018).

Hard-to-cook defect in common bean is caused by variation in ability to achieve cell separation. Pectin methyl-esterase (PME) enzyme during hydration influences cell separation. Comparison of fast- and slow-cooking beans showed minor differences in PME enzyme activity. Gene expression reveals a total of 2132 differentially expressed genes (DEGs), with genes encoding cell-wall-related enzymes predominantly expressed. *Phvul010g080300* and *Phvul005g007600* were highly expressed in fast- and slow-cooking genotypes, respectively. Additionally, 8% of DEGs were upregulated in cell periphery of the slow-cooking bean. Thus, unlocking the role of pectin methyl-esterase-related genes and novel cell wall processes may provide new insights to the occurrence of the hard-to-cook defect (Toili et al., 2022).

The textural hardening during postharvest storage and soaking could be due to pectin-cation-phytate mechanism in common bean (Chen et al., 2023a). The texture exhibited after cooking of a single bean seed before ageing and its texture after ageing are correlated. The cell-wall-associated Ca concentration is positively and significantly correlated with the texture of both fresh and aged cooked red haricot bean cotyledons. Ageing results in a significant enrichment of Ca at the cell wall due to intracellular phytate hydrolysis, affecting the texture distribution of aged beans during cooking significantly. This relocation of the mineral cations from the cell interior to the cell wall occurs mainly during storage rather than subsequent soaking of the cotyledons. The demethylesterification of the cell wall pectin and increased pectin-Ca interactions upon ageing of the cotyledons further confirm the pectin-cation-phytate hypothesis at work to cause hard-to-cook defect in common bean (Chen et al., 2023b).

Gene expression during soaking influences CT. Genes related to cell wall growth and development as well as hypoxic stress (a condition where seed experience oxygen deficiency) are differentially expressed between the fast- and slow-cooking beans due to soaking. The candidate genes in slow-cooking beans increase intracellular calcium concentrations and cell wall modification enzymes. The expression of cell-wall-strengthening enzymes in the slow-cooking beans may increase their CT and ability to resist osmotic stress by preventing cell separation and water uptake in the cotyledon (Jeffery et al., 2023).

A major QTL on chromosome 7 and two novel QTL on chromosomes 1 and 2 affect seed hydration, while a major QTL affecting the visual appeal of cooked bean mapped on chromosome 4 (Sandhu et al., 2018). Cooked beans are used to assess for flavor and texture among common bean genotypes. Several SNPs were significantly associated with total and individual flavors, seed coat perception, and cotyledon texture—for example, six SNPs were associated with total flavor intensity; five SNPs with beany, three SNPs with earthy, one SNP with starchy, and one SNP with bitter flavors; and three and two SNPs associated with seed coat perception and cotyledon texture (Bassett et al., 2021b).

Farmers and consumers prefer fresh pod fragrance. *BADH* is the key gene associated with fragrance. *LpBADH2* in lablab bean shows a high degree of similarity with those of soybean homolog *GmBADH2*. A single nsSNP and three sSNPs were detected in *LpBADH2*. The substitution of the amino acid tyrosine in fragrant accessions with phenylalanine in nonfragrant accessions in protein coded by *LpBADH2* causes switch over from fragrance to nonfragrance in lablab bean (Basanagouda et al., 2023).

Hardness of cooked beans also affects the processing of soybean-based products. Hardness in soybean is primarily controlled by the pectin methylesterase gene *Glyma03g03360*, while Ca content has a secondary effect on it. Amplification refractory mutation system-polymerase chain reaction (ARMS-PCR), instead of cleaved amplified polymorphic sequence (CAPS) method, detects SNP polymorphism of *Glyma03g03360* and correctly classified cooked beans. ARMS-PCR could be useful for a marker-assisted selection of soybean with soft-cooked beans, while calcium content may be used as additional selection criterion (Toda et al., 2020).

Snap beans (or green, string, or French beans) are a form of common bean, of which immature pods and seeds are consumed as a vegetable. They have lower protein and total carbohydrate levels but possess vitamins and carotenoids, deficient or only in trace amounts in dry beans. Two major loci with epistatic effect, mapped on chromosomes Pv01 and Pv06, are involved in the genetic control of edible pod characteristics (García-Fernández et al., 2023). Pod size, weight, and color (green and purple or purple with stripes) are important quality traits in snap beans. SNPs associated with pod length and pod weight were successfully converted into KASP markers and validated for use in snap bean breeding (Li et al., 2023).

The edible use of snap bean pods is mainly affected by their colors (green, purple, or yellow). Several SNPs and candidate genes are associated either with greenness regulating chlorophyll and photosynthesis activity or affecting colors other than green regulating flavonoid biosynthetic pathway expressed in pods and seeds in snap beans (Celebioglu et al., 2023). Three anthocyanin biosynthesis pathway genes are differentially expressed in A18–1 (a cultivar with green stem and yellow pod) and its purple mutant pv-pur. The wild and its mutant lines differ in ten metabolites. In the mutant, *F3′5′H* expression was increased while *F3′H* and *ANS* were downregulated. The purple phenotype in the mutant is caused by the increased expression of F3′5′H, the key enzyme in the transformation from dihydroflavanol to dihydromyricetone in the anthocyanin biosynthetic pathway. The purple mutant in relation to its wild type produced significantly higher total anthocyanin levels (Liu et al., 2022b). The phenotypic characterization of snap bean purple mutant (*pv-pur*) has purple cotyledon, hypocotyl, stem, leaf vein, flower, and pod tissues. Fine mapping revealed that *pv-pur* is in the 243.9 kb region of chromosome 06, and a probable candidate gene *Phvul.006g018800.3* encoding *F3'5'H.* The coding region of this gene contains six single-base mutations that altered protein structure. The transgenic *Arabidopsis*, relative to wild type, containing *PV-PUR* or *pv-pur*, had a purple leaf base and internode, demonstrating that *PV-PUR* is a crucial gene for anthocyanin biosynthesis in snap beans (Liu et al., 2023). A single recessive gene, *pv-ye*, mapped on chromosome Pv 02, regulates the golden pods. The gene is localized to an interval of 182.9 kb between SSR markers, SSR77 and SSR93. *Phvul.002G006200* is the probable candidate gene for *pv-ye*. *Phvul.002G006200* sequencing unfolded a single-nucleotide polymorphism (SNP) in *pv-ye* that co-segregate with yellow phenotype (Yang et al., 2021).

7.5.2 Seed Coat Darkening

Seed coat in common seed darkens with age. Darker beans at harvest time and after storage are subject to decreased market values. The darkened seeds are associated with lower nutritional quality, increased cooking time, and decreased palatability. A major QTL for the non-darkening trait after seed storage, flanked by SNP 715646341 and SNP 715646348 on chromosome Pv10 and delimited within 13.2 cM and 40 candidate genes, explained 48% of the phenotypic variation for seed coat darkening (Erfatpour et al., 2018). Three QTL, two mapped on chromosome Pv04 and one on Pv10, and several candidate genes control variation for seed coat lightness at harvest and after 30 days storage (de Almeida et al., 2021). Postharvest seed coat darkening is attributed to proanthocyanidin accumulation and its subsequent oxidation in the seed coat. *Phvul.010G130600* is tightly linked with seed coat darkening in common bean. A single-nucleotide deletion (c.703delG) in the *Phvul.010G130600* caused non-darkening seeds. A gene-based dominant marker has been developed for the selection of dominant allele of *Phvul.010G130600* which can be used for marker-assisted selection of non-darkening beans (Erfatpour and Pauls, 2020).

Pinto beans are most affected by postharvest seed coat darkening. Some cultivars of pinto bean darken faster after harvest and accumulate higher level of proanthocyanidins compared to slow darkening cultivars. A novel allele of the *P* gene, P^{sd}, is associated with slow darkening seed coat, and gene-based marker developed to select for seed-coat non-darkening character in pinto beans (Islam et al., 2020). The vacuolar transporters play a key role in the accumulation of proanthocyanidins (PA). A vacuolar transporter *PvMATE8* is expressed at a higher level in regular-darkening pinto bean compared to slow-darkening pinto bean, which may be useful in unlocking the mechanism of PA accumulation in common bean (Islam et al., 2022). The synthesis and accumulation of proanthocyanidin is reduced in slow darkening pinto beans (Duwadi et al., 2018).

Both prolong darkening (grains placed in a room at a controlled temperature and humidity for 90 days simulating the supermarket shelf) or accelerated darkening (grains exposed to UV-C light at 254 nm for 96 h) and monitored by NMR spectroscopy may be used to screen for variation in slow darkening trait in beans (Salvador et al., 2022).

7.5.3 Seed Size, Color, Oil, and Oil Quality in Peanut

Multiple hub genes (i.e., genes that are highly connected to other genes regulating network as a whole) and candidate genes are associated with seed size, weight, and/or oil content in peanut (Miao et al., 2023; Yang et al., 2023). Roasted peanut flavor is a complex trait, affected by inherent genetic variation for total sugars and sucrose of the peanut cultivars and processing technology. A few QTLs and candidate genes across environments are associated with total sugars and sucrose (Zhang et al., 2023b). *Arahy.Y2LWD9* regulates sucrose content in peanut (Guo et al., 2023). Two mutant alleles, *ahFAD2A* mapped on LG a09 and *ahFAD2B* on LG b09, control fatty acid composition for higher oleic/linoleic ratio in peanut (Bera et al., 2019). Normal-oleic peanuts, however, show stronger characteristic flavor than high-oleic peanut oil. Methylpyrazine, 2,5-dimethylpyrazine, 2-ethyl-5-methylpyrazine and benzaldehyde are key volatiles that contribute to dark roast, roast peanutty, and sweet aroma of peanut oil (Hu et al., 2021). There is a sizable market for boiled and salted peanuts in Africa, Asia, and the USA. This segment of consumers prefers red-colored 3–4-seeded Valenica peanuts. The seeds are very sweet and have better flavor (Variath and Janila, 2017).

7.5.4 Edamame (Vegetable) Soybean

A codominant PCR-based marker, based on the 2-bp deletion in *GmAMADH2* gene, truly co-segregate with aromatic soybean germplasm as well as the aromatic segregants of the mapping population involving aromatic and nonaromatic soybeans, which could be efficiently deployed for molecular breeding of soybeans for aroma (Arikit et al., 2011). A functional marker based on sequence variation between fragrant and nonfragrant accessions and its further validation shows perfect association between marker genotype and aroma phenotype suggesting that *GmBADH2* is a major aroma (pandan-like aroma) conferring gene in soybean (Qian et al., 2022). Furthermore, the key biosynthesis enzymes are differentially expressed during seed development and may contribute to variations in the content of quality traits, including aroma at the vegetable and mature stages (Agyenim-Boateng et al., 2023).

Seed size and sweetness are equally important for vegetable soybeans. Several candidate genes (four sucrose synthase, one sucrose-phosphate synthase, and four sugar transport genes) associates with seed sweetness and seed size in vegetable soybean (Liu et al., 2022a). *Glyma.02G294000* is one of candidate gene associated with total soluble sugar content (Lu et al., 2022b). Sweetness flavor of edamame-derived products depend on variation in sucrose and alanine contents. Four genes with known effects on sucrose biosynthesis and three genes regulating alanine content, in addition to several novel sucrose- and alanine-related genes (Wang et al., 2023) and *GSTT1*, *GL3*, and *GSTL3* as candidate genes associated with the isoflavone content (Yang et al., 2022), are useful genomic resource in edamame beans. Consumption of isoflavone provides protective health benefits to minimize the risk of age-related diseases in humans (Křížová et al., 2019).

TABLE 7.4
Genes and Markers Associated with Aroma, Culinary, Sensory, Textural Traits, and Pod Darkening in Food Legumes from 2011 to 2023

SNPs associated with specific culinary, sensory, and textural traits	Reference
Common bean (*Phaseolus vulgaris*)	
Several SNPs associated with flavor attributes	Bassett et al. (2021b)
A major QTL for the non-darkening, flanked by SNP 715646341 and SNP 715646348 on chromosome Pv10, contributed 48% of the phenotypic variation for seed coat darkening	Erfatpour et al. (2018)
Lablab bean (*Lablab purpureus*)	
LpBADH2 associated with fresh pod fragrance	Basangouda et al. (2023)
Mung bean (*Vigna radiata*)	
2-acetyl-1-pyrroline (2AP) associated with characteristics aroma	Attar et al. (2017)
Peanut (*Arachis hypogaea*)	
A major QTL, *qSUCA06*, mapped on chromosome A06, across environments accounted for 32–41% variation for sucrose content	Guo et al. (2023)
Four and eight candidate genes associated with total sugars and sucrose	Zhang et al. (2023a)
37 volatile compounds detected in roasted peanut	Eker et al. (2023)
Multiple hub genes with potential roles in seed size/weight and/or oil content	Yang et al. (2023)
Arahy.D52S1Z, *Arahy.IBM9RL*, *Arahy.W18Y25*, *Arahy.CPLC2W*, and *Arahy.14EF4H* associated with pod and seed weight	Miao et al. (2023)
Two stable QTL regions on chromosomes 2 and 16 associated with four seed related traits	Zhang et al. (2019)

(*Continued*)

Table 7.4 *(Continued)*
Genes and Markers Associated with Aroma, Culinary, Sensory, Textural Traits, and Pod Darkening in Food Legumes from 2011 to 2023

SNPs associated with specific culinary, sensory, and textural traits	Reference
AHGA44686 associated with seed length and hundred-seed weight in multiple environments with large phenotypic variance	Zhao et al. (2017)
Soybean (*Glycine max*)	
SNPs and genes associated with sucrose and alanine contents in edamame beans; accessions with high sucrose and Ala contents, useful genetic resource	Wang et al. (2023)
GmBADH2, a major aroma-conferring gene	Qian et al. (2022)
Glyma.02G294000 associated with total soluble sugar content	Lu et al. (2022b)
GSTT1, *GL3*, and *GSTL3* associated with the isoflavone content	Yang et al. (2022)
4, 1, and 4 gene for sucrose synthase, sucrose-phosphate synthase, and sugar transport genes associated with seed sweetness and seed size in vegetable soybean	Liu et al. (2022a)
A codominant PCR-based marker, designed based on the 2-bp deletion in *GmAMADH2*, associated with aroma	Arikit et al. (2011)

7.6 REDUCING OFF-FLAVOR IN SOYBEAN PRODUCTS

Soybeans in general are poorly accepted as food by consumers due to their off-flavor. The genistein/daidzein mass ratio is an important factor in regulating off-flavor, and *isoflavone synthase2* (*IFS2*) is the key gene to potentially lower off-flavor formation in soybean (Kumar et al., 2023). Light-induced off-odor is the major problem in soybean oils. Furan fatty acids in the triacylglycerol of soybean oil is the major source of such off-odor, which produce 3-methyl-2,4-nonanedione (3-MND) upon photooxidation. Reduction in furan fatty acids leads to a decrease in 3-MND, which minimizes off-odor induced by light exposure. Mutant lines low or no furan fatty acid levels in soybean seeds are reported. *Glyma.20G201400* and *Glyma.04G054100* are associated with furan fatty acid biosynthesis in soybean. The ectopic expression of both genes produces furan fatty acids in transgenic soybean hairy roots. Homologs of *Glyma.20G201400* and *Glyma.04G054100* are widely conserved in the plant kingdom. Seed oil produced from these mutants in relation to wild type shows lower amounts of 3-MND and reduced off-odor when exposed to light (Watanabe et al., 2023).

7.7 CONCLUSION

Human perception and appreciation of food depends on how culinary and sensory matrix of edible-grade raw materials, in addition to other nutritional traits, are integrated with other ingredients into the food systems. Assessing and exploiting functional diversity for culinary and sensory traits together with other equally important traits like yield, nutrition, and stress tolerance is a significant breeding challenge.

Multiple-fold increase in productivity of staple food crops has been achieved to date via improved seed technology and optimum crop management practices and policy support. The proof of concept has been demonstrated, at least in some crops, that there exists variability in gene pools for culinary and sensory traits that is heritable, and in some cases, genes and associated markers have been used to develop next-generation consumer-driven improvements in diverse food commodities.

REFERENCES

Agyenim-Boateng, K.G., Zhang, S., Zhang, S., Khattak, A.M., Shaibu, A., Abdelghany, A.M., et al. (2023). The nutritional composition of the vegetable soybean (maodou) and its potential in combatting malnutrition. *Front. Nutr.* 9, 1034115. https://doi.org/10.3389/fnut.2022.1034115

Ahomondji, S.E., Agoyi, E.E., Agrangba, C.E., Lokonon, B.E., Agbahoungba, S., Assogbadjo, A.E., et al. (2023). Sensory preference criteria and willingness to adopt vegetable soybean "Edamame" in Benin (West Africa). *J. Sens. Stud.* 38, e12797. https://doi.org/10.1111/joss.12797

Akwero, A., Ocani, D., Akechi, W., Lamo, J., Ochwo-Ssemakula, M. (2020). Allelic variations in aroma gene in cultivated rice varieties. *African Crop Sci. J.* 28, 241–254. https://doi.org/10.4314/acsj.v28i2.9

Amongi, W., Nkalubo, S.T., Ochwo-Ssemakula, M., Badji, A., Dramadri, I.O., Odongo, T.L., et al. (2023). Phenotype based clustering, and diversity of common bean genotypes in seed iron concentration and cooking time. *PLoS ONE* 18, e0284976. https://doi.org/10.1371/journal.pone.0284976

Arikit, S., Yoshihashi, T., Wanchana, S., Tanya, P., Juwattanasomran, R., Srinives, P., et al. (2011). A PCR-based marker for a locus conferring aroma in vegetable soybean (*Glycine max* L.). *Theor. Appl. Genet.* 122, 311–316. https://doi.org/10.1007/s00122-010-1446-y

Attar, U., Hinge, V., Zanan, R., Adhav, R., and Nadaf, A. (2017). Identification of aroma volatiles and understanding 2-acetyl-1-pyrroline biosynthetic mechanism in aromatic mung bean (*Vigna radiata* (L.) Wilczek). *Physiol. Mol. Biol. Plants* 23, 443–451. https://doi.org/10.1007/s12298-017-0414-2

Basanagouda, G., Ramesh, S., Siddu, C.B., Chandana, B.R., Kalpana, M.P., Rotti, K., et al. (2023). A non-synonymous SNP in homolog of *BADH2* gene is associated with fresh pod fragrance in dolichos bean (*Lablab purpureus* var. lignosus (Prain) Kumari). *Genet. Resour. Crop Evol.* 70, 373–380. https://doi.org/10.1007/s10722-022-01535-y

Bassett, A., Hooper, S., and Cichy, K. (2021a). Genetic variability of cooking time in dry beans (*Phaseolus vulgaris* L.) related to seed coat thickness and the cotyledon cell wall. *Food Res. Int.* 141, 109886. https://doi.org/10.1016/j.foodres.2020.109886

Bassett, A., Kamfwa, K., Ambachew, D., and Cichy, K. (2021b). Genetic variability and genome-wide association analysis of flavor and texture in cooked beans (*Phaseolus vulgaris* L.). *Theor. Appl. Genet.* 134, 959–978. https://doi.org/10.1007/s00122-020-03745-3

Beecher, B., Bowman, J., Martin, J.M., Bettge, A.D., Morris, C.F., Blake, T.K., et al. (2002). Hordoindolines are associated with a major endosperm-texture QTL in barley (*Hordeum vulgare*). *Genome.* 45, 584–591. https://doi.org/10.1139/g02–008

Bera, S.K., Kamdar, J.H., Kasundra, S.V., Patel, S.V., Jasani, M.D., Maurya, A.K., et al. (2019). Steady expression of high oleic acid in peanut bred by marker-assisted backcrossing for fatty acid desaturase mutant alleles and its effect on seed germination along with other seedling traits. *PLoS ONE* 14, e0226252. https://doi.org/10.1371/journal.pone.0226252

Bi, S., Xu, X., Luo, D., Lao, F., Pang, X., Shen, Q., et al. (2020). Characterization of key aroma compounds in raw and roasted peas (*Pisum sativum* L.) by application of instrumental and sensory techniques. *J. Agric. Food Chem.* 68, 2718–2727. https://doi.org/10.1021/acs.jafc.9b07711

Bindusree, G., Natarajan, P., Kalva, S., and Madasamy, P. (2017). Whole genome sequencing of Oryza sativa L. cv. Seeragasamba identifies a new fragrance allele in rice. *PLoS One* 12, e0188920. https://doi.org/10.1371/journal.pone.0188920

Bradbury, L., Henry, R., Jin, Q., Reinke, R., and Waters, D. (2005b). A perfect marker for fragrance genotyping in rice. *Mol. Breed.* 16, 279–283. https://doi.org/10.1007/s11032-005-0776-y

Bradbury, L.M.T., Fitzgerald, T.L., Henry, R.J., Jin, Q., and Waters, D.L.E. (2005a). The gene for fragrance in rice. *Plant Biotechnol. J.* 3, 363–370. https://doi.org/10.1111/j.1467–7652.2005.00131.x

Carneiro, R., Adie, K., Yu, D., Beverly, M., Neill, C., Zhang, B., et al. (2022). Understanding the role of overall appearance and color in consumers' acceptability of edamame. *Front. Sustain. Food Syst.* 6, 738453. https://doi.org/10.3389/fsufs.2022.738453

Celebioglu, B., Hart, J.P., Porch, T., Griffiths, P., and Myers, J.R. (2023). Genome-wide association study to identify possible candidate genes of snap bean leaf and pod color. *Genes* 14, 2234. https://doi.org/10.3390/genes14122234

Chen, D., Bernaerts, T., Debon, S., Oduah, C.O., Zhu, L., Wallecan, J., et al. (2023b). Novel insights into the role of the pectin-cation-phytate mechanism in ageing induced cooking texture changes of red haricot beans through a texture-based classification and in situ cell wall associated mineral quantification. *Food Res. Intl.* 163, 112216. https://doi.org/10.1016/j.foodres.2022.11221

Chen, D., Ding, A., Zhu, L., Grauwet, T., Loey, A.V., Hendrickx, M., et al. (2023a). Phytate and mineral profile evolutions to explain the textural hardening of common beans (*Phaseolus vulgaris* L.) during postharvest storage and soaking: Insights obtained through a texture-based classification approach. *Food Chem.* 404 (Pt A), 134531. https://doi.org/10.1016/j.foodchem.2022.134531

Çilesiz, Y., Nadeem, M.A., Gürsoy, N., Kul, R., and Karaköy, T. (2023). Assessing the cooking and quality traits diversity in the seeds of faba bean germplasm. *Turk. J. Agric. For.* 47, 448–466. https://doi.org/10.55730/1300–011X.3101

de Almeida, C.P., Santos, I.L., de Carvalho Paulino, J.F., Barbosa, C.C.F., Pereira, C.C.A., Carvalho, C.R.L., et al. (2021). Genome-wide association mapping reveals new *loci* associated with light-colored seed coat at harvest and slow darkening in carioca beans. *BMC Plant Biol.* 21, 343 (2021). https://doi.org/10.1186/s12870-021-03122-2

Ding, B., Wang, F., Zhang, B., Feng, M., Chang, L., Shao, Y., et al. (2023). Flavor characteristics of ten peanut varieties from China. *Foods* 12, 4380. https://doi.org/10.3390/foods12244380

Duwadi, K., Austin, R.S., Mainali, H.R., Bett, K., Marsolais, F., and Dhaubhadel, S. (2018). Slow darkening of pinto bean seed coat is associated with significant metabolite and transcript differences related to proanthocyanidin biosynthesis. *BMC Genomics* 19, 260. https://doi.org/10.1186/s12864-018-4550-z

Dwivedi, S.L., Puppala, N., Maleki, S., Ozias-Akins, P., and Ortiz, R. (2014). Peanut improvement for human health. *Plant Breed. Rev.* 38, 143–185.

Eker, T., Darici, M., and Cabaroglu, T. (2023). The impact of seed size on volatile composition in roasted peanuts extracted by the Purge and Trap method. *J. Food Meas. Charact.* 17, 211–223. https://doi.org/10.1007/s11694-022-01583-z

Erfatpour, M., Navabi, A., and Pauls, K.P. (2018). Mapping the non-darkening trait from 'Wit-rood boontje' in bean (*Phaseolus vulgaris*). *Theor. Appl. Genet.* 131, 1331–1343. https://doi.org/10.1007/s00122-018-3081-y

Erfatpour, M., and Pauls, K.P. (2020). A R2R3-MYB gene-based marker for the non-darkening seed coat trait in pinto and cranberry beans (*Phaseolus vulgaris* L.) derived from 'Wit-rood boontje'. *Theor. Appl. Genet.* 133, 1977–1994. https://doi.org/10.1007/s00122-020-03571-7

Erkinbaev, C., Derksen, K., and Paliwal, J. (2019). Single kernel wheat hardness estimation using near infrared hyperspectral imaging. *Infrared Phys. Technol.* 98, 250–255. https://doi.org/10.1016/j.infrared.2019.03.033

Forde, C.G., and de Graaf, K. (2022). Influence of sensory properties in moderating eating behaviors and food intake. *Front. Nutr.* 9, 841444. https://doi.org/10.3389/fnut.2022.841444

Frankin, S., Cnáani, A., Bonfil, D.J., Tzin, V., Nashef, K., Degen, D., et al. (2023). New flavors from old wheats: Exploring the aroma profiles and sensory attributes of local Mediterranean wheat landraces. *Front. Nutr.* 10, 1059078. https://doi.org/10.3389/fnut.2023.1059078

Fritz, K.R., Dean, L.L., Hendrix, K.W., Andres, R.J., Newman, C.S., Oakley, A.T., et al. (2022). Flavor quality and composition of accession resources in the North Carolina State University peanut breeding program. *Crop Sci.* 62, 1880–1890. https://doi.org/10.1002/csc2.20774

Gama, A.P., and Adhikari, K. (2019). Sensory characterization of dominant Malawi peanut varieties after roasting. *J. Food Sci.* 84, 1554–1562. https://doi.org/10.1111/1750-3841.14641

García-Fernández, C., Jurado, M., Campa, A., Bitocchi, E., Papa, R., and Ferreira, J.J. (2023). Genetic control of pod morphological traits and pod edibility in a common bean RIL population. *Theor. Appl. Genet.* 137, 6. https://doi.org/10.1007/s00122-023-04516-6

Guo, J., Qi, F., Qin, L., Zang, M., Sun, Z., Li, H., et al. (2023). Mapping of a QTL associated with sucrose content in peanut kernels using BSA-seq. *Front. Genet.* 13, 1089389. https://doi.org/10.3389/fgene.2022.1089389

Guo, L., Huang, L., Cheng, X., Gao, Y., Zhang, X., Yuan, X., et al. (2022). Volatile flavor profile and sensory properties of vegetable soybean. *Molecules* 27, 939. https://doi.org/10.3390/molecules27030939

Guzmán, C., Ibba, M.I., Álvarez, J.B., and Morris, C. (2022). Wheat quality. In: Reynolds, M.P., and Braun, H.J. (eds.) *Wheat Improvement.* Springer, Cham. https://doi.org/10.1007/978-3-030-90673-3_11

Han, E., Ahn, H., Lee, S., and Lee, K.-G. (2022). Analysis of volatile compounds of black bean, mung bean, and soybean extracts prepared with distillation under reduced pressure–continuous liquid–liquid extraction and hot water extraction. *Chem. Biol. Technol. Agric.* 9, 95. https://doi.org/10.1186/s40538-022-00369-1

He, Q., and Park, Y.J. (2015). Discovery of a novel fragrant allele and development of functional markers for fragrance in rice. *Mol. Breed.* 35, 217. https://doi.org/10.1007/s11032-015-0412-4

Hu, H., Shi, A., Liu, H., Liu, L., Fauconnier, M.L., and Wang, Q. (2021). Study on key aroma compounds and its precursors of peanut oil prepared with normal- and high-oleic peanuts. *Foods* 10, 3036. https://doi.org/10.3390/foods10123036

Ibba, M.I., Kiszonas, A.M., See, D.R., Skinner, D.J., and Morris, C.F. (2019). Mapping kernel texture in a soft durum (*Triticum turgidum* subsp. *durum*) wheat population. *J. Cereal Sci.* 85, 20–26. https://doi.org/10.1016/j.jcs.2018.10.006

Islam, N.S., Bett, K.E., Pauls, K.P., Marsolais, F., and Dhaubhadel, S. (2020). Postharvest seed coat darkening in pinto bean (*Phaseolus vulgaris*) is regulated by *Psd*, an allele of the basic helix-loop-helix transcription factor *P. Plants People Planet* 2, 663–677. https://doi.org/10.1002/ppp3.10132

Islam, N.S., Duwadi, K., Chen, L., Pajak, A., McDowel, T., Marsolais, F., and Dhaubhadel, S. (2022). Global analysis of common bean multidrug and toxic compound extrusion transporters (PvMATEs): PvMATE8 and pinto bean seed coat darkening. *Front. Plant Sci.* 13, 1046597. https://doi.org/10.3389/fpls.2022.1046597

Isleib, T.G., Pattee, H.E., Gorbet, D.W., and Gierbrech, F. (2000). Genotypic variation in roasted peanut flavor quality across 60 years of breeding. *Peanut Sci.* 27, 92–98. https://doi.org/10.3146/i0095-3679-27-2-11

Jeffery, H.R., Mudukuti, N., Buell, C.R., Childs, K.L., and Cichy, K. (2023). Gene expression profiling of soaked dry beans (*Phaseolus vulgaris* L.) reveals cell wall modification plays a role in cooking time. *Plant Genome* 16, e20364. https://doi.org/10.1002/tpg2.20364

Jiang, S., Zhang, X., Yang, X., Liu, C., Wang, L., Ma, B., et al. (2023). A chromosome-level genome assembly of an early matured aromatic *Japonica* rice variety Qigeng10 to accelerate rice breeding for high grain quality in Northeast China. *Front. Plant Sci.* 14, 1134308. https://doi.org/10.3389/fpls.2023.1134308

Jin, W., Cai, W., Zhao, S., Gao, R., and Jiang, P. (2023a). Uncovering the differences in flavor volatiles of different colored foxtail millets based on gas chromatography-ion migration spectrometry and chemometrics. *Curr. Res. Food Sci.* 7, 100585. https://doi.org/10.1016/j.crfs.2023.100585

Jin, W., Zhao, S., Sun, H., Pei, J., Gao, R., and Jiang, P. (2023b). Characterization and discrimination of flavor volatiles of different colored wheat grains after cooking based on GC-IMS and chemometrics. *Curr. Res. Food Sci.* 7, 100583. https://doi.org/10.1016/j.crfs.2023.100583

Kang, M., Guo, Y., Ren, Z., Ma, W., Luo, Y., Zhao, K., et al. (2023). Volatile fingerprint and differences in volatile compounds of different foxtail millet (*Setaria italica* Beauv.) varieties. *Foods* 12, 4273. https://doi.org/10.3390/foods12234273

Karolkowski, A., Guichard, E., Briand, L., and Salles, C. (2021). Volatile compounds in pulses: A review. 120, 3140. https://doi.org/10.3390/foods10123140

Klevorn, C.M., Dean, L.L., and Johanningsmeier, S.D. (2019). Metabolite profiles of raw peanut seeds reveal differences between market-types. *J. Food Sci.* 84, 397–405. https://doi.org/10.1111/1750–3841.14450

Kottur, G.L. (1919). Classification and description of the jowars of the Bombay and Karnataka. In: *Bulletin Number 92 of 1919.* Department of Agriculture, The Yeravda Prison Press, Poona, p. 16.

Kovach, M.J., Calingacion, M.N., Fitzgerald, M.A., and McCouch, S.R. (2009). The origin and evolution of fragrance in rice (*Oryza sativa* L.). *Proc. Natl. Acad. Sci. U. S. A.* 106, 14444–14449. https://doi.org/10.1073/pnas.0904077106

Křížová, L., Dadáková, K., Kašparovská, J., and Kašparovský, T. (2019). Isoflavones. *Molecules* 24, 1076. https://doi.org/10.3390/molecules24061076

Kumar, S., Banerjee, S., Kaur, A., Sasi, M., Kumari, S., Sachdev, S., et al. (2023). Isoflavones play a potential role in off-flavour scavenging, with a key role of *IFS2* in isoflavone accumulation in soybean seeds. *Food Technol. Biotechnol.* 61, 514–522. https://doi.org/10.17113/ftb.61.04.23.8231

Kumar, V., Rani, A., Goel, L., Pratap, D., Billore, S.D., and Chauhan, G.S. (2011). Evaluation of vegetable-type soybean for sucrose, taste-related amino acids, and isoflavones contents. *Int. J. Food Prop.* 14, 1142–1151. https://doi.org/10.1080/10942911003592761

Li, M., Wu, X., Wang, B., Wu, X., Wang, Y., Wang, J., et al. (2023). Genome-wide association analysis reveals the optimal genomic regions for pod size in bean. *Front. Plant Sci.* 14, 1138988. https://doi.org/10.3389/fpls.2023.1138988

Li, S., Zhao, W., Liu, S., Li, P., Zhang, A., Zhang, J., et al. (2021). Characterization of nutritional properties and aroma compounds in different colored kernel varieties of foxtail millet (*Setaria italica*). *J. Cereal Sci.* 100, 103248. https://doi.org/10.1016/j.jcs.2021.103248

Li, W., Yoo, E., Lee, S.-K., Sung, J., Noh, H.J., Hwang, S.J., et al. (2022). Seed weight and genotype influence the total oil content and fatty acid composition of peanut seeds. *Foods* 11, 3463. https://doi.org/10.3390/foods11213463

Li, W., Zeng, X., Li, S., Chen, F., and Gao, J. (2020). Development and application of two novel functional molecular markers of BADH2 in rice. *Electronic J. Biotechnol.* 46, 1–7. https://doi.org/10.1016/j.ejbt.2020.04.004.

Liu, C., Yang, X., He, Y., Chen, Q., Huang, Y., Yan, Z., et al. (2023). Fine mapping and characterisation of a *PV-PUR* mediating anthocyanin synthesis in snap bean (*Phaseolus vulgaris* L.). *Mol. Breed.* 43, 15. https://doi.org/10.1007/s11032-023-01362-8

Liu, C., Yang, X.X., Yan, Z.S., Liu, D.J., and Feng, G.J. (2022b). Identification and characterization of a mutant PV-PUR gene responsible for the purple phenotype of snap bean (*Phaseolus vulgaris* L.) *Int. J. Mol. Sci.* 23, 1265. https://doi.org/10.3390/ijms23031265

Liu, N., Niu, Y., Zhang, G., Feng, Z., Bo, Y., Lian, J., et al. (2022a). Genome sequencing and population resequencing provide insights into the genetic basis of domestication and diversity of vegetable soybean. *Hort. Res.* 9, uhab052. https://doi.org/10.1093/hr/uhab052

Longin, F.H., Beck, H., Gütler, H., and Heilig, W. (2020). Aroma and quality of breads baked from old and modern wheat varieties and their prediction from genomic and flour-based metabolite profiles. *Food Res. Intl.* 129, 108748. https://doi.org/10.1016/j.fodres.2019.108748

Lu, R., Liu, J., Wang, X., Song, Z., Ji, X., Li, N., et al. (2022a). Chromosome-level genome assembly of a fragrant Japonica rice cultivar 'Changxianggeng 1813' provides insights into genomic variations between fragrant and non-fragrant Japonica rice. *Int. J. Mol. Sci.* 23, 9705. https://doi.org/10.3390/ijms23179705

Lu, W., Sui, M., Zhao, X., Jia, H., Han, D., Yan, X., et al. (2022b). Genome-wide identification of candidate genes underlying soluble sugar content in vegetable soybean (*Glycine max* L.) *via* association and expression analysis. *Front. Plant Sci.* 13, 930639. https://doi.org/10.3389/fpls.2022.930639

Ma, X., Sajjad, M., Wang, J., Yang, W., Sun, J., Li, X., et al. (2017). Diversity, distribution of puroindoline genes and their effect on kernel hardness in a diverse panel of Chinese wheat germplasm. *BMC Plant Biol.* 17, 158. https://doi.org/10.1186/s12870-017-1101-8

Mendoza, F.A., Wiesinger, J.A., Lu, R., Nchimbi-Msolla, S., Miklas, P.N., Kelly, J.D., et al. (2018). Prediction of cooking time for soaked and unsoaked dry beans (*Phaseolus vulgaris* L.) using hyperspectral imaging technology. *Plant Phenome J.* 1, 1–9. https://doi.org/10.2135/tppj2018.01.0001

Miao, P., Meng, X., Li, Z., Sun, S., Chen, C.Y., and Yang, X. (2023). Mapping quantitative trait loci (QTLs) for hundred-pod and hundred-seed weight under seven environments in a recombinant inbred line population of cultivated peanut (*Arachis hypogaea* L.). *Genes (Basel)* 14, 1792. https://doi.org/10.3390/genes14091792

Mishra, P.K., Tripathi, J., Gupta, S., and Variyar, P.S. (2019). GC-MS olfactometric characterization of odor active compounds in cooked red kidney beans (*Phaseolus vulgaris*). *Heliyon* 5, e02459. https://doi.org/10.1016/j.heliyon.2019.e02459

Misra, G., Badoni, S., Domingo, C.J., Cuevas, R.P.O., Llorente, C., Mbanjo, E.G.N., et al. (2018). Deciphering the genetic architecture of cooked rice texture. *Front. Plant Sci.* 9, 1405. https://doi.org/10.3389/fpls.2018.01405

Monkhan, T., Chen, X., and Somta, P. (2021). *BADH1* is associated with fragrance in sorghum (*Sorghum bicolor* (L.) Moench) cultivar 'Ambemohor'. *J. Genet.* 100, 3. https://doi.org/10.1007/s12041-020-01256-0

Nair, R.M., Boddepalli, V.N., Yan, M.-R., Kumar, V., Pan, R.S., Wang, C., et al. (2023). Global status of vegetable soybean. *Plants* 12, 609. https://doi.org/10.3390/plants12030609

Nawade, B., Mishra, G.P., Radhakrishnan, T., Dodia. S.M., Ahmad, S., Kumar, A., et al. (2018). High oleic peanut breeding: Achievements, perspectives, and prospects. *Trends Food Sci. Technol.* 78, 107–119. https://doi.org/10.1016/j.tifs.2018.05.022

Ndikuryayo, C., Ndayiragije, A., Kilasi, N., and Kusolwa, P. (2023). Screening for aroma in rice genotypes with African and Asian backgrounds. *Urban Agric. Reg. Food Syst.* 8, e20045. https://doi.org/10.1002/uar2.20045

Nordon, A.J., Gorbet, D.W., Knauft, D.A., and Young, C.T. (1987). Variability in oil quality among peanut genotypes in the Florida breeding program. *Peanut Sci.* 36, 126–132. https://doi.org/10.3146/i0095-3679-14-1-3

Nucia, A., Okoń, S., Tomczyńska-Mleko, M., and Nawrocka, A. (2021). Molecular and physical characterization of grain hardness in European spring common wheat (*Triticum aestivum* L.). *3 Biotech.* 11, 345. https://doi.org/10.1007/s13205-021-02897-3

Okpala, N.E., Mo, Z., Duan, M., and Tang, X. (2019). The genetics and biosynthesis of 2-acetyl-1-pyrroline in fragrant rice. *Plant Physiol. Biochem.* 135, 272–276. https://doi.org/10.1016/j.plaphy.2018.12.012

Pattee, H.E., Isleib, T.G., Gorbet, D.W., and Giesbrecht, F.G. (2001). Parent selection in breeding for roasted peanut flavor quality 1. *Peanut Sci.* 28, 51–58. https://doi.org/10.3146/i0095-3679-28-2-3.

Pattee, H.E., Isleib, T.G., Gorbet, D.W., and Giesbrecht, F.G. (2002). Selection of alternative genetic sources of large-seed size in Virginia-type peanut: Evaluation of sensory, composition, and agronomic characteristics. *J. Agric. Food Chem.* 50, 4885–4889. https://doi.org/10.1021/jf025601j

Phitaktansakul, R., Kim, K.-W., Aung, K.M., Maung, T.Z., Min, M.-H., Somsri, A., et al. (2022). Multi-omics analysis reveals the genetic basis of rice fragrance mediated by *betaine aldehyde dehydrogenase 2*. *J. Adv. Res.* 40, 303–314. https://doi.org/10.1016/j.jare.2021.12.004

Qian, L., Jin, H., Yang, Q., Zhu, L., Yu, X., Fu, X., et al. (2022).Sequence variation in *GmBADH2* enhances soybean aroma and is a functional marker for improving soybean flavor. *Int. J. Mol. Sci.* 23, 4116. https://doi.org/10.3390/ijms23084116

Rao, M.S.S., Bhagsari, A.S., and Mohamed, A.I. (2002). Fresh green seed yield and seed nutritional traits of vegetable soybean genotypes. *Crop Sci.* 42, 1950–1958. https://doi.org/10.2135/cropsci2002.1950

Ribeiro, N.D., and Maziero, S.M. (2023). Indirect selection for culinary quality and minerals in beans based on genotype × environment interaction. *Rev. Ciênc. Agron.* 54. https://doi.org/10.5935/1806-6690.20230044

Ritter, S.W., Enssil, S., Gastl, M.I., and Becker, T.M. (2024). Identification of key aroma compounds of faba beans (*Vicia faba*) and their development during germination—a SENSOMICS approach. *Food Chem.* 435, 137610. https://doi.org/10.1016/j.foodchem.2023.137610

Rivera, A., Casquero, P.A., Prieto, S.M., and Almirall, A. (2016). Culinary and sensory traits diversity in the Spanish Core Collection of common beans (*Phaseolus vulgaris* L.). *Spanish J. Agric. Res.* 14, e0701. https://doi.org/10.5424/sjar/2016141-7726

Rivera, A., Plans, M., Sabaté, J., Casañas, F., Casals, J., Rull, A., et al. (2018). The Spanish core collection of common beans (*Phaseolus vulgaris* L.): An important source of variability for breeding chemical composition. *Front. Plant Sci.* 9, 1642. https://doi.org/10.3389/fpls.2018.01642

Sadohara, R., Izquierdo, P., Couto Alves, F., Porch, T., Beaver, J., Urrea, C., et al. (2022). The *Phaseolus vulgaris* L. Yellow bean collection: Genetic diversity and characterization for cooking time. *Genet. Resour. Crop Evol.* 69, 1627–1648. https://doi.org/10.1007/s10722-021-01323-0

Sadohara, R., Long, Y., Izquierdo, P., Urrea, C.A., Morris, D., and Cinchy, C. (2021). Seed coat color genetics and G × E in a yellow bean collection via image analysis paired with machine-learning and GWAS. *Plant Genome* e20173. https://doi.org/10.1002/tpg2.20173

Salvador, M.V., Kock, F.V.C., Santos, I.L., Paulino, J.F.C., de Almeida, C.P., and Dos Santos Garcia, R.H. (2022). Monitoring stimulated darkening from UV-C light on different bean genotypes by NMR spectroscopy. *Molecules* 27, 2060. https://doi.org/10.3390/molecules27072060

Sandhu, K.S., You, F.M., Conner, R.L., Balasubramanian, P.M., and Hou, A. (2018). Genetic analysis and QTL mapping of the seed hardness trait in a black common bean (*Phaseolus vulgaris*) recombinant inbred line (RIL) population. *Mol. Breed.* 38, 34. https://doi.org/10.1007/s11032-018-0789-y

Schifferstein, H.N.J., Kudrowitz, B.M., and Breuer, C. (2020). Food perception and aesthetics—Linking sensory science to culinary practice. *J. Culin. Sci. Technol.* 20, 293–335. https://doi.org/10.1080/15428052.2020.1824833

Shabrangy, A., Roustan, V., Reipert, S., Weidinger, M., Roustan, P.-J., Stoger, E., et al. (2018). Using RT-qPCR, proteomics, and microscopy to unravel the spatio-temporal expression and subcellular localization of hordoindolines across development in barley endosperm. *Front. Plant Sci.* 9, 775. https://doi.org/10.3389/fpls.2018.00775

Shao, G., Tang, S., Chen, M., Wei, X., He, J., and Luo, J. (2013). Haplotype variation at *Badh2*, the gene determining fragrance in rice. *Genomics* 101, 157–162. https://doi.org/10.1016/j.ygeno.2012.11.010

Shao, G.N., Tang, A., Tang, S.Q., Luo, J., Jiao, G.A., Wu, J.L., et al. (2011). A new deletion mutation of fragrant gene and the development of three molecular markers for fragrance in rice. *Plant Breed.* 130, 172–176. https://doi.org/10.1111/j.1439–0523.2009.01764.x

Shi, W., Yang, Y., Chen, S., and Xu, M. (2008). Discovery of a new fragrance allele and the development of functional markers for the breeding of fragrant rice varieties. *Mol. Breed.* 22, 185–192.

Takahashi, A., Ikeda, T.M., Takayama, T., and Yanagisawa, T. (2010). A barley hordoindoline mutation resulted in an increase in grain hardness. *Theor. Appl. Genet.* 120, 519–526. https://doi.org/10.1007/s00122-009-1172-5

Tallury, S., Pattee, H., Isleib, T., and Stalker, H. (2008). Flavor profiles of wild species-derived peanut breeding lines. *Peanut Sci.* 35, 81–85. https://doi.org/10.3146/PS07-113.1

Terasawa, Y., Rahman, S.M., Takata, K., and Ikeda, T.M. (2012). Distribution of hordoindoline genes in the genus *Hordeum*. *Theor. Appl. Genet.* 124, 143–151. https://doi.org/10.1007/s00122-011-1693-6

Toda, K., Kato, S., Hirata, K., Kikuchi, A., Nihei, Y., and Hajika, M. (2020). DNA marker-assisted evaluation of cooked bean hardness of three soybean progeny lines. *Breed. Sci.* 70, 487–493. https://doi.org/10.1270/jsbbs.19171

Toili, M.E.M., de Koning, R., Kiekens, R., Ndumba, N., Wahome, S., Anami, S., et al. (2022). Transcriptome-guided identification of pectin methyl-esterase-related enzymes and novel molecular processes effectuating the hard-to-cook defect in common bean (*Phaseolus vulgaris* L.). *Foods* 11, 1692. https://doi.org/10.3390/foods11121692

Variath, M.T., and Janila, P. (2017). Economic and academic importance of peanut. In: Varshney, R.K., Pandey, M.K., and Puppala, N. (eds.) *The Peanut Genome, Compendium of Plant Genomes*. Springer International Publishing. https://doi.org/10.1007/978-3-319-63935-2_2

Wafula, E.N., Wainaina, I.N., Buvé, C., Nguyen, N.-D.-T., Kinyanjui, P.K., Saeya, W., et al. (2020), Application of near-infrared spectroscopy to predict the cooking times of aged common beans (*Phaseolus vulgaris* L.). *J. Food Eng.* 284, 110056. https://doi.org/10.1016/j.jfoodeng.2020.110056

Wang, M.L., Tonnis, B.D., Chen, C.Y., Li, X., Pinnow, D.L., Tallury, S., et al. (2022). Evaluation of variability in seed coat color, weight, oil content, and fatty acid composition within the entire USDA-cultivated peanut germplasm collection. *Crop Sci.* 62, 2332–2346. https://doi.org/10.1002/csc2.20830

Wang, Z., Yu, D., Morota, G., Dhakal, K., Singer, W., Lord, N., et al. (2023). Genome-wide association analysis of sucrose and alanine contents in edamame beans. *Front. Plant Sci.* 13, 1086007. https://doi.org/10.3389/fpls.2022.1086007

Watanabe, S., Omagari, A., Yamada, R., Matsumoto, A., Kimura, Y., Makita, N., et al. (2023). Mutations in the genes responsible for the synthesis of furan fatty acids resolve the light-induced off-odor in soybean oil. *Plant J.* https://doi.org/10.1111/tpj.16560

Wiesinger, J.A., Cichy, K.A., Glahn, R.P., Grusak, M.A., Brick, M.A., Thompson, H.J., et al. (2016). Demonstrating a nutritional advantage to the fast-cooking dry bean (*Phaseolus vulgaris* L.). *J. Agric. Food Chem.* 64, 8592–8603. https://doi.org/10.1021/acs.jafc.6b03100

Yang, C., Yan, J., Jiang, S., Li, X., Min, H., Wang, X., et al. (2022). Resequencing 250 soybean accessions: New insights into genes associated with agronomic traits and genetic networks. *GPB* 20, 29–41. https://doi.org/10.1016/j.gpb.2021.02.009

Yang, L., Yang, L., Ding, Y., Chen, Y., Liu, N., Zhou, X., et al. (2023). Global transcriptome and co-expression network analyses revealed hub genes controlling seed size/weight and/or oil content in peanut. *Plants (Basel)* 12, 3144. https://doi.org/10.3390/plants12173144

Yang, X., Liu, C., Li, Y., Yan, Z., Liu, D., and Feng, G. (2021). Identification and fine genetic mapping of the golden pod gene (pv-ye) from the snap bean (*Phaseolus vulgaris* L.). *Theor. Appl. Genet.* 134, 3773–3784. https://doi.org/10.1007/s00122-021-03928-6

Yuan, F., Fu, X., Yu, X., Yang, Q., Jin, H., and Zhu, L. (2021). Comparative analysis and development of a flavor fingerprint for volatile compounds of vegetable soybean seeds sased on headspace-gas chromatography-ion mobility spectrometry. *Front. Plant Sci.* 12, 768675. https://doi.org/10.3389/fpls.2021.768675

Yundaeng, C., Somta, P., Tangphatsornruang, S., Wongportchai, S., and Srinives, P. (2013). Gene discovery and functional marker development for fragrance in sorghum (*Sorghum bicolor* (L.) Moench). *Theor. Appl. Genet.* 126, 2897–906. https://doi.org/10.1007/s00122-013-2180-z

Zanan, R., Khandagale, K., Hinge, V., Elangovan, M., Henry, R.J., and Nadaf, A. (2016). Characterization of fragrance in sorghum (*Sorghum bicolor* (L.) Moench) grain and development of a gene-based marker for selection in breeding. *Mol. Breed.* 36, 146. https://doi.org/10.1007/s11032-016-0582-8.

Zhang, H., Dean, L., Wang, M.L., Dang, P., Lamb, M., and Chen, C. (2023a). GWAS with principal component analysis identify QTLs associated with main peanut flavor-related traits. *Front. Plant Sci.* 14, 1204415. https://doi.org/10.3389/fpls.2023.1204415

Zhang, K., He, Y., Lu, X., Shi, Y., Zhao, H., Li, X., et al. (2023b). Comparative and population genomics of buckwheat species reveal key determinants of flavor and fertility. *Mol. Plants* 16, 1427–1444. https://doi.org/10.1016/j.molp.2023.08.013

Zhang, S., Hu, X., Miao, H., Chu, Y., Cui, F., Yang, W., et al. (2019). QTL identification for seed weight and size based on a high-density SLAF-seq genetic map in peanut (*Arachis hypogaea* L.). *BMC Plant Biol.* 19, 537. https://doi.org/10.1186/s12870-019-2164-5

Zhao, J., Huang, L., Ren, X., Pandey, M.K., Wu, B., Chen, Y., et al. (2017). Genetic variation and association mapping of seed-related traits in cultivated peanut (*Arachis hypogaea* L.) using single-locus simple sequence repeat markers. *Front. Plant Sci.* 8, 2105. https://doi.org/10.3389/fpls.2017.02105

Zhao, M., Qian, L., Chi, Z., Jia, X., Qi, F., Yuan, F., et al. (2022). Combined metabolomic and quantitative RT-PCR analyses revealed the synthetic differences of 2-Acetyl-1-pyrroline in aromatic and non-aromatic vegetable soybeans. *Int. J. Mol. Sci.* 23, 14529. https://doi.org/10.3390/ijms232314529

Zhogoleva, A., Alas, M., and Rosenvald, S. (2023). Characterization of odor-active compounds of various pea preparations by GC-MS, GC-O, and their correlation with sensory attributes. *Future Foods* 8, 100243. https://doi.org/10.1016/j.fufo.2023.100243

8 Exploiting Natural and CRISPR/Cas9-Induced Allelic Variations in *Betaine Aldehyde Dehydrogenase (BADH)* to Develop Fragrant Grains, Fruits, and Vegetables

Sangam L. Dwivedi and Autar K. Mattoo

8.1 INTRODUCTION

Crops with appealing appearance, texture, flavor, and aroma have greater marketability. Today's consumers emphasis is on plant-based foods that not only have novel sensory attributes but also are nutritionally superior and offer therapeutic potential. Both conventional and modern breeding approaches have been effectively deployed to breed such crop attributes. Naturally occurring betaine aldehyde dehydrogenase 2 (BADH2) plays a key role in the accumulation of a fragrant compound, 2-acetyl-1-pyrroline (2AP). The demand for fragrant edible produce is gaining worldwide popularity among consumers. Fragrant rice fetches much higher premium than non-fragrant rice. 2AP is the master regulator of aroma/fragrance in jasmine and basmati rice (Bhattacharjee et al., 2002).

Both genetics and agronomy, in addition to the environment, influence the 2-AP contents. Cultivation practices, soil micro- and macronutrients, amino acid precursors, growth regulators, and environmental factors (e.g., drought, salinity, light, and temperature) influence the 2-AP biosynthesis in fragrant rice grains (Imran et al., 2023). Altitude, soil nitrogen content, and temperature also impact aroma in rice. Analysis of aromatic rice variety (Mushk Budiji), sampled from 5000 to 7000 feet above mean sea level from the highlands of the Himalayas, indicated 35 aromatic compounds. Fatty acid degradation and linoleic acid metabolic genes were found to be overexpressed at higher altitudes, while lipid biosynthesis was negatively influenced (Fayaz et al., 2024). Spring vs. summer sowing, depending on weather conditions, may also influence aroma. For example, foxtail millet grains produced

DOI: 10.1201/9781003467335-8

from spring season have higher aroma than those from summer produced grains. The aroma composition and characteristics also vary significantly depending on the sowing period and production region. Aldehydes were the key characteristic aroma components in foxtail millet grains, (*E,E*)-2,4-decadienal, (*E*)-2-nonenal, (*E,E*)-2,4-nonadienaldehyde, and nonanal, respectively, being the predominant aldehyde components in the same order (Li et al., 2025).

The use of CRISPR-Cas technologies (e.g., base editing, prime editing, multiplex gene editing, epigenome editing, gene delivery methods, etc.) are reliable, with precise gene editing to knock out, replace, and knock in the desired gene fragments at targeted locations in the genome with heritable mutations in plants (Verma et al., 2023).

8.2 MOLECULAR AND GENETIC BASIS OF FRAGRANCE

8.2.1 Grain Crops

Fragrant grains have been reported in rice, sorghum, and soybean. 2-acetyle 1- pyrroline (2AP) is a major volatile compound causing fragrance in these crops. A major gene for fragrance (*badh2*) mapped on chromosome 8 regulates fragrance in rice. An 8-bp deletion in exon 7 of this gene results in truncation of betaine aldehyde dehydrogenase enzyme whose loss of function leads to the accumulation of 2AP. Nonetheless, there may be other genetic loci controlling fragrance in rice (Sakthivel et al., 2009). Re-sequencing of large collection of Korean rice germplasm revealed novel functional haplotypes in the *BADH2* coding region, and an array of quantitative trait loci (eQTLs) expression and *trans*-protein quantitative trait loci (pQTLs) associated with *BADH2* expression and protein accumulation are likely regulators mediating 2AP variation in fragrant rice. Fifteen volatile compounds discriminated against fragrant haplotypes, which along with 2AP can be used as potential biomarkers for rice fragrance (Phitaktansakul et al., 2022).

A continuous 1,444 bp deletion causes a premature stop codon in *SbBADH2* in sorghum. Non-function of *SbBADH2* causes fragrance in sorghum accession IS19912 (Yundaeng et al., 2013), while *SbBADH1* regulates coconut-juice-like fragrance in another sorghum accession, ambemohar (Monkhan et al., 2021).

A nonsynonymous SNP (nsSNP) in *GmBADH2* causes fragrance in vegetable soybean variety Kaori (Juwattanasomran et al., 2011). The characteristic pandan-like aroma in some varieties of vegetable soybeans is also due to the presence of 2AP compound. A sequence comparison of *GmBADH1/2* between aromatic and non-aromatic soybean germplasm revealed that a mutation in *GmBADH2* is a major aroma-conferring gene in soybean (Qian et al., 2022).

Metabolome and gene expressions involving aromatic and nonaromatic vegetable grain soybeans harvested at 35 and 40 days after anthesis showed differentially accumulated metabolites, classified as flavonoids, lipids, amino acids, and derivatives, saccharides and alcohols, organic acids, nucleotides and derivatives, phenolic acids, alkaloids, and vitamin, which contribute to phenotypic differences in the 2AP synthesis between aromatic and nonaromatic soybeans. The accumulation of 2AP mainly occurs from 1-pyrroline-5-carboxylate (P5C) and the gene involved were *P5CDH*, *P5CS*, *PRODH*, and *P5CR* (Zhao et al., 2022).

8.2.2 Fresh Pod Fragrance

Fresh pods of lablab bean, also known as dolichos bean, (*Lablab purpureus* var. *lignosus*) are known for their characteristic fragrance due to the presence of two fatty acids: trans-2-dodecenoic acid and trans-2-tetradecenoic acid. The pods release oily substances that emit the characteristic scent. Fragrant pods fetch a premium price in the market. Betaine aldehyde dehydrogenase (BADH) causes fragrance in vegetable soybean and mung bean. *LpBADH2* isolated from lablab beans shows 97.3% homology with soybean homolog *GmBADH2* and houses a single nsSNPs and three sSNPs in *LpBADH2*. The substitution of the amino acid tyrosine in (fragrant genotypes) with phenylalanine (nonfragrant genotypes) in protein coded by *LpBADH2* caused switch over from fragrance to nonfragrance in lablab bean (Basanagouda et al., 2023).

8.2.3 Fragrance in Vegetables

A cucumber (*Cucumis sativus*) accession, PK2011T202, emits characteristic fragrance, caused by 2AP. An SNP (A1855G) in *CsBADH* gene, mapped on chromosome 1, is associated with fragrance in cucumber. Mutation(s) in *BADH2* gene is associated with biosynthesis of 2AP. A pandan-like fragrance was recently discovered from a Thailand cucumber cultivar. Sequence comparison of CsBADH involving fragrant (PK2011T202) and nonfragrant (301176) accessions revealed a single base substitution (A1855G) in exon 5, which causes an amino acid change in a highly conserved motif of BADH, Y163C. SNP polymorphism between the two accessions successfully differentiated fragrance alleles. A major QTL for fragrance, *qFgr*, co-localized with the *CsBADH* (Yundaeng et al., 2015).

LcBADH is associated with aroma in the sponge gourd (*Luffa cylindrica*). A single SNP (A > G) in exon 5 of *LcBADH* causes an amino acid change from tyrosine to cysteine at position 163, forming substrate binding pocket of the BADH enzyme. SNP marker, AroLuff, perfectly co-segregated with the aroma phenotypes in an F_2 population involving aromatic and nonaromatic sponge gourd population, supporting Mendelian's simple recessive inheritance. AroLuff marker was also effective in selection of aromatic ridge gourd (*Luffa acutangula*) progenies from a cross between aromatic sponge gourd and nonaromatic ridge gourd, suggesting that AroLuff is useful to introduce fragrance in both the species. The fragrance trait is not reported among ridge gourd accessions (Chaubey et al., 2022; Saensuk et al., 2022). The taro-like aroma, pleasant flavor, and value-added trait are also reported in pumpkin species. 2-acetyl-1-pyrroline is the key contributor to the taro-like aroma in pumpkin (Junxing et al., 2022).

8.3 HARNESSING AND EXPLOITING ALLELIC VARIATION FOR FRAGRANCE

8.3.1 Grain Crops

An 8-bp deletion in the exon of *badh2*, mapped on chromosome 8, results in truncation of betaine aldehyde dehydrogenase (BADH) enzyme. Loss of function of

BADH leads to the accumulation of a major aromatic compound, 2-acetyle 1-pyrroline (2AP), in fragrant rice. Exception to this mutation also indicates that there exist some other genetic loci in controlling grain fragrance in rice (Fitzgerald et al., 2008; Sakthivel et al., 2009). To date, 19 alleles of *badh2* gene accounted for unique spectrum of fragrance in aromatic rice (Akwero et al., 2020). *badh2.1* is the most prevalent allele detected among traditional basmati and jasmine type rice, while *bdh2.7* having a G insertion in the 14th exon is the second most prevalent allele in Asian rice germplasm. A novel allele with a T deletion in 9th exon has been reported among the Thai rice accessions. Both *badh2.1* and *badh2.7* alleles were identified with DNA markers (Kovach et al., 2009; Withana et al., 2020).

The traditional basmati rice cultivars are tall. They are prone to lodging, sensitive to photoperiod and temperature, and are low-yielding, hence the need to introduce short stature and stress tolerance to enhance productivity without loss of grain quality and aroma. BDAH2 functional markers have been developed to differentiate fragrant rice germplasm from nonfragrant types (Addision et al., 2020; Bradbury et al., 2005, Giang et al., 2023, Li et al., 2020, Shao et al., 2011, Singh et al., 2022, Okpala et al., 2019). A novel, efficient, high-throughput, and low-cost KASP (kompetitive allele-specific PCR) functional marker FM-E14 accurately and rapidly identifies fragrant homozygous, nonfragrant homozygous, and heterozygous genotypes (Pingyong et al., 2023).

Both traditional and marker-assisted breeding have led to development and release of several semi-dwarf, lodging resistant, and high-yielding rice cultivars, maintaining grain quality and aroma similar to traditional basmati types in Southeast Asian countries (Carsono et al., 2023; Grover et al., 2020; Kumar et al., 2023; Singh et al., 2018; Wang et al., 2021a). Pusa Basmati 1121 (PB 1121) is a landmark basmati rice in India with exceptional quality (extra-long slender milled grains, pleasant aroma, with exceptionally high cooked kernel elongation ratio of 2.5 with a cooked kernel length of up to 22 mm, volume expansion more than four times, appealing taste, good mouth feel and easy digestibility). It has brought revenue to the tune of US$ 20.8 billion to the country since 2008 and prosperity to millions of basmati farmers. Currently, it occupies ~70% of the total area under basmati rice cultivation in India (Singh et al., 2018). More recently, high-quality aromatic and anthocyanin rice (F25) has been developed using CRISPR/Cas9 technology, hybridization, and marker-assisted selection, to accelerate multi-trait improvement and meet the growing demand of premium quality rice in China (Shi et al., 2023).

Aromatic sorghum grain germplasm and functional markers, associated with characteristic pandan-like flavor, have been reported to effect marker-assisted breeding for development of aromatic grain sorghums (Yundaeng et al., 2013; Zanan et al., 2016).

8.3.2 Fruits and Vegetables

Most commercial cultivars of winter melon lack the pandan-like aroma, while it is present in some landraces. This type of aroma adds higher economic value. It is thus an attractive breeding target. A sequence comparison of *BhAMADH* between the aromatic and non-aromatic accessions revealed an 804-bp deletion encompassing exons

11–13 in the aromatic accession. A functional marker, based on the 804-bp deletion, differentiated aromatic and non-aromatic segregants in 237 F_2 progenies. A perfect association of the marker genotypes and aroma phenotypes suggests *BhAMADH*, a major gene conferring aroma in winter melon (Ruangnam et al., 2017).

8.4 CRISPR/CAS9-INDUCED FRAGRANCE IN CEREAL GRAINS

Conventional selection via crossbreeding has led to a limited genetic improvement in the quality of staple food crops. Advances in gene-editing technology such as CRISPR/Cas9 have opened the door to significant breakthroughs by editing specific genes of crops due to their accuracy and efficiency. Precise editing of key genes by CRISPR/Cas9 technology has improved quality and biomass yield of staple food crops (Guo et al., 2023; Wei et al., 2023). Maize genome harbors two functional *BADH2* homologs, *ZmBADH2a* and *ZmBADH2b*. CRISPR/Cas9 system was used to generate single and double mutant in four maize inbred lines. Double mutant (*zmbadh2a-zmbadh2b*) seeds produced a characteristic popcorn-like aroma but not from either single mutant or the wild type (Wang et al., 2021b). Knockout of SbBADH2 via CRISPR/Cas9 generated sorghum lines with aromatic smell in both seeds and leaves compared to wild type (Zhang et al., 2022). More recently, a sweet sorghum line was successfully created by editing *SbBADH2* (Cheng et al., 2024). CRISPR/Cas9 generated novel alleles of OsBADH2 with characteristic aroma into an elite non-aromatic rice cultivar ASD16. Sequence analysis of putative aromatic transformant (T_0) lines detected 22 different types of mutations, of which those with -1/-2 bp or -8/-5 bp deletions produced strong aroma that is stably inherited in the T_1 generation (Ashokkumar et al., 2020). Further, a two-line hybrid combining glutinous and aromatic traits or a three-line aromatic hybrid via CRISPR/Cas9 system have been reported (Hui et al., 2022; Zhang et al., 2023a). Popcorn aroma is a valuable trait in cereals. Millet grains lack this trait. Using CRISPR/Cas9 technology and *SiBADH2* gene in foxtail millet, a genetic variant successfully created an aromatic foxtail millet variety with popcorn-like fragrance in both seeds and leaves. Thus, it is a valuable resource to enhance aroma in foxtail millet grains (Zhang et al., 2023b). Consumers will soon have food products derived from aromatic foxtail millet, maize, and sorghum (including sweet sorghum) on their plates.

8.5 GEOGRAPHICAL IDENTIFICATION (GI) TAGS AND ACHIEVING AUTHENTICITY OF AROMATIC GRAINS

The agricultural products of high value, such as aromatic basmati rice grains from India, are characterized by the environmental conditions of the production region, especially climatic and soil conditions, and are often protected by GI tags. Over 100 countries have introduced this GI protection system for their unique products. Stable isotope (δD, $\delta^{13}C$, $\delta^{15}N$, $\delta^{18}O$, and $\delta^{34}S$) analysis is often used to trace the geographical origin of food materials (Suzuki, 2021). Multielement fingerprinting, including trace element ratio, combined with appropriate statistical modeling and analysis, can also be used to identify the geographic origin of the produce (Bitter et al., 2020; Rocha et al., 2019; Zhao et al., 2020).

The growing demand for such unique food products worldwide has increased food fraud opportunities (mislabeling and adulteration) in the supply chain, delivering poor-quality food in the name of branded products. Various techniques, including the use of DNA markers, have been developed to check adulteration in grain lots. A two-tiered testing regime of rapid screening using portable near infrared technology (NIR) supported by mass-spectrometry-based analysis effectively differentiated high-value basmati rice grains from a specific geographic region and their potential adulterants (McGrath et al., 2021). However, a tailor-made solution is needed since rice fraud manifests differently in different countries. NIR analysis supported by headspace–gas chromatography–mass spectrometry (HS-GC-MS) correctly predicted nearly 100% of Pusa 1121 and Taraori, the two high-value types of Indian basmati rice grains, from potential adulterants. Fingerprinting of rice volatile organic compounds (VOC) indicated >0.99 accuracy for differentiating high-quality basmati varieties from potential adulterant varieties, with aldehydes as key VOC marker compounds (Shannon et al., 2021).

Paw San rice (or Myanmar pearl rice) produced in the Shwe Bo District in the Sagaing region of Myanmar is the most valued and highly priced kind of Paw San rice. Fourier-transform near-infrared (FT-NIR) spectroscopy and headspace–gas chromatography–ion mobility spectrometry (HS-GC-IMS), coupled with chemometric modeling, effectively differentiated the geographic origin of Paw San rice (Thantar et al., 2024). Hom Mali rice is of highest quality rice in Thailand. Fourier-transform infrared spectroscopy with attenuated total reflection (FTIR-ATR) and NIR spectroscopy effectively differentiated Hom Mali rice from two geographical regions of Thailand (Srinuttrakul et al., 2021). A combination of stable isotope and multielement analytical techniques with chemometrics aided by partial least squares discriminant analysis (PLS-DA) and linear discriminant analysis (LDA) effectively differentiated authenticity of rice samples from eight production systems, with isotope of δ^{2}H, δ^{13}C, Ba, Al, Mg, δ^{34}S, Pb, and δ^{18}O as most important variables for testing rice origin traceability (Zhi et al., 2023).

8.6 CONCLUSIONS

Aromatic edible products are in demand worldwide. Allelic diversity and associated marker technology are routinely used to develop rice cultivars with aromatic grains, marketed as basmati or jasmine rice from the Asia region. There is also aroma/fragrance in other grain crops (foxtail millet, sorghum, soybean), the lablab bean, and in fruits (winter melon) and vegetables (cucumber, sponge gourd). *BADH* or its variants control fragrance. CRISPR/Cas9 system is equally effective in inducing allelic variation associated with aroma in maize, rice, and sorghum. A new rice cultivar combining grain aroma and anthocyanins demonstrates that multi-trait improvement could be achieved to meet the growing demand of quality food produce. Regular consumption of anthocyanin-rich food minimizes the risk of noncommunicable diseases in humans. Many countries have adapted the use of GI tags to their unique agricultural products, bringing prosperity to millions of farmers cultivating these unique products. Product uniqueness could be identified either by analysis of stable isotope (δD, δ^{13}C, δ^{15}N, δ^{18}O, and δ^{34}S) or by multielement fingerprinting, including trace element ratios.

REFERENCES

Addision, C.K., Angira, B., Kongchum, M., Harrell, D.L., Baisakh, N., Linscombe, S.D., et al. (2020). Characterization of haplotype diversity in the *BADH2* aroma gene and development of a KASP SNP assay for predicting aroma in U.S. rice. *Rice (N Y)* 13, 47. https://doi.org/10.1186/s12284-020-00410-7

Akwero, A., Ocan, D., Akech, W., Lamo, J., Ochwo-Ssemakula, M., and Rubaihayo, P. (2020). Allelic variations in aroma gene in cultivated rice varieties. *African Crop Sci. J.* 28, 241–254. https://doi.org/10.4314/acsj.v28i2.9

Ashokkumar, S., Jaganathan, D., Ramanathan, V., Rahman, H., Palaniswamy, R., Kambale, R., et al. (2020). Creation of novel alleles of fragrance gene *OsBADH2* in rice through CRISPR/Cas9 mediated gene editing. *PLoS ONE* 15, e0237018. https://doi.org/10.1371/journal.pone.0237018

Basanagouda, G., Ramesh, S., Siddu, C.B., Chandana, B., Kalpana, M.P., Rotti, K., et al. (2023). A non-synonymous SNP in homolog of *BADH2* gene is associated with fresh pod fragrance in dolichos bean (*Lablab purpureus* var. lignosus (Prain) Kumari). *Genet. Resour. Crop Evol.* 70, 373–380. https://doi.org/10.1007/s10722-022-01535-y

Bhattacharjee, P., Singhal, R.S., and Kulkarni, P.R. (2002). Basmati rice: A review. *Int. J. Food Sci. + Technol.* 37, 1–12. https://doi.org/10.1046/j.1365–2621.2002.00541.x

Bitter, N.Q., Fernandez, D.P., Driscoll, A.W., Howa, J.D., and Ehleringer, J.R. (2020). Distinguishing the region-of-origin of roasted coffee beans with trace element ratios. *Food Chem.* 320, 126602. https://doi.org/10.1016/j.foodchem.2020.126602

Bradbury, L.M.T., Fitzgerald, T.L., Henry, R.J., Jin, Q., and Waters, D.L.E. (2005). The gene for fragrance in rice. *Plant Biotechnol. J.* 3, 363–370. https://doi.org/10.1111/j.1467–7652.2005.00131.x

Carsono, N., Tambunam, R.R., Sari, S., and Wicaksana, N. (2023). Molecular and phenotypic markers for pyramiding multiple traits in rice. *Open Agric.* 8, 20220187. https://doi.org/10.1515/opag-2022–0187

Chaubey, T., Sagar, V., Singh, R.K., Chanotyia, C.S., Pndey, S., Singh, P.M., et al. (2022). Volatile compounds governed by single recessive gene impart aroma in sponge gourd (*Luffa cylindrica* L. Roem). *Plants (Basel)* 11, 2881. https://doi.org/10.3390/plants11212881

Cheng, Z., Li, K., Liu, H., Wei, X., Yin, T., Xing, X., et al. (2024). Establishment of a genome-editing system to create fragrant germplasm in sweet sorghum. *aBIOTECH.* https://doi.org/10.1007/s42994-024-00180-6

Fayaz, U., Hussain, S.Z., Naseer, B., Mahdi, S.S., Mir, J.I., Ghosh, A., et al. (2024). Flavor profiling and gene expression studies of indigenous aromatic rice variety (*Mushk Budiji)* grown at different altitudes of Highland Himalayan regions. *Sci. Rep.* 14, 1010. https://doi.org/10.1038/s41598-024-51467-z

Fitzgerald, M.A., Hamilton, N.R.S., Calingacion, M.N., Verhoeven, H.A., and Butardo, V.M. (2008). Is there a second fragrance gene in rice? *Plant Biotechnol. J.* 6, 416–423. https://doi.org/10.1111/j.1467–7652.2008.00327.x

Giang, V.Q., Ky, H., Tung, N.C.T., Hien, N.L., van Manh, N., Thanh, N.N., et al. (2023). Novel deletion in exon 7 of *Betaine Aldehyde Dehydrogenase 2* (*BADH2*). *Rice Sci.* 30, 104–112. http://doi.org/10.1016/j.rsci.2023.01.00

Grover, N., Kumar, A., Yadav, A.K., Krishnan, S.G., Ellur, R.K., Bhowmick, P.K., et al. (2020). Marker assisted development and characterization of herbicide tolerant near isogenic lines of a mega basmati rice variety, "Pusa Basmati 1121". *Rice* 13, 68. https://doi.org/10.1186/s12284-020-00423-2

Guo, Y., Zhao, G., Gao, X., Zhang, L., Zhang, Y., Cai, X., et al. (2023). CRISPR/Cas9 gene editing technology: A precise and efficient tool for crop quality improvement. *Planta* 258, 36. https://doi.org/10.1007/s00425-023-04187-z

Hui, S., Li, H., Mawia, A.M., Zhou, L., Cai, J., Ahmad, S., et al. (2022). Production of aromatic three-line hybrid rice using novel alleles of *BADH2*. *Plant Biotechnol. J.* 20, 59–74. https://doi.org/10.1111/pbi.13695

Imran, M., Shafiq, S., Ashraf, U., Qi, J., Mo, Z., and Tang, X. (2023). Biosynthesis of 2-Acetyl-1-pyrroline in fragrant rice: Recent insights into agro-management, environmental factors, and functional genomics. *J. Agric. Food Chem.* 71, 4201–4215. https://doi.org/10.1021/acs.jafc.2c07934

Junxing, L.I., Aiqing, M., Gangjun, Z., Xiaoxi, L., Haibin, W., Jianning, L., et al. (2022). Assessment of the 'taro-like' aroma of pumpkin fruit (*Cucurbita moschata* D.) via E-nose, GC-MS and GC-O analysis. *Food Chem. X* 15, 100435. https://doi.org/10.1016/j.fochx.2022.100435

Juwattanasomran, R., Somta, P., Chankaew, S., Shimizu, T., Wongpornchai, S., and Kaga, A. (2011). A SNP in *GmBADH2* gene associates with fragrance in vegetable soybean variety "Kaori" and SNAP marker development for the fragrance. *Theor. Appl. Genet.* 122, 533–541. https://doi.org/10.1007/s00122-010-1467-6

Kovach, M.J., Calingacion, M.N., Fitzgerald, M.A., and McCouch, S.R. (2009). The origin and evolution of fragrance in rice (*Oryza sativa* L.). *Proc. Natl. Acad. Sci. U. S. A.* 106, 14444–14449. https://doi.org/10.1073/pnas.0904077106

Kumar, M., Singh, R.P., Jena, D., Singh, V., Rout, D., Arsode, P.B., et al. (2023). Marker-assisted improvement for durable bacterial blight resistance in aromatic rice cultivar HUR 917 popular in eastern parts of India. *Plants (Basel)* 12, 1363. https://doi.org/10.3390/plants12061363

Li, P., Li, S., Zhao, W., Zhang, A., Liu, J., Wang, Y., et al. (2025). Characteristic odor of foxtail millet from different area with different sowing time based on gas chromatography-mass spectrometry. *J. Future Foods* 5, 50–56. https://doi.org/10.1016/j.jfutfo.2024.01.004

Li, W., Zeng, X., Li, S., Chen, F., and Gao, J. (2020). Development and application of two novel functional molecular markers of *BADH2* in rice. *Electr. J. Biotechnol.* 46, 1–7. https://doi.org/10.1016/j.ejbt.2020.04.004

McGrath, T.F., Shannon, M., Chevallier, O.P., Ch, R., Xu, F., Kong, F., et al. (2021). Food fingerprinting: Using a two-tiered approach to monitor and mitigate food fraud in rice. *J AOAC Int.* 104, 16–28. https://doi.org/10.1093/jaoacint/qsaa109

Monkhan, T., Chen, X., and Somta, P. (2021). *BADH1* is associated with fragrance in sorghum (*Sorghum bicolor* (L.) Moench) cultivar 'Ambemohor'. *J. Genet.* 100, 3. https://doi.org/10.1007/s12041-020-01256-0

Okpala, N.E., Mo, Z., Duan, M., and Tang, X. (2019). The genetics and biosynthesis of 2-acetyl-1-pyrroline in fragrant rice. *Plant Physiol. Biochem.* 135, 272–276. https://doi.org/10.1016/j.plaphy.2018.12.012

Phitaktansakul, R., Kim, K.-W., Aung, K.M., Maung, T.Z., Min, M.-H., Somsri, A., et al. (2022). Multi-omics analysis reveals the genetic basis of rice fragrance mediated by *betaine aldehyde dehydrogenase 2*. *J. Adv. Res.* 42, 303–314. https://doi.org/10.1016/j.jare.2021.12.004

Pingyong, S., Wuhan, Z., Li, Z., Fu, S., Qiang, H., Na, X., et al. (2023). Development and application of a novel functional marker for fragrance in rice. *Rice Sci.* 30, 176–180. https://doi.org/10.1016/j.rsci.2023.03.002

Qian, L., Jin, H., Yang, Q., Zhu, L., Yu, X., Fu, X., et al. (2022). A sequence variation in *GmBADH2* enhances soybean aroma and is a functional marker for improving soybean flavor. *Int. J. Mol. Sci.* 23, 4116. https://doi.org/10.3390/ijms23084116

Rocha, S., Pinto, E., Almeida, A., and Fernandes, E. (2019). Multi-elemental analysis as a tool for characterization and differentiation of Portuguese wines according to their Protected Geographical Indication. *Food Control* 103, 27–35. https://doi.org/10.1016/j.foodcont.2019.03.034

Ruangnam, S., Wanchana, S., Phoka, N., Saeansuk, C., Mahatheeranont, S., de Hoop, S.J., et al. (2017). A deletion of the gene encoding amino aldehyde dehydrogenase enhances the "pandan-like" aroma of winter melon (*Benincasa hispida*) and is a functional marker for the development of the aroma. *Theor. Appl. Genet.* 130, 2557–2565. https://doi.org/10.1007/s00122-017-2976-3

Saensuk, C., Ruangnam, S., Pitloka, M.K., Dumhai, R., Mahatheeranont, S., de Hoop, S.J., et al. (2022). A SNP of betaine aldehyde dehydrogenase (BADH) enhances an aroma (2-acetyl-1-pyrroline) in sponge gourd (*Luffa cylindrica*) and ridge gourd (*Luffa acutangula*). *Sci. Rep.* 12, 3718. https://doi.org/10.1038/s41598-022-07478-9

Sakthivel, K., Sundaram, R.M., Sobha Rani, N., Balachandran, S.M., and Neeraja, C.N. (2009). Genetic and molecular basis of fragrance in rice. *Biotechnol. Adv.* 27, 468–473. https://doi.org/10.1016/j.biotechadv.2009.04.001

Shannon, M., Ratnasekhar, C.H., McGrath, T.F., Kapil, A.P., and Elliott, C.T. (2021). A two-tiered system of analysis to tackle rice fraud: The Indian Basmati study. *Talanta* 225, 122038. https://doi.org/10.1016/j.talanta.2020.122038

Shao, D.N., Tang, A., Tang, S.Q., Luo, J., Jiao, G.A., Wu, J.L., et al. (2011). A new deletion mutation of fragrant gene and the development of three molecular markers for fragrance in rice. *Plant Breed.* 130, 172–176. https://doi.org/10.1111/j.1439–0523.2009.01764.x

Shi, T., Gao, Y., Xu, A., Wang, R., Lyu, M., Sun, Y., et al. (2023). A fast-breeding strategy creates fragrance- and anthocyanin-enriched rice lines by marker-free gene-editing and hybridization. *Mol. Breed.* 43, 23. https://doi.org/10.1007/s11032-023-01369-1

Singh, G., Krishnan, S.G., Kumar, A., Vinod, K.K., Bollinedi, H., Ellur, R.K., et al. (2022). Molecular profiling of *BADH2* locus reveals distinct functional allelic polymorphism associated with fragrance variation in Indian aromatic rice germplasm. *Physiol. Mol. Biol. Plants* 28, 1013–1027. https://doi.org/10.1007/s12298-022-01181-9

Singh, V., Singh, A.K., Mohapatra, T., Krishnan, G., and Ellur, R.K. (2018). Pusa Basmati 1121—a rice variety with exceptional kernel elongation and volume expansion after cooking. *Rice* 11, 19. https://doi.org/10.1186/s12284-018-0213-6

Srinuttrakul, W., Mihailova, A., Islam, M.D., Liebisch, B., Maxwell, F., Kelly, S.D., et al. (2021). Geographical differentiation of Hom Mali rice cultivated in different regions of Thailand using FTIR-ATR and NIR spectroscopy. *Foods* 10, 1951. https://doi.org/10.3390/foods10081951

Suzuki, Y. (2021). Achieving food authenticity and traceability using an analytical method focusing on stable isotope analysis. *Anal. Sci.* 37, 189–199. https://doi.org/10.2116/analsci.20SAR14

Thantar, S., Mihailova, A., Islam, M.D., Maxwell, F., Hamed, I., Vlachou, C., et al., (2024). Geographical discrimination of Paw San rice cultivated in different regions of Myanmar using near-infrared spectroscopy, headspace-gas chromatography-ion mobility spectrometry and chemometrics. *Talanta* 273, 125910. https://doi.org/10.1016/j.talanta.2024.125910

Verma, V., Kumar, A., Partap, M., Thakur, M., and Bhargava, B. (2023). CRISPR-Cas: A robust technology for enhancing consumer-preferred commercial traits in crops. *Front. Plant Sci.* 14, 1122940. https://doi.org/10.3389/fpls.2023.1122940

Wang, X., Guo, X., Ma, X., Luo, L., Fang, Y., Zhao, N., et al. (2021a). Development of new rice (*Oryza. sativa* L.) breeding lines through marker-assisted introgression and pyramiding of brown planthopper, blast, bacterial leaf blight resistance, and aroma genes. *Agronomy* 11, 2525. https://doi.org/10.3390/agronomy11122525

Wang, Y., Liu, X., Zheng, X., Wang, W., Yin, X., Liu, H., et al. (2021b). Creation of aromatic maize by CRISPR/Cas. *J. Integr. Plant Biol.* 63, 1664–1670. https://doi.org/10.1111/jipb.13105

Wei, X., Long, Y., Yi, C., Pu, A., Hou, Q., Liu, C., et al. (2023). Bibliometric analysis of functional crops and nutritional quality: Identification of gene resources to improve crop nutritional quality through gene editing technology. *Nutrients* 15, 373. https://doi.org/10.3390/nu15020373

Withana, W.V.E., Kularathna, R.M.R.E., Kottearachchi, N.S., Kekulandra, D.S., Weerasena, J., and Steele, K. (2020). In silico analysis of the fragrance gene (badh2) in Asian rice (*Oryza sativa* L.) germplasm and validation of allele specific markers. *Plant Genet. Resour.* 18,71–80. https://doi.org/10.1017/S1479262120000015

Yundaeng, C., Somta, P., Tangphatsornruang, S., Chankaew, S., and Srinives, P. (2015). A single base substitution in *BADH/AMADH* is responsible for fragrance in cucumber (*Cucumis sativus* L.), and development of SNAP markers for the fragrance. *Theor. Appl. Genet.* 128(9), 1881–1892. https://doi.org/10.1007/s00122-015-2554-5

Yundaeng, C., Somta, P., Tangphatsornruang, S., Wongpornchai, S., and Srinivas, P. (2013). Gene discovery and functional marker development for fragrance in sorghum *Sorghum bicolor* (L.) Moench. *Theor. Appl. Genet.* 126, 2897–2906. https://doi.org/10.1007/s00122-013-2180-z

Zanan, R., Khadagale, K., Hinge, V., Elangovan, M., Henry, R.I., and Nadaf, A. (2016). Characterization of fragrance in sorghum *Sorghum bicolor* (L.) Moench grain and development of a gene-based marker for selection in breeding. *Mol Breed.* 36, 146. https://doi.org/10.1007/s11032-016-0582-8

Zhang, C., Yun, P., Xia, J., Zhou, K., Wang, L., Zhang, J., Bo, Z., et al. (2023a). CRISPR/Cas9-mediated editing of *Wx* and *BADH2* genes created glutinous and aromatic two-line hybrid rice. *Mol. Breed.* 43, 24. https://doi.org/10.1007/s11032-023-01368-2

Zhang, D., Tang, S., Xie, P., Yang, D., Wu, Y., Cheng, S., et al. (2022). Creation of fragrant sorghum by CRISPR/Cas9. *J. Integr. Plant Biol.* 64, 961–964. https://doi.org/10.1111/jipb.13232

Zhang, Y., He, Q., Zhang, S., Man, X., Sui, Y., Jia, G., et al. (2023b). *De novo* creation of popcorn-like fragrant foxtail millet. *J. Integr. Plant Biol.* 65, 2412–2415. https://doi.org/10.1111/jipb.13556

Zhao, H., Wang, F., and Yang, Q. (2020). Origin traceability of peanut kernels based on multi-element fingerprinting combined with multivariate data analysis. *J. Sci. Food Agric.* 100, 4040–4048. https://doi.org/10.1002/jsfa.10449

Zhao, M., Qian, L., Chi, Z., Jia, X., Qi, F., Yuan, F., et al. (2022). Combined metabolomic and quantitative RT-PCR analyses revealed the synthetic differences of 2-acetyl-1-pyrroline in aromatic and non-aromatic vegetable soybeans. *Int. J. Mol. Sci.* 23, 14529. https://doi.org/10.3390/ijms232314529

Zhi, L., Yuan, W., Yudi, H., Wei, L., Bin, L., and Guiyuan, M. (2023). Multi-stable isotope and multi-element origin traceability of rice from the main producing regions in Asia: A long-term investigation during 2017–2020. *Food Chem.* 412, 135417. https://doi.org/10.1016/j.foodchem.2023.135417

9 Biofortification

Transforming Staple Food Crops for Enhanced Aroma, Flavonoids, and Micronutrients with Reduced Seed Phytate

Rodomiro Ortiz and Sangam L. Dwivedi

9.1 INTRODUCTION

Biofortification, the process of enhancing the nutritional content of food crops through crossbreeding or genetic engineering, has gained momentum as a sustainable strategy to address malnutrition and improve public health. Furthermore, products derived from whole grains improve human health and safety due to their unique blend of bioactive compounds (e.g., resistant starch, vitamins, minerals, phytochemicals, and antioxidants), which provide benefits (Rawat et al., 2023). This chapter delves into the fascinating realm of biofortification, focusing on the development of staple food crops enriched with fragrance, flavonoids, and essential micronutrients while concurrently minimizing seed phytate content.

Biofortification represents a promising approach to combat widespread nutrient deficiencies, particularly in regions where access to diverse diets is limited. Birol and Bouis (2023) indicated that socioeconomics help identify priority micronutrient-crop-geography combinations for maximum impact. Moreover, randomized controlled bio-efficacy trials and farmer-adoption models assist in assessing the potential health impacts of biofortification, projecting outcomes through empirical results obtained by nutritionists. Similarly, surveys of growers and end-users, along with monitoring systems, provide necessary information for estimating farmer adoption rates and understanding input/output markets, as well as farmer and consumer preferences. These integrated approaches ensure that biofortification efforts are targeted effectively and have a tangible impact on improving public health and nutrition outcomes. Furthermore, as noted by Hall et al. (2008), enhancing the comprehension of the pathways responsible for biosynthesizing nutritionally relevant metabolites is essential for gaining more precise control over the presence or absence and levels of such components in food. This increased understanding enables us to manipulate these pathways more effectively, thereby facilitating the

 DOI: 10.1201/9781003467335-9

development of crops with desired nutritional profiles and ensuring the availability of essential nutrients in our diets.

By fortifying staple crops with essential vitamins and minerals, such as iron, zinc, and vitamin A, biofortification offers a sustainable solution to improve the nutritional status of vulnerable populations. Additionally, the incorporation of aromatic compounds and flavonoids not only enhances the sensory attributes of food but also contributes to overall health and well-being through their antioxidant and anti-inflammatory properties.

Crop husbandry techniques may modestly enhance the nutritional value of certain foods, yet the burgeoning field of molecular biology is swiftly being utilized to engineer crops with significantly enhanced key nutrients. These nutritional enhancements target various aspects, including elevated mineral content, optimized fatty acid composition, augmented levels of amino acids, and heightened antioxidant levels. By leveraging advancements in molecular biology, researchers aim to revolutionize agricultural practices, creating crops that not only yield higher yields but also offer improved nutritional profiles, thereby addressing global health challenges related to nutrient deficiencies more effectively (Hirschi, 2009). Through innovative crossbreeding techniques and advances in genetic engineering, significant strides in developing biofortified cultivars of staple crops continue. As highlighted by Zhu et al. (2020), advancements in plant synthetic metabolic engineering, particularly regarding the development of research strategies for multigene stacking tools and the engineering of complex metabolic pathways, have proven effective and precise in enhancing traits associated with micronutrients, phytonutrients, and bioactive components. These breakthroughs underscore the potential of plant synthetic metabolic engineering in precisely modifying key nutritional aspects of crops. These efforts hold great promises for addressing malnutrition on a global scale and promoting sustainable food systems that prioritize both human health and environmental stewardship. By harnessing these advances and innovations, we can accelerate the development of nutrient-enriched crops tailored to meet the diverse nutritional needs of humanity, thereby contributing to improved public health and nutrition outcomes globally.

9.2 FRAGRANCE-ENRICHED STAPLE CROPS

The aroma of staple foods significantly influences culinary preferences and enriches the overall dining experience. Aromas are integral to the flavor and quality management of fruits and vegetables, as highlighted by Abbas et al. (2023). Subtle variations in aroma compounds can have a profound impact on the taste and texture of these crops, ultimately shaping their consumer appeal. Leveraging biofortification techniques, plant breeders and other scientists are actively working to enhance the fragrance of staple crops, elevating the act of eating to a sensory delight. For example, fragrant rice cultivars like basmati and jasmine rice have been meticulously bred, captivating consumers with their distinct aromas and enhancing the gastronomic allure of these essential grains. Through such advancements, biofortification not only addresses nutritional needs but also enhances the sensory pleasure associated with consuming staple foods.

9.3 FLAVONOID-RICH STAPLE CROPS

Flavonoids, a class of polyphenolic compounds, have gained considerable attention due to their potent antioxidant properties and potential health-promoting effects. Biofortification initiatives aim to elevate flavonoid levels in staple crops, not only conferring nutritional benefits but also enhancing the vibrant colors and distinct flavors associated with certain varieties. Dwivedi et al. (2022) provide insights into the strides made in characterizing genes involved in anthocyanin biosynthesis across cereal and other crops. Through breeding efforts, grain cultivars rich in anthocyanins have been developed and introduced in regions spanning Europe, America, and Asia. Notably, advancements in genetic engineering techniques, including the targeted manipulation of transcription factors and precise gene editing, have facilitated the development of anthocyanin-enriched genetic variants without compromising yield. This technological progress has paved the way for a diverse array of food products derived from colored grains or flours, now readily available in grocery stores worldwide. Furthermore, biofortification extends to crops like purple and red-fleshed potatoes, which are enriched with anthocyanins, a subclass of flavonoids. These biofortified potatoes not only offer visual appeal but also boast enhanced nutritional value, aligning with consumer preferences for healthier food choices. As such, the integration of flavonoid-rich crops into our diets not only enhances nutritional intake but also contributes to the sensory enjoyment of meals, ultimately promoting overall well-being.

9.4 MICRONUTRIENT-ENHANCED STAPLE CROPS

Micronutrient deficiencies, often referred to as hidden hunger, remain a global health challenge. Biofortification addresses this issue by developing staple crops with elevated levels of essential micronutrients such as iron, zinc, and vitamin A. Golden rice enriched with beta-carotene (a precursor to vitamin A) exemplifies the potential of biofortification to combat vitamin A deficiency, particularly in regions where rice is a dietary staple, e.g., South Asia.

In the era of post-genomics and computational systems biology, the integration of high-throughput genomics and robust statistical analysis, particularly in mapping quantitative trait loci (QTL), has significantly enhanced our understanding of the molecular basis of natural diversity for complex quantitative traits (Garcia-Oliveira et al., 2018). Recent molecular mapping studies have illuminated the co-localization of QTL associated with iron (Fe) and zinc (Zn) with those linked to other potentially toxic metals such as cadmium (Cd), lead (Pb), and arsenic (As). This knowledge offers valuable insights into designing targeted crosses for marker-assisted breeding (MAB), aiming to develop cereal cultivars with elevated levels of Fe and Zn. Nevertheless, while QTL specific to Fe or Zn have been identified, revealing plants' ability to discriminate between essential nutrients and chemically analogous toxic ions, concerns persist regarding inadvertent breeding for nonessential metals harmful to both plants and animals, even at low concentrations. Moreover, research on mineral nutrients often overlooks modifications in the accumulation of these toxic elements crucial for food safety. Hence, harnessing natural genetic variation for mineral nutrients through molecular breeding holds promise for the future. Additionally, the significant genetic variability observed in Fe and Zn bioavailability inhibitors and promoters presents opportunities for enhancing the bioavailability of these minerals in cereal grains. However, genes responsible for this variability remain largely undiscovered and underutilized in breeding programs, representing a potential avenue for further exploration. In conclusion, improving Fe and Zn bioavailability from cereal grains through breeding strategies, whether by accumulating anti-nutrient agents or prebiotics, shows potential for enhancing the nutritional quality of staple crops and addressing micronutrient deficiencies in human populations.

9.5 REDUCED SEED PHYTATE CONTENT

Phytate, while present in seeds, acts as an anti-nutrient by binding essential minerals and inhibiting their absorption during digestion. The reduction of seed phytate content is another crucial aspect of biofortification efforts (Dwivedi and Ortiz, 2021). Phytates, while present in seeds and grains, can bind essential minerals, rendering them less available for absorption in the human body. By minimizing phytate levels in biofortified crops, the bioavailability of key nutrients is increased, maximizing the nutritional impact of these foods. Hence, biofortification strategies include reducing seed phytate levels in staple crops to enhance mineral bioavailability. Low-phytate soybeans, for instance, have been developed to improve the absorption of minerals like iron and zinc, addressing malnutrition issues related to these micronutrients.

9.6 EXAMPLES OF BIOFORTIFIED STAPLE CROPS

9.6.1 Aromatic Rice

Basmati rice, jasmine rice, and other aromatic cultivars have been biofortified to enhance fragrance, making them sought-after choices in global culinary traditions (Behera and Panda, 2023). Hence, aromatic rice holds a significant position in the global market due to its distinctive aroma and superior eating and cooking qualities. Leading the trade in aromatic rice are Asian countries like India and Pakistan, renowned for their basmati rice, and Thailand, a major supplier of jasmine rice on the international stage (Giraud, 2013). Their captivating aroma is attributed to a complex interplay of more than 300 volatile compounds. Among these compounds, 2-acetyl-1-pyrroline (2-AP) stands out as the principal component responsible for the characteristic fragrance. Notably, the presence of 2-AP is linked to spontaneous mutations in the recessive gene *OsBadh2* or *Badh2*, underscoring the genetic basis of aroma development in rice cultivars. The allure of aromatic rice extends beyond its sensory appeal, as it commands premium prices in both domestic and international markets. Basmati rice, with its long grains and delicate aroma, is favored in gourmet cuisines and commands a loyal consumer base worldwide. Similarly, jasmine rice, known for its subtly sweet aroma and tender texture, is sought after by discerning chefs and consumers for its exceptional culinary qualities. The cultivation and trade of aromatic rice not only contributes significantly to the agricultural economies of producing countries but also enriches global culinary diversity. As consumer preferences continue to evolve and demand for high-quality rice varieties grows, the allure of aromatic rice remains undiminished, cementing its status as a prized commodity in the global marketplace.

9.6.2 Anthocyanin-Rich Potatoes

Such potatoes are characterized by their vibrant purple and red flesh, boast elevated levels of anthocyanins, adding not only nutritional value but also unique colors and flavors to this staple tuber crop. Mattoo et al. (2022) highlighted that advancements in genetics have led to the development and release of anthocyanin-rich potato cultivars, such as the purple potatoes Kufri Neelkanth and KufriJamunia' in northern India. These cultivars represent a significant advancement in biofortification efforts, offering consumers not only visually appealing options but also potatoes with enhanced health benefits due to their rich anthocyanin content.

Kufri Neelkanth

Kufri Jamunia

9.6.3 Golden (β-Carotene) Rice

Genetically engineered through a transgenic approach to produce β-carotene, golden rice addresses vitamin A deficiency, particularly in regions such as South Asia where rice consumption is prevalent. This innovative rice germplasm owes its name to its distinctive golden hue, which is a result of the high levels of beta-carotene, a precursor of vitamin A, accumulated in its grains. The development of golden rice represents a milestone in the field of biofortification, as it offers a sustainable solution to combat vitamin A deficiency, a condition that can lead to serious health issues, including impaired vision and compromised immune function (De Steur et al., 2022). By incorporating genes responsible for beta-carotene synthesis from other organisms into rice plants, golden rice provides a means of increasing the vitamin A content of this staple crop. Despite its potential to address a pressing public health issue, the adoption and cultivation of golden rice has faced various challenges, including regulatory hurdles and concerns surrounding its safety and acceptance by consumers. Nevertheless, ongoing research and efforts continue to refine and promote this biofortified rice variety as part of a comprehensive approach to improving global nutrition and reducing the prevalence of vitamin A deficiency-related diseases.

9.6.4 Low-Phytate Soybeans

Such germplasm was produced from breeding efforts aimed at reducing phytate levels, thereby enhancing mineral absorption and contributing to improved nutritional outcomes. In this regard, Kuerban et al. (2020) identified significant genotypic variation in seed phytate-P concentration across a diverse core subset of 256 soybean accessions from China, ranging from 0.69 to 5.49 mg P per gram of dry weight. Interestingly, the geographical distribution of seed phytate-P concentration revealed the highest levels in genotypes from Hainan Province, while the lowest levels were observed in genotypes from Inner Mongolia. Surprisingly, seed phytate-P concentration showed no correlation with desirable root traits associated with enhanced phosphorus acquisition. In pursuit of soybean cultivars with both low phytate concentrations and desirable phosphorus acquisition traits, two accessions, namely, Siliyuan and Diliuhuangdou-2, were selected for further breeding efforts. Likewise, Punjabi et al. (2018) introduced an inositol polyphosphate 6-/3-/5-kinase gene-specific RNAi construct into the seeds of the Pusa-16 soybean cultivar. This intervention resulted in the development of soybean lines with very low phytic acid levels, moderate accumulation of inorganic phosphate, and elevated mineral content. Importantly, these low-phytic-acid lines exhibited robust agronomic performance, including normal seedling emergence, indicating their suitability for widespread cultivation. Overall, the development of low-phytate soybean varieties holds promise for addressing mineral deficiencies and improving nutritional outcomes, while maintaining excellent agronomic performance. These advancements underscore the potential of genetic engineering and breeding strategies to enhance the nutritional quality of staple crops and contribute to global food security.

9.7 CHALLENGES AND CONSIDERATIONS

While biofortification holds immense promise, challenges such as public acceptance, regulatory frameworks (Mejia et al., 2017), and ethical concerns surrounding genetically modified organisms (GMOs) warrant careful consideration. Engaging with communities, promoting awareness, and addressing these concerns are crucial for the successful implementation of biofortification programs.

While there is existing data on farming households engaged in cultivating biofortified crops, there remains a dearth of information regarding the extent of coverage of biofortified foods within the broader population. This gap in knowledge is pivotal for evaluating the efficacy of biofortification initiatives and informing decisions pertaining to program execution, all the while gauging strides made toward attaining anticipated outcomes. Obtaining insights into the penetration of biofortified foods among the general populace is crucial for accurately assessing the impact of biofortification programs and refining strategies to maximize their reach and effectiveness.

Despite a relatively high level of awareness regarding iron-biofortified beans (IBBs) among the households surveyed, the consumption of IBBs remains low, underscoring the necessity of exploring strategies to promote their uptake. Further investigation is imperative to shed light on the factors impeding the consumption of IBBs (Dusingizimana et al., 2023). In this context, the development of test methods and indicators for evaluating household coverage of biofortified foods is paramount. Petry et al. (2020) have devised five indicators to assess population-wide household coverage, drawing upon methodologies utilized in the evaluation of large-scale food fortification programs. These indicators encompass the consumption of the food, awareness of the biofortified food, availability of the biofortified food, consumption of the biofortified food (ever), and consumption of the biofortified food (current). These metrics have proven effective in evaluating the coverage of IBBs (as well as sweet potatoes with orange flesh) in Rwanda, offering valuable insights into the reach and impact of biofortification efforts. Such indicators serve as invaluable tools for monitoring and optimizing the implementation of biofortification programs, ultimately facilitating the realization of their intended nutritional and public health benefits.

Vaiknoras et al. (2019) revealed that approximately 29% of households in Rwanda have cultivated an IBB cultivar, facilitated by a formal delivery approach that accelerated the adoption process. The rapid spread of IBBs was primarily attributed to diffusion through social networks, highlighting the influential role of community connections in driving adoption rates. Interestingly, the study observed that women farmers tend to dis-adopt IBB cultivars at a slower pace than their male counterparts. This finding underscores the importance of considering gender dynamics in agricultural interventions, as it suggests that women may exhibit greater resilience or attachment to new crop husbandry practices or technologies once adopted. Overall, these insights shed light on the complex dynamics influencing the adoption and retention of biofortified crop varieties, informing strategies for promoting sustainable agricultural practices and improving food security and nutrition outcomes in Rwanda and similar contexts.

Effective biofortification programs rely on robust monitoring, evaluation, and learning (MEL) systems to inform their management and implementation. These systems provide valuable insights into program progress, impact, and areas for improvement. However, despite the presence of MEL frameworks for biofortification programs, challenges persist in harmonizing indicators, metrics, methods, and tools (IMMT), hindering the accurate tracking of program outcomes (Rodas-Moya et al., 2022). Harmonization of IMMT is essential for ensuring consistency and comparability across different biofortification initiatives. Without standardized IMMT, it becomes difficult to measure and compare the effectiveness of various interventions, assess program impact, and allocate resources efficiently. To address this issue, there is a need for concerted efforts to develop and adopt common IMMT standards within the biofortification community. This could involve the establishment of guidelines, protocols, and training programs to facilitate the implementation of standardized MEL practices. Additionally, collaboration between stakeholders, including government agencies, research institutions, non-governmental organizations, and donors, is crucial for promoting the adoption of harmonized IMMT across biofortification programs. By harmonizing IMMT, biofortification programs can enhance their accountability, transparency, and effectiveness. This, in turn, will facilitate evidence-based decision-making, resource allocation, and program scaling, ultimately contributing to the successful implementation and sustainability of biofortification efforts worldwide.

9.8 IMPACTS

Dwivedi et al. (2023) indicate a significant milestone in agricultural biofortification efforts, with over 400 mineral-enriched (such as iron and zinc) and provitamin-A-rich cultivars released in the Global South. This represents a concerted effort to address widespread nutrient deficiencies prevalent in these regions. Presently, the impact of these biofortified cultivars is palpable, with approximately 4.6 million households cultivating zinc-rich rice and wheat. Furthermore, around 3 million households across sub-Saharan Africa and Latin America are benefiting from iron-rich beans, a crucial source of this essential mineral for combating anemia and related health issues. Additionally, the adoption of provitamin-A-rich cassava is making a tangible difference in the nutritional landscape, with approximately 2.6 million people in sub-Saharan Africa and Brazil incorporating this staple crop into their diets. This highlights the diverse range of biofortified crops available and their potential to improve the nutritional status and overall well-being of millions of individuals in regions where malnutrition is a pressing concern. These statistics underscore the success and scalability of biofortification initiatives, demonstrating their effectiveness in addressing nutritional deficiencies and promoting sustainable food systems in the Global South.

Research on farmer adoption and consumer acceptance consistently indicates that both groups favor the production and consumption characteristics of biofortified cultivars (Saltzman et al., 2017). Surprisingly, this preference often extends to biofortified varieties as much as, if not more than, popular conventional varieties, even in situations where nutritional information is not readily available. Farmers are attracted to biofortified cultivars due to their favorable agronomic traits, such as higher yields, disease resistance, and environmental resilience. Additionally, biofortified crops may offer advantages in terms of reduced input requirements, making them appealing options for cultivation. Similarly, consumers are drawn to biofortified cultivars for their enhanced sensory attributes, including taste, texture, and appearance. Despite the absence of explicit nutritional information, consumers perceive biofortified foods as high-quality products that meet their expectations for flavor and overall eating experience. This widespread acceptance of biofortified cultivars among farmers and consumers underscores the intrinsic value of these crops beyond their nutritional benefits. By prioritizing traits that align with end-user preferences, biofortification initiatives stand to achieve greater adoption and impact, ultimately contributing to improved nutrition and food security on a global scale.

9.8.1 Iron-Rich Beans

The analysis of the impact of adopting RWR2245, a popular iron-biofortified bean (IBB) cultivar released in Rwanda, reveals compelling outcomes (Vaiknoras and Larochelle, 2021). With a yield gain of 20–49% over local bush bean cultivars, households adopting RWR2245 experience notable shifts in consumption and economic behavior. First, there was a significant increase in household consumption of beans sourced from their own production. This suggests that the high grain yields of RWR2245 contributed to greater self-sufficiency in bean consumption, thus reducing reliance on external sources. Second, adopting households also decreased their bean purchases. The improved productivity of RWR2245 likely led to surplus beans, thereby reducing the need to buy beans from the market. Moreover, those cultivating RWR2245 increased their bean sales by 12%. This indicates that the surplus beans generated by adopting RWR2245 can be sold, providing an additional source of income for these households. Furthermore, the adoption of RWR2245 leads to an increase in iron intake among adopting households. This is a significant health benefit, particularly in regions where iron deficiency is prevalent. Overall, the adoption of RWR2245 not only enhances food security and economic well-being but also contributes to improved nutritional outcomes, demonstrating the multifaceted benefits of biofortified cultivars in addressing key challenges faced by agricultural communities.

The production and marketing of biofortified crops have often proceeded without a comprehensive regulatory framework or adequate government oversight (Mejia et al., 2017). In many cases, the absence of regulatory guidance has led to decentralized approaches to biofortification initiatives, with various stakeholders independently pursuing the development and dissemination of biofortified crops. Despite these challenges, several countries have recognized the potential of biofortified crops to address malnutrition and have incorporated them into their nutrition agendas. These countries have taken proactive steps to promote the adoption and consumption of biofortified foods through policy measures, public awareness campaigns, and nutrition programs. However, the lack of uniform regulatory standards across regions poses challenges for the international trade and harmonization of biofortified crops. Without consistent regulatory frameworks, the movement of biofortified products across borders may be hindered, limiting their global impact and accessibility. Moving forward, there is a growing need for governments and international organizations to collaborate in establishing clear regulatory frameworks that ensure the safety, quality, and efficacy of biofortified crops. By fostering a supportive regulatory environment, policymakers can facilitate the widespread adoption and acceptance of biofortified foods, ultimately contributing to improved nutrition and public health outcomes on a global scale.

9.9 CONCLUSIONS

Biofortification stands at the forefront of agricultural innovation, offering a transformative approach to address malnutrition and enhance the nutritional quality of staple food crops. By focusing on fragrance, flavonoids, and micronutrients while concurrently reducing seed phytate content, biofortification not only improves the health outcomes of populations but also contributes to the sensory pleasure derived from staple foods. Fernández-Tomë et al. (2023) also emphasizes that plant proteins obtained from grains and seeds offer nutritionally balanced dietary options, promoting improved health outcomes, poverty reduction, and enhanced food security. These protein sources also harbor a plethora of functional compounds, further augmenting their nutritional value. Moreover, proteins serve as a novel reservoir of bioactive peptides, which confer various health benefits, thus underscoring the multifaceted advantages of incorporating plant-based proteins into diets for both human health and socioeconomic well-being. As we navigate the complexity of modern agriculture, the biofortification of staple crops emerges as a beacon of hope, thereby promising a future where our daily sustenance is not only nutritionally rich but also a source of delight for our senses.

REFERENCES

Abbas, F., Zhou, Y., O'Neill Rothenberg, D., Alam, I., Ke, Y., and Wang, H.C. (2023). Aroma components in horticultural crops: Chemical diversity and usage of metabolic engineering for industrial applications. *Plants (Basel)* 12, 1748. https://doi.org/10.3390/plants12091748

Behera, P.K., and Panda, D. (2023). Germplasm resources, genes and perspective for aromatic rice. *Rice* 30, 294–305. https://doi.org/10.1016/j.rsci.2023.03.011

Birol, E., and Bouis, H.E. (2023). Role of socio-economic research in developing, delivering and scaling new crop varieties: The case of staple crop biofortification. *Front. Plant Sci.* 14, 1099496. https://doi.org/10.3389/fpls.2023.1099496.

De Steur, H., Stein, A.J., and Demont, M. (2022). From Golden Rice to Golden Diets: How to turn its recent approval into practice. *Global Food Sec.* 32, 100596. https://doi.org/10.1016/j.gfs.2021.100596.

Dusingizimana, T., Jones, A., Vasanthakaalam, H., and Kjellqvist, T. (2023). Household-level coverage of iron-biofortified beans in the Northern Province of Rwanda. *Curr. Dev. Nutr.* 7, 100106. https://doi.org/10.1016/j.cdnut.2023.100106.

Dwivedi, S.L., Garcia-Oliveira, A.L., Govindaraj, M., and Ortiz, R. (2023). Biofortification to avoid malnutrition in humans in a changing climate: Enhancing micronutrient bioavailability in seed, tuber, and storage roots. *Front. Plant Sci.* 14, 1119148. https://doi.org/10.3389/fpls.2023.1119148

Dwivedi, S.L., Mattoo, A.K., Garg, M., Dutt, S., Singh, B., and Ortiz, R. (2022). Developing germplasm and promoting consumption of anthocyanin-rich grains for health benefits. *Front. Sust. Food Syst.* 6, 867897. https://doi.org/10.3389/fsufs.2022.86789

Dwivedi, S.L., and Ortiz, R. (2021). Food crops low in seed-phytate improve human and other non-ruminant animals' health. *Crop Breed. Genet. Genom.* 3, e210009. https://doi.org/10.20900/cbgg20210009

Fernández-Tomé, S., Ashaolu, T.J., and Hernández-Ledesma, B. (2023). Exploration of the nutritional and functional properties of underutilized grains as an alternative source for the research of food-derived bioactive peptides. *Nutrients* 15, 351. https://doi.org/10.3390/nu15020351

Garcia-Oliveira, A.L., Chander, S., Ortiz, R., Menkir, A., and Gedil, M. (2018). Genetic basis and breeding perspectives of grain iron and zinc enrichment in cereals. *Front. Plant Sci.* 9, 937. https://doi.org/10.3389/fpls.2018.00937.

Giraud, G. (2013). The world market of fragrant rice, main issues and perspectives. *Int. Food Agribus. Manag. Rev.* 16, 1–20. https://doi.org/10.22004/ag.econ.148577

Hall, R.D., Brouwer, I.D., and Fitzgerald, M.A. (2008). Plant metabolomics and its potential application for human nutrition. *Physiol. Plant.* 132, 162–175. https://doi.org/10.1111/j.1399–3054.2007.00989.x

Hirschi, K.D. (2009). Nutrient biofortification of food crops. *Ann. Rev. Nutr.* 29, 401–421. https://doi.org/10.1146/annurev-nutr-080508–141143

Kuerban, M., Jiao, W., Pang, J., Jing, J., Qiu, L.J., Ding, W., et al. (2020). Targeting low-phytate soybean genotypes without compromising desirable phosphorus-acquisition traits. *Front. Genet.* 14, 11, 574547. https://doi.org/10.3389/fgene.2020.574547.

Mattoo, A.K., Dwivedi, S.L., Dutt, S., Singh, B., Garg, M., and Ortiz, R. (2022). Anthocyanin-rich vegetables for human consumption—focus on potato, sweetpotato and tomato. *Int. J. Mol. Sci.* 23, 2634. https://doi.org/10.3390/i,jms23052634

Mejía, L.A., Dary, O., and Boukerdenna, H. (2017). Global regulatory framework for production and marketing of crops biofortified with vitamins and minerals. *Ann. New York Acad. Sci.* 1390, 47–58. https://doi.org/10.1111/nyas.13275

Petry, N., Wirth, J.P., Friesen, V.M., Rohner, F., Nkundineza, A., Chanzu, E., et al. (2020). Assessing the coverage of biofortified foods: Development and testing of methods and indicators in Musanze, Rwanda. *Curr. Dev. Nutr.* 4, nzaa107. https://doi.org/10.1093/cdn/nzaa107.

Punjabi, M., Bharadvaja, N., Jolly, M., Dahuja, A., and Sachdev, A. (2018). Development and evaluation of low phytic acid soybean by sirna triggered seed specific silencing of *Inositol Polyphosphate 6-/3-/5-Kinase* gene. *Front. Plant Sci.* 9, 804. https://doi.org/10.3389/fpls.2018.00804.

Rawat, M., Varshney, A., Rai, M., Chikara, A., Pohty, A.L., Joshi, A., et al. (2023). A comprehensive review on nutraceutical potential of underutilized cereals and cereal-based products. *J. Agric. Food Res.* 12, 100619. https://doi.org/10.1016/j.jafr.2023.100619

Rodas-Moya, S., Giudici, F.M., Mudyahoto, B., Birol, E., Kodish, S.R., Lachat, C., et al. (2022). Critical review of indicators, metrics, methods, and tools for monitoring and evaluation of biofortification programs at scale. *Front. Nutr.* 9, 963748. https://doi.org/10.3389/fnut.2022.963748

Saltzman, A., Birol, E., Oparinde, A., Andersson, M.S., Asare-Marfo, D., Diressie, M.T., et al. (2017). Availability, production, and consumption of crops biofortified by plant breeding: Current evidence and future potential. *Ann. New York Acad. Sci.* 1390, 104–114. https://doi.org/10.1111/nyas.13314

Vaiknoras, K., and Larochelle, C. (2021). The impact of iron-biofortified bean adoption on bean productivity, consumption, purchases and sales. *World Dev.* 139, 105260. https://doi.org/10.1016/j.worlddev.2020.105260.

Vaiknoras, K., Larochelle, C., Birol, E., Asare-Marfo, D., and Herrington, C. (2019). Promoting rapid and sustained adoption of biofortified crops: What we learned from iron-biofortified bean delivery approaches in Rwanda. *Food Policy* 83, 271–284. https://doi.org/10.1016/j.foodpol.2018.11.003.

Zhu, Q., Wang, B., Tan, J., Liu, T., Li, L., and Liu, Y.-G. (2020). Plant synthetic metabolic engineering for enhancing crop nutritional quality. *Plant Commun.* 1, 100017. https://doi.org/10.1016/j.xplc.2019.100017

10 Minimizing the Risk of Toxins in Staple Food Crops via Crossbreeding and Biotechnological Interventions

H.V. Veerendrakumar, D.S. Shailaja, U. Nikhil Sagar, Deekshitha Bomireddy, Khaja Moinuddin, Yogesh Dashrath Naik, Hari Kishan Sudini, Baozhu Guo, Boshou Liao, Sunil S. Gangurde, and Manish K. Pandey

10.1 INTRODUCTION

The occurrence of food contaminants is a critical threat worldwide for consumers. Toxins produced by fungi are known as mycotoxins, the secondary metabolites predominantly produced by *Penicillium, Fusarium,* or *Aspergillus*, adversely impacting human and animal health (Umereweneza et al., 2018). Exposure (acute and chronic) to mycotoxins may cause deleterious effects on health, like reduced immune response (AFB1, ochratoxin A), retarded growth, abdominal pain (deoxynivalenol), vomiting, infertility, and carcinogenic diseases (Taheur et al., 2019). It keeps the food and feed crops at high risk globally (Darwish et al., 2014; Pleadin et al., 2019; Rodrigues and Naehrer, 2012). Mycotoxins are present in various products from raw agricultural products to their processed/commercial value-added products. Contamination of mycotoxins may occur at any stage of food chain, like at the time of harvesting, industry-level processing, transportation, and storage (Schatzmayr et al., 2006). It is affecting the health of consumers by decreasing the food grain availability for consumption due to high levels of contamination; it is reported that mycotoxins are the responsible for food wastage of 1.3 billion metric tons annually (Mesterházy et al., 2020). On average 400 types of mycotoxins have been identified, but 30 are reported as harmful. Long-term exposure to toxins may lead to synergistic impact on health (Sun et al., 2015; Zhou et al., 2017). Therefore, it is crucial to identify methods for detection and measurement of mycotoxin contamination. Chromatographic methods are widely adopted to determine mycotoxin. However, these methodologies

 DOI: 10.1201/9781003467335-10

are less feasible or efficient for quick analysis of samples due to higher cost and need for highly trained manpower to operate the equipment (Awuchi et al., 2021; Marc, 2022). To overcome this, many innovative methods have been developed for detection of toxins, such as enzyme-linked immunosorbent assay (ELISA), lateral flow assays (LFAs), or biosensors for the detection and rapid measurement of mycotoxins (Janik et al., 2021).

Developing strategies for reducing, preventing, and eliminating mycotoxins in food products is a significant challenge to those involved in agri-food business. Traditionally, few physical, chemical, and biological methods are used to reduce fungal growth and minimize the risk of mycotoxin contamination in food products. A few mycotoxins, however, are stable in nature, meaning that such toxins are not detoxified even after applying traditional heat treatments (Peng et al., 2018). Since there is an increasing trend in favor of utilizing non-heat-based methodologies in reducing the level of mycotoxins in foods during sensory quality improvement. Hence, thorough analysis of novel techniques is imperative in their efficacy enhancement in mycotoxins mitigation and ensuring safe food to consumers. For example, probiotics promote digestion, improve intestinal function, and have the potential to detoxify aflatoxins. *Bacillus velezensis* NWPZ-8 (A43) not only detoxify 75% aflatoxin B1 but also increase protein content and amino acid composition in peanut meals. Extracellular proteins in A43 strain are the main source of aflatoxin detoxification, indicating probiotic strains could be used to detoxify aflatoxins in peanut meal (Gu et al., 2024). *B. amyloliquefaciens* (WF2020) is another probiotic strain efficient in detoxifying aflatoxin B1 in peanuts. WF2020 suppresses the expression of ten aflatoxin pathway genes and two transcription factors (*alfR* and *alfS*), suggesting that might inhibit aflatoxin B1 synthesis in *A. flavus* (Chen et al., 2022). It is beyond the scope of this chapter to detail the progress in all staple food crops, but limited, as proof of concept, to maize (*Zea mays*) and peanut (*Arachis hypogaea*, known as groundnut in South Asia). This chapter, therefore, discusses in detail (i) the vulnerability of maize and peanuts to aflatoxin contamination due to climate change, (ii) the adverse impact of aflatoxin on human and animal health, (iii) management strategies to minimize the risk of contamination during postharvest and storage, (iv) the genetic and molecular basis of aflatoxin contamination and the genes and DNA markers associated with aflatoxins, and (v) progress to date realized in developing improved germplasm/cultivars resistant to aflatoxins in maize and peanuts.

10.2 TYPE OF MYCOTOXINS DETECTED IN MAIZE AND PEANUTS

Maize is known to naturally biosynthesize aflatoxins G1, G2, B1, and B2 (Table 10.1). *Aspergillus* and *Penicillium* species, such as *A. glaucus*, *A. ochraceus*, *A. carbonarius*, *A. alliaceus*, *A. niger*, *A. auricomus*, *A. melleus*, *P. verrucosum*, and *P. nordicum*, among others, are known to produce ochratoxins (OTs) (Patriarca and Pinto, 2017) and *P. verrucosum* (Kuiper-Goodman and Scott, 1989). Recently a pangenome showing species diversity in 346 *A. flavus* isolates from 12 states in the USA was developed, and toxin-producing genes from isolates collected from various crops are identified (Gangurde et al., 2024). The primary OT forms are ochratoxin A, B, and C, that vary in toxicity and chemical makeup. The non-chlorinated form of OTA is called ochratoxin B (OTB), and the ethyl ester of OTA that forms when rumen fluid is present is called ochratoxin C (OTC) (Ruyck et al., 2015). The most common and toxic of these ochratoxins is OTA, which is followed by OTC and OTB (Heussner and Bingle, 2015).

Genus *Fusarium*, primarily *F. graminearum* and *F. culmorum* but also *F. acuminatum F. equiseti*, *F. semitectum*, *F. crookwellense*, *F. verticillioides*, *F. oxysporum* and *F. sporotrichioides*, generate zearalenone (ZEA) (Zhang et al., 2016a). It has been identified that maize is most susceptible to ZEA infection. ZEA is xenogenic, hepatotoxic, and immunotoxic (Ropejko and Twaruzek, 2021). Fungal species belonging to the *Fusarium* genus produce a variety of mycotoxins (A, B, C, and D) known as trichothecenes (Wu et al., 2013). The primary components of the type A group include neosolaniol (NEO), monoacetoxy (MAS), diacetoxy-scirpenol (DAS), and HT-2 and T-2 toxins. Fusarenone X, Nivalenol (NIV), and DON and its derivatives 15Ac-DON and 3Ac-DON are the principal constituents of the B group (Patriarca and Pinto, 2017; Ferrigo et al., 2016). The most prevalent types of trichothecenes are found in maize. Fortunately, type A trichothecenes are found in much lower concentrations in contaminated cereals than type B trichothecenes, despite investigations showing that trichothecenes of type A are more harmful than those of type B (Patriarca and Pinto, 2017).

TABLE 10.1
Toxins Produced in Maize and Peanut, Their Pathogens, and Their Effects on Human Health

Toxins associated with maize	Pathogens that produce toxins	Toxicity	References
Aflatoxins M1, M2, B1, B2, G1, and G2	*A. flavus* *A. pseudotamarii* *A. nomius* *A. parasiticus* *A. ochraceoroseus* *A. bombycis*	AFM1's has mutagenic and teratogenic effects, leading to hepatocellular carcinoma, and difficulties with embryonic growth.	Frisvad et al. (2019) Chu and Ueno (1977) El-Sayed et al. (2022)
Ochratoxins (OTA, OTB, and OTC)	*A. ochraceus* *A. niger* *A. carbonarius* *P. nordicum* *P. viridicatum* *P. verrucosum*	OTA exhibits hepatotoxic, neurotoxic, teratogenic, and nephrotoxic properties, bringing on pig nephropathy. In humans, OTA has been linked to urothelial malignancies, BEN, renal failures, and interstitial nephropathy.	Ismaiel and Papenbrock (2015) Tao et al. (2018), Pinotti et al. (2016) Bayman et al. (2002)

(*Continued*)

Table 10.1 *(Continued)*
Toxins Produced in Maize and Peanut, Their Pathogens, and Their Effects on Human Health

Toxins associated with maize	Pathogens that produce toxins	Toxicity	References
Trichothecenes-DON, T-2/HT-2 toxins	*F. graminearum* *F. poae* *F. culmorum* *F. sporotrichoides* *F. langsethiae* *Trichoderma* *Stachybotrys* *Trichothecium*	DON and T-2 toxins infiltrate cellular membranes and interact with ribosomes to hamper translation; exposure to T-2 toxins results in ATA in humans and also prevents the synthesis of DNA, RNA, and proteins, which in turn causes apoptosis, lipid peroxidation, and cytokinesis.	Nathanail et al. (2015) Bhat et al. (2010) Foroud et al. (2019)
Zearalenone	*F. cerealis* *F. graminearum* *F. equiseti,* *F. culmorum* *F. vertillioides*	ZEN induces liver cancer, tumors in the uterus, and thyroid adenoma in rats. ZEN causes testicular atrophy, nephropathy, retinopathy, and cataracts in rats. ZEN is involved in regulation of enzyme expression in the biosynthetic pathways.	Conkov et al. (2003) Ostry et al. (2017) Mazaheri et al. (2021)

10.3 IMPACT OF MYCOTOXINS ON ANIMALS AND HUMAN HEALTH

Mycotoxins are dangerous to animals and human health (Awuchi et al., 2021; Ruyck et al., 2015). Large-scale consumption of severely contaminated food or feed can cause acute poisoning. Chronic illnesses including liver cancer caused by AFs have been linked to long-term exposure to mycotoxins at concentrations slightly over regulatory limits in China and Africa (Wu et al., 2013). In Africa and certain regions of China, FUMs induce esophageal cancer, whereas OTA causes kidney-related illnesses in Europe and Egypt (Marasas et al., 1988; Chu and Li, 1994; Alizadeh et al., 2012). Oral ingestion of mycotoxin-contaminated foods/feeds has resulted in various levels of harmful consequences in both animals and humans, including immunosuppression, genotoxicity, teratogenicity, and malignant mutations (Janik et al., 2020).

Acute mycotoxin poisoning is often caused by aflatoxins (AFs), which show up as epidemics (Table 10.2). AF symptoms might include vomiting, convulsions, edema, jaundice, stomach discomfort, liver failure, and even death. Twenty-six of the 1967 cases in Taiwan had eaten rice having approximately 200 μg/kg of AFs leading to the death of three people (Pitt, 1989). Hours before they died, they have consumed Chinese noodles massively contaminated with high levels of AFs that were

TABLE 10.2
Toxin-Induced Health Hazards Reported in Maize between 1988 and 2018 in African Countries

Year	Location	Substance	Cases	Deaths	Fatality rate	Reference
1981	Kenya	Maize	20	12	60%	Probst et al. (2007)
2004	Kenya	Maize	>125	>125	>100%	Kamala et al. (2018)
2016	Tanzania	Maize	68	20	29.41%	Marasas et al. (1988)

being manufactured in Hong Kong, causing acute hepatic encephalopathy in a city in northwestern Malaysia, where thirteen Chinese children died in 1988 (Lye et al., 1995). Gastroenteritis, initially typical vomiting, is seen due to a shot of massive amounts of Deoxynivalenol (DON) (Ekwomadu et al., 2021). T-2 toxin is famously known for causing alimentary toxic aleukia outbreaks in Russia, caused the death of thousands (Dey et al., 2023). AFB1 has shown to exert adverse effects in several organs such as liver, kidney, lung, intestine, and heart (Chen et al., 2021).

The impact of mycotoxicosis have on human health are influenced by age, weight, gender, food type and quantity consumed, presence of other mycotoxins and bioactive compounds, infections already present, and so forth. The facts indicated that minor levels of AFs becoming experienced by taking in, breathing, or hit through the skin, may lead to cancer, liver illness, teratogenicity, and genetic mutations (Alvito and Pereira-Da-Silva, 2022). According to a recent overview, AFs are the most ubiquitous and have been associated with well-documented toxic effects on fetuses and children (Reddy and Raghavender, 2007). AFs cause severe growth retardation in pups from AF-exposed dams reared under cold conditions, and they increase the incidence of neural tube defects in embryos affected by fumonisins (FUMs) (Wen et al., 2016). The liver carcinogenicity of long-term AF exposure has been recognized for decades (Atkins and Norman, 1998; Alvito et al., 2022). In addition, individuals with hepatitis B and hepatitis C infection in China and Africa were at higher risk for liver cancer when exposed to AFs compared with those not infected with hepatitis (Wu et al., 2013, Christopher et al., 2015). Among 155,000 instances of AF-associated liver cancer reported annually worldwide, over 40%, or 59,900, occur in Africa (Wagacha and Muthomi, 2008). Liver cancer has high mortality in China, and it ranks first in morbidity among various cancers. Given that maize is the predominant staple food in such populations, there is probably a connection between this and the high prevalence of liver cancer. It has been reported that besides liver carcinogenicity, human and animal studies identified teratogenic and mutagenic effects of AFs at low concentrations due to at least part of many mechanisms of AF toxicity not yet fully understood (Theumer et al., 2018; Claeys et al., 2020). Multiple tests have confirmed that AFs are mutagenic and genotoxic in bacteria and might have an embryotoxic effect, inducing birth defects in offspring and immunosuppressive properties that reduce resistance to infections such as HIV/AIDS and tuberculosis, as evidenced in Gambian and Ghanaian studies (Omotayo et al., 2019).

10.4 NON-CROP IMPROVEMENT METHODS FOR TOXIN DETOXIFICATION

Mycotoxin contamination can arise at any stage of the food chain, leading to considerable economic losses and endangering consumer health. Urgent implementation of effective mycotoxin removal methods is imperative to safeguard both consumers and the food industry. While breeding and biotechnology offer effective avenues to mitigate toxin levels in crops, alternative strategies are available. Prevention emerges as a pivotal strategy, applicable before harvesting, in processed foods and even in raw materials. During food processing, a variety of decontamination techniques, such as physical, chemical, or biological processes, might be used. While these methods can partially destroy or inactivate mycotoxins, complete elimination remains rare (Khaneghah et al., 2017, 2018; Ostry et al., 2013; Zhu et al., 2017). Preventive measures involve practices, such as ensuring timely harvesting, avoiding harvesting when crops have excessive moisture, and drying crops before storage if needed (Zhang et al., 2019). When mycotoxins are already present in food, controlling or detoxifying the food becomes necessary to lessen the negative effects of the mycotoxins (Awuchi et al., 2021; Zhang et al., 2019).

Physical detoxification of mycotoxins involves processes that physically alter the structure or remove mycotoxins from contaminated materials without chemically altering the toxins themselves. There are mainly three types of physical detoxification approaches for mycotoxins—thermal processing, irradiation, and adsorption (Figure 10.1). Chemical detoxification of mycotoxins involves the use of chemical agents to modify or degrade the toxins, rendering them less harmful or non-toxic. Chemical approaches include ozonation and acid and alkali treatment (Sipos et al., 2021). Biological detoxification of mycotoxins involves the use of living organisms or biological agents to degrade or transform mycotoxins into less toxic forms. This process harnesses the metabolic activities of microorganisms, enzymes, or other biological entities to break down mycotoxins effectively. These biological detoxification methods (microbial detoxification, and enzymatic detoxification) (Liu et al., 2022) hold promise for mitigating mycotoxin contamination in specific crops and enhancing food safety and quality in agriculture.

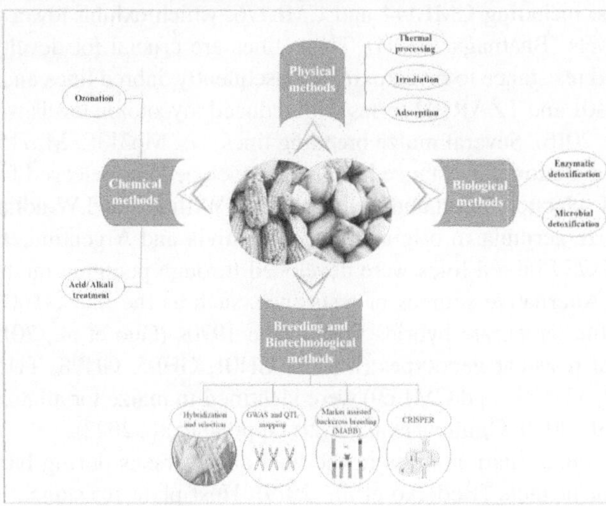

10.5 BREEDING AND BIOTECHNOLOGICAL TOOLS IN LIMITING MYCOTOXIN PRODUCTION

Over the past three decades, extensive research has been conducted to unravel genetic behavior, molecular mechanisms, and intricate biology underlying host-pathogen interactions. Advances in various omics approaches have greatly enhanced our understanding of resistance mechanisms and the pathways involved in these interactions.

10.5.1 Traditional Breeding Methods for Reducing Aflatoxin Contamination

In conventional breeding, plants with desirable features, including resistance to fungal infection, are chosen and crossed in a methodical hybridization process with the goal of creating superior genotypes. To enhance the resistance of commercial cultivars, resistant lines need to be developed by involving multiple breeding cycles and extensive screening for aflatoxin resistance traits and then incorporate them into breeding programs. As aflatoxin susceptibility is a highly complex, quantitative trait regulated by multiple genes, selecting the complete source of resistance is quite challenging (Mayfield et al., 2011). As a result, selecting lines with tighter husks and harder kernels that are less conducive to fungal growth were used to develop maize cultivars resistant to *Aspergillus* (Brown et al., 2013). The complexity of this pathogen is further increased by the strong host-by-pathogen interactions affecting both colonization and aflatoxin production, making breeding for aflatoxin resistance a challenging task (Amaike and Keller, 2011; Christensen and Kolomiets, 2011). Six recombinant inbred lines (RILs) obtained from a cross between Mp715 and Va35 had comparable aflatoxin resistance as of the resistant parent Mp715 (Williams, 2006). Three hybrids (GH01, GH05, and GH08) showed stable resistance to aflatoxin levels with 11.9, 14.3, and 3.8 ppb, respectively. It also highlighted the non-significance of genotype × environment interaction (Oppong et al., 2022). Maize breeding has produced several resistant lines, including CML144 and CML176, which exhibit lower aflatoxin and fumonisin levels (Bhatnagar, 2004). These lines are crucial for developing hybrids with enhanced resistance to mycotoxins. Consequently, inbred lines and hybrids such as the TZAR101 and TZAR102 series with reduced mycotoxin levels were developed (Brown et al., 2016). Several maize breeding lines, *viz*. Mp313E, Mp715, and Mp717, with enhanced aflatoxin resistance have been developed and released following multiple cycles of selection (Scott and Zummo, 1990; Williams and Windham, 2012). By using the maize germplasm originating from Bolivia and Argentina, Tx736, Tx740, Tx772, and Tx739 inbred lines were developed through pedigree method (Mayfield et al., 2012). Alternative sources of resistance, such as the line GT603, were identified from elite temperate hybrids bred in the 1970s (Guo et al., 2011). Similarly, the number of resistant genotypes (Mp719, GH01, GH05, GH08, TUN15, TUN61, TUN37, CY2, TUN49, and CML69) were identified in maize for aflatoxin resistance (Womack et al., 2020; Ogunola et al., 2021; Oppong et al., 2022).

Aflatoxin contamination starts in the field and worsens during harvest, drying, and storage of peanuts (Bediako et al., 2019). Host plant resistance is recognized as a primary approach for improving resistance to *Aspergillus* infection. Breeding resistance to *Aspergillus* infection or aflatoxin production is pivotal for management, requiring genetic variability, stable resistance sources, and effective screening techniques. Extensive work in peanut has identified resistant lines, such as ICG 1471, NC3033, PI 337394F, PI 337409, ICGV 87084, ICGV 87094, ICG 12625, F1334, and F1344, which exhibit minimal infection and aflatoxin production and have been utilized in breeding programs (Pandey et al., 2019). Recently, the Zh.h1312 genotype was identified in peanut that exhibited resistance to both seed infection and aflatoxin production, harboring all identified resistance alleles (Ding et al., 2023). These reported genotypes can be used for breeding programs or to understand the complexity of aflatoxin resistance through genetic approaches such as trait mapping.

10.5.2 Genomic-Assisted Breeding to Reduce Aflatoxin Contamination

Maize: The integration of molecular markers, high-throughput sequencing, and web-based genomic data has significantly advanced back-crossbreeding for aflatoxin resistance. These technological advancements enable more precise selection and facilitate the timely introgression of genes and quantitative trait loci (QTLs). *Fusarium* (FER) and *Gibberella* ear rots (GER) are significant maize diseases that lead to yield losses and mycotoxin contamination. Several QTLs associated with fumonisin and FER resistance in maize have been identified using both linkage and association mapping approaches. Extensive research over the past decade has focused on identifying QTLs linked to reduced deoxynivalenol (DON) accumulation and enhanced GER resistance. For instance, analysis of two biparental Canadian maize populations revealed eight shared QTLs for FER and fumonisin resistance (Maschietto et al., 2017). In three separate environments, a major QTL across environments for GER resistance was identified in a Chinese biparental maize population (Yin et al., 2014). Seven chitinase genes associated with *Aspergillus* resistance were reported through QTL mapping (Hawkins et al., 2015). QTLs linked to *Fusarium graminearum* resistance have been detected on all ten chromosomes of elite maize double haploid populations (Han et al., 2016, 2018). Additionally, significant QTLs for ear rot resistance were observed in some biparental maize populations (Giomi et al., 2016). A total of 12 QTLs for aflatoxin resistance were identified in a linkage mapping study, collectively explaining 46% of the phenotypic variation (Warburton et al., 2011).

High-density single-nucleotide polymorphism (SNP) arrays, applied in genome-wide association studies (GWAS), offer an effective approach for identifying QTLs and candidate genes associated with reduced mycotoxin contamination. Using association mapping in 130 elite European dent and flint lines, eight significant SNPs associated with DON accumulation were identified (Han et al., 2018). Moreover, 15 QTLs were detected on all maize chromosomes, except chromosomes 6, 7, and 8, using QTL analyses of four biparental tropical maize populations for FER resistance, which validated the results of GWAS (Chen et al., 2016). A recent GWAS study involving 188 maize lines reported 14 significant SNPs and four genes (*GRMZM2G066153*, *GRMZM2G131254*, *GRMZM2G060993*, and *GRMZM2G153359*) encoding for G proteins, gibberellic acid synthesis, and flavonoid production, which are involved in FER resistance (Coan et al., 2018). Previous research indicates that QTL mapping for fumonisin contamination is limited, primarily due to the strong correlation between fumonisin levels and FER infection, as well as the high costs associated with mycotoxin testing in large maize populations. Since *Fusarium verticillioides* is the predominant ear rot pathogen globally, most GWAS studies have focused on identifying candidate genes for resistance to *F. verticillioides* rather than *Aspergillus flavus* or *F. graminearum* (Chilaka et al., 2016; Wang et al., 2016).

A meta-QTL analysis incorporating 224 QTLs from 15 studies identified 40 stable meta-QTLs associated with FER and GER resistance traits (Akohoue and Miedaner, 2022). Candidate genes within these meta-QTLs, such as terpene synthase 21 and flavonoid O-methyltransferase 2, exhibit potential for enhancing resistance and reducing mycotoxin accumulation through genomics-assisted breeding strategies. Linkage mapping and association mapping approaches were utilized by Zhang et al. (2016b) to pinpoint loci for reduced aflatoxin deposition, identifying significant SNPs on chromosomes 5, 8, and 9, and a shared QTL, qAA8, on chromosome 8. Kompetitive allele-specific PCR (KASP) genotyping assays were subsequently developed using six crucial SNPs discovered through GWAS to facilitate marker-assisted selection in *A. flavus* resistance breeding programs (Warburton et al., 2015). Integrating linkage and association mapping methodologies adjusted to specific population types can enhance the detection of QTLs. Additionally, advanced phenotyping techniques that minimize background noise are critical for reducing false positives in genomic selection.

Peanut: QTL mapping using RILs detected two major QTLs (*qAFB1A07* and *qAFB1B06.1*) for aflatoxin resistance (Yu et al., 2019), whereas, another study reported 11 QTLs for aflatoxin resistance, with three major QTLs (*qAFTB06.3*, *qAFTB05.2*, and *qAFTA05.1*) (Jin et al., 2023). A candidate gene *AhAftr1* for aflatoxin production resistance in peanut seed was reported, which detected 36 peanut germplasms, with over 77% reduction in aflatoxin production than susceptible control, Zhonghua12 (Yu et al., 2024). Hub genes (18 genes encoding pathogenesis-related proteins, 1-aminocyclopropane-1-carboxylate oxidase, MAPK kinase, serine/threonine kinase, pattern recognition receptors, cytochrome P450, SNARE protein SYP121, pectinesterase, phosphatidylinositol transfer protein, and pentatricopeptide repeat protein) were found positively associated with resistance to *A. flavus*, which may be exploited to breed peanuts with *A. flavus* resistance by genomic-assisted breeding (Cui et al., 2022).

10.6 APPLYING BIOTECHNOLOGICAL TOOLS (RNAI AND/OR GENOME EDITING) FOR REDUCING AFLATOXIN CONTAMINATION

Maize: The development of advanced biotechnological techniques, such as ribonucleic acid interference (RNAi), has opened new ways for genetic transformation to enhance mycotoxin resistance in crops (Majumdar et al., 2017). RNAi technology can silence specific genes involved in mycotoxin biosynthesis, thereby reducing or eliminating the production of these toxic compounds. During RNAi, double-stranded RNA molecules are processed into small interfering RNAs (siRNAs), which are short RNA duplexes that bind to complementary mRNA sequences, leading to mRNA cleavage and subsequent gene silencing or regulation (Nunes and Dean, 2012; Koch and Kogel, 2014). For example, the *aflC* gene, which encodes an enzyme crucial for aflatoxin biosynthesis, was successfully silenced in maize using a kernel-specific RNAi gene cassette. This resulted in undetectable levels of aflatoxin in transgenic maize kernels, while non-transgenic maize kernels accumulated aflatoxin levels in thousands of parts per billion (Thakare et al., 2017). The absence of significant differences in transcript profiles between aflatoxin-free transgenic kernels and non-transgenic kernels further highlights the potential of RNAi for mitigating mycotoxin contamination. Continuous research and development are essential to optimize RNAi-based approaches for practical agricultural applications.

Genome-editing techniques, such as clustered regularly interspaced short palindromic repeats and CRISPR-associated protein 9 (CRISPR/Cas9), offer additional benefits over traditional breeding and mutagenesis by enabling precise gene substitutions, knock-ins, knockouts, and targeted DNA repair, thereby accelerating crop improvement (Chapman et al., 2012). Among the currently available genome-editing tools, CRISPR/Cas9, an endonuclease associated with clustered regularly interspaced short palindromic repeats (CRISPR), shows the greatest promise for enhancing mycotoxin resistance. The Cas9-mediated system allows for the creation of marker-free mutants. Gardiner and Kazan (2018) utilized a dual ribozyme-based single-guide RNA (sgRNA) expression system along with a codon-optimized Cas9 nuclease to induce mutations in the *FgOs1* gene of *Fusarium graminearum*. The mutations were then screened using a fungicide resistance phenotypic assay, demonstrating the utility of this approach for generating mycotoxin-resistant lines in cereals. Several research groups have successfully developed transgenic and mutant lines exhibiting resistance to *F. graminearum* infection and reduced accumulation of DON (deoxynivalenol) (Alexander et al., 2009; Cao et al., 2021; Liu et al., 2019; Maier et al., 2006; Takeshita et al., 2008; Urban et al., 2003). To effectively develop and deploy mycotoxin-resistant cultivars for agricultural use, greater collaboration between public and private seed companies and research institutes is crucial. This partnership will support the dissemination of resistant varieties to farmers, ultimately contributing to food safety and security by reducing mycotoxin contamination in crops.

Genome sequencing of hosts and pathogens has revealed insights into complex genetic architectures and novel genes. Additionally, transcriptomic, proteomic, and metabolomic studies have enhanced our understanding of gene expression and pathway enrichment that determine phenotypes. Genome-wide gene expression profiling in maize under *A. flavus* inoculation revealed that the gene encoding glycine-rich RNA-binding protein 2 is associated with host hypersensitivity and susceptibility in the Va35 genotype (Kelley et al., 2012).

Peanut: Few studies have been reported in peanuts as well, 23 genes encoding glycine-rich RNA-binding proteins (Ah.GR-RBPs) from peanuts were identified. Furthermore, expression analysis revealed that *Ah.GR-RBP.1, Ah.GR-RBP.3, Ah.GR-RBP.12* and *Ah.GR-RBP.15* exhibited higher expression levels in the susceptible genotype (Jose et al., 2024). RNA-seq was used to study host-pathogen interactions in peanut, identifying 578 differentially expressed genes (DEGs) in *A. flavus* involved in fungal growth, aflatoxin biosynthesis, and various cellular processes during compatible interactions in peanut lines differing in resistance or susceptibility

(Nayak et al., 2017). Metabolite profiling in peanuts using liquid chromatography–high-resolution mass spectrometry revealed key compounds like phenylpropanoids and lignin. These metabolites, particularly hydroxycinnamic acid amides and lignins, were significantly higher in the resistant genotype, reinforcing the cell wall to block pathogen entry (Avuthu et al., 2024). Transcriptomic and metabolomic analysis of resistant peanut seeds infected with *A. flavus* showed high expression of DEGs, including those for transcription factors, disease-related proteins, peroxidase, chitinase and phenylalanine ammonia-lyase (Wang et al., 2023).

To sum up, multi-omics approaches are effective in identifying key genes and pathways involved in resistance to *A. flavus* and aflatoxin contamination. High expression of specific key genes, disease-related proteins, peroxidase, chitinase, and phenylalanine ammonia-lyase, in resistant genotypes, highlights their importance in developing aflatoxin-resistant crop varieties. These findings highlight the potential of integrating multi-omics data to enhance breeding programs, providing valuable insights for the selection and development of aflatoxin-resistant lines and mitigating aflatoxin contamination in crops.

10.7 CONCLUSIONS

The objective to minimize the risk of toxins in staple food crops through breeding and biotechnology represents a vital intersection of agricultural science, public health, and sustainable development. By integrating traditional crossbreeding techniques with advanced biotechnological methods, significant progress can be achieved in reducing the presence of harmful substances like mycotoxins, cyanogenic glycosides, and other naturally occurring toxins in essential food crops. Crossbreeding, with its long history and proven track record, provides a foundational approach to enhancing crop resilience and toxin resistance. Selective breeding programs have successfully developed crops resistant to diseases, thereby indirectly reducing the incidence of toxin accumulation. Identification of genomic regions and genes helping in resisting the fungal pathogen or toxin formation should be identified and introduced into elite lines. Biotechnological interventions, particularly genetic engineering and CRISPR-based genome editing, offer precise and accelerated methods to address the issue of food crop toxins. These technologies enabled the targeted manipulation of specific genes responsible for toxin production or susceptibility, leading to crops that are inherently safer for consumption. Moreover, advancements in omics technologies (genomics, proteomics, metabolomics) provide comprehensive insights into plant metabolic pathways, facilitating the identification of novel targets for intervention. A synergistic approach that combines the strengths of traditional breeding and modern biotechnology will be paramount in our quest to ensure food security and safety.

REFERENCES

Akohoue, F., and Miedaner, T. (2022). Meta-analysis and co-expression analysis revealed stable QTL and candidate genes conferring resistances to *Fusarium* and *Gibberella* ear rots while reducing mycotoxin contamination in maize. *Front. Plant Sci.* 13, 1050891. https://doi.org/10.3389/fpls.2022.1050891

Alexander, N.J., Proctor, R.H., and McCormick, S.P. (2009). Genes, gene clusters, and biosynthesis of trichothecenes and fumonisins in *Fusarium. Toxin Rev.* 28, 198–215. https://doi.org/10.1080/15569540903092142

Alizadeh, A.M., Roshandel, G., Roudbarmohammadi, S., Roudbary, M., Sohanaki, H., Ghiasian, S.A., and Aghasi, M. (2012). Fumonisin B1 contamination of cereals and risk of esophageal cancer in a high risk area in northeastern Iran. *Asian Pac. J. Cancer Prev.* 13(6), 2625–2628.

Alvito, P., Assunção, R.M., Bajard, L., Martins, C., Mengelers, M.J., Mol, H., and Silva, M.J. (2022). Current advances, research needs and gaps in mycotoxins biomonitoring under the HBM4EU—lessons learned and future trends. *Toxins* 14(12), 826.

Alvito, P., and Pereira-da-Silva, L. (2022). Mycotoxin exposure during the first 1000 days of life and its impact on children's health: A clinical overview. *Toxins* 14(3), 189.

Amaike, S., and Keller, N.P. (2011). *Aspergillus flavus. Ann. Rev. Phytopathol.* 49,107–133. PMID:21513456. https://doi.org/10.1146/annurev-phyto-072910-095221

Atkins, D., and Norman, J. (1998). Mycotoxins and food safety. *Nutr. Food Sci.* 98, 260–266. https://doi.org/10.1108/00346659810224172

Avuthu, T., Sanivarapu, H., Prasad, K., Sharma, N., Sudini, H.K., and Yogendra, K. (2024). Comparative metabolomics analysis reveals secondary cell wall thickening as a barrier to resist *Aspergillus flavus* infection in groundnut. *Physiol. Plant.* 176, e14169. https://doi.org/10.1111/ppl.14169

Awuchi, C.G., Ondari, E.N., Ogbonna, C.U., Upadhyay, A.K., Baran, K., and Okpola, C.O.R. (2021). Mycotoxins affecting animals, foods, humans, and plants: Types, occurrence, toxicities, action mechanisms, prevention, and detoxification strategies—A revisit. *Foods* 10, 1279. https://doi.org/10.3390/foods10061279

Bayman, P., Baker, J.L., Doster, M.A., Michailides, T.J., and Mahoney, N.E. (2002). Ochratoxin production by the *Aspergillus ochraceus* group and *Aspergillus alliaceus. Appl. Environ. Microbiol.* 68(5), 2326–2329.

Bediako, K.A., Dzidzienyo, D., Ofori, K., Offei, S.K., Asibuo, J.Y., Amoah, R.A., and Obeng, J. (2019). Prevalence of fungi and aflatoxin contamination in stored groundnut in Ghana. *Food Control*, 104, 152–156.

Bhat, R., Rai, R.V., and Karim, A.A. (2010). Mycotoxins in food and feed: Present status and future concerns, Compr. *Rev. Food Sci. Food Saf.* 9, 57–81. https://doi.org/10.1111/j.1541-4337.2009.00094.x

Bhatnagar, S. (2004). *Phenotypic and Genotypic Characterization of High Lysine Maize.* Texas A&M University.

Brown, R.L., Chen, Z.Y., Cleveland, T.E., and Russin, J.S. (2013). Advances in the development of host resistance in corn to aflatoxin contamination by *Aspergillus flavus. Phytopathology* 93, 1044–1050. https://doi.org/10.1094/PHYTO.1999.89.2.113

Brown, R.L., Williams, W.P., Windham, G.L., Menkir, A., and Chen, Z.Y. (2016). Evaluation of African-bred maize germplasm lines for resistance to aflatoxin accumulation. *Agronomy* 6, 24. https://doi.org/10.3390/agronomy6020024

Cao, S., Li, W., Li, C., Wang, G., Jiang, W., and Sun, H. (2021). The CHY-type zinc finger protein FgChy1 regulates polarized growth, pathogenicity, and microtubule assembly in *Fusarium graminearum. Mol. Plant-Microbe Interact.* 34, 362–375. https://doi.org/10.1094/MPMI-07-20-0206-R

Chapman, J.R., Taylor, M.R., and Boulton, S.J. (2012). Playing the end game: DNA double-strand break repair pathway choice. *Mol. Cell* 47, 497–510. https://doi.org/10.1016/j.molcel.2012.07.029

Chen, G., Fang, Q., Liao, Z., Xu, C., Liang, Z., Liu, T., et al. (2022). Detoxification of aflatoxin B1 by a potential probiotic *Bacillus amyloliquefaciens* WF2020. *Front. Microbiol.* 13, 891091. https://doi.org/10.3389/fmicb.2022.891091

Chen, J., Shrestha, R., Ding, J., Zheng, H., Mu, C., Wu, J., et al. (2016). Genome-wide association study and QTL mapping reveal genomic loci associated with *Fusarium* ear rot resistance in tropical maize germplasm. *G3* 6, 3803–3815. https://doi.org/10.1534/g3.116.034561

Chen, J., Wen, J., Tang, Y., Shi, J., Mu, G., and Yan, R. (2021). Research progress on fumonisin B1 contamination and toxicity: A review. *Molecules* 26, 5238. https://doi.org/10.3390/molecules26175238

Chilaka, A.C., De Boevre, M., Atanda, O.O., and Saeger, D.S. (2016). Occurrence of *Fusarium* mycotoxins in cereal crops and processed products (Ogi) from Nigeria. *Toxins (Basel)* 8, 1–18. https://doi.org/10.3390/toxins8110342

Christensen, S.A., and Kolomiets, M.V. (2011). The lipid language of plant–fungal interactions. *Fungal Genet. Biol.* 48, 4–14. PMID:20519150. https://doi.org/10.1016/j.fgb.2010.05.005

Christopher, P., Wild, J., Miller, D., John, D., and Groopman, J.D. (2015). Mycotoxin control in low and middle-income countries. *IARC Working Group Reports, No. 9.* International Agency for Research on Cancer: Lyon, France, pp. 13–16.

Chu, F.S., and Li, G.Y. (1994). Simultaneous occurrence of fumonisin B1 and other mycotoxins in moldy corn collected from the People's Republic of China in regions with high incidences of esophageal cancer. *Appl. Environ. Microbiol.* 60(3), 847–852.

Chu, F.S., and Ueno, I. (1977). Production of antibody against aflatoxin B1. *Applied and Environmental Microbiology*, 33(5), 1125–1128.

Claeys, L., Romano, C., De Ruyck, K., Wilson, H., Fervers, B., and Korenjak, M. (2020). Mycotoxin exposure and human cancer risk: A systematic review of epidemiological studies. *Compr. Rev. Food Sci. Food Saf.* 19, 1449–1464. https://doi.org/10.1111/1541-4337.12567

Coan, M.M.D., Senhorinho, H.J.C., and Pinto, R.J.B. (2018). Genome-wide association study of resistance to ear rot by *Fusarium verticillioides* in a tropical field maize and popcorn core collection. *Crop Sci.* 58, 564–578. https://doi.org/10.2135/cropsci2017.05.0322

Conkov, E., Laciakova, A., Kovac, G., and Seidel, H. (2003). Fusarial toxins and their role in animal diseases. *Vet. J.* 165, 214–220. https://doi.org/10.1016/S1090-0233(02)00127-2

Cui, M., Han, S., Wang, D., Haider, M.S., Guo, J., and Zhao, Q. (2022). Gene co-expression network analysis of the comparative transcriptome identifies hub genes associated with resistance to *Aspergillus flavus* L. in cultivated peanut (*Arachis hypogaea* L.). *Front. Plant Sci.* 13, 899177. https://doi.org/10.3389/fpls.2022.899177

Darwish, W.S., Ikenaka, Y., Nakayama, S.M.M., and Ishizuka, M. (2014). An overview on mycotoxin contamination of foods in Africa. *J. Vet. Med. Sci.* 76, 789–797. https://doi.org/10.1292/jvms.13-0563

Dey, D.K., Kang, J.I., Bajpai, V.K., Kim, K., Lee, H., and Sonwal, S. (2023). Mycotoxins in food and feed: Toxicity, preventive challenges, and advanced detection techniques for associated diseases. *Crit. Rev. Food Sci. Nutr.* 63, 8489–8510. https://doi.org/10.1080/10408398.2022.2059650

Ding, C., Tian, T., Liu, Q., Zhao, S., Tao, T., Wu, H., and Guo, L. (2023). Transcriptomics and metabolomic profiling identify molecular mechanism for *Aspergillus flavus* infection in grain. *Food Front.* 4(4), 1845–1858.

Ekwomadu, T.I., Akinola, S.A., and Mwanza, M. (2021). *Fusarium* mycotoxins, their metabolites (Free, Emerging, and Masked), food safety concerns, and health impacts. *Int. J. Environ. Res. Public Health* 18, 11741. https://doi.org/10.3390/ijerph182211741

El-Sayed, R.A., Jebur, A.B., Kang, W., and El-Demerdash, F.M. (2022). An overview on the major mycotoxins in food products: Characteristics, toxicity, and analysis. *J. Future Foods* 2(2), 91–102.

Ferrigo, D., Raiola, A., and Causin, R. (2016). *Fusarium* toxins in cereals: Occurrence, legislation, factors promoting appearance and their management. *Molecules* 21, 627. https://doi.org/10.3390/molecules21050627

Foroud, N.A., Baines, D., Gagkaeva, T.Y., Thakor, N., Badea, A., and Steiner, B. (2019). Trichothecenes in cereal grains–an update. *Toxins* 11, 634. https://doi.org/10.3390/toxins11110634

Frisvad, J.C., Hubka, V., Ezekiel, C.N., Hong, S.B., NovßkovB, A., and Chen, A.J. (2019). Taxonomy of *Aspergillus* section Flavi and their production of aflatoxins, ochratoxins and other mycotoxins. *Stud. Mycol.* 93, 1–63. https://doi.org/10.1016/j.simyco.2018.06.001

Gangurde, S.S., Korani, W., Bajaj, P., Wang, H., Fountain, J.C., and Agarwal, G. (2024). *Aspergillus flavus* pangenome (AflaPan) uncovers novel aflatoxin and secondary metabolite associated gene clusters. *BMC Plant Biol.* 24, 354. https://doi.org/10.1186/s12870-024-04950-8

Gardiner, D.M., and Kazan, K. (2018). Selection is required for efficient Cas9-mediated genome editing in *Fusarium graminearum. Fungal Biol.* 122, 131–137. https://doi.org/10.1016/j.funbio.2017.11.006

Giomi, G.M., Kref, E.D., Iglesias, J., Fauguel, C.M., Fernandez, M., and Oviedo, M.S. (2016). Quantitative trait loci for *Fusarium* and *Gibberella* ear rot resistance in Argentinian maize germplasm. *Euphytica* 211, 287. https://doi.org/10.1007/s10681-016-1725-z

Gu, R., Yang, X., Zhang, Q., Li, P., and Mao, J. (2024). Screening, identification, and application of *Bacillus velezensis* NWPZ-8 probiotics to aflatoxin detoxification in peanut meal. *ACS Food Sci. Technol.* 4, 2155–2166. https://doi.org/10.1021/acsfoodscitech.4c00371

Guo, B.Z., Krakowsky, M.D., Ni, X., Scully, B.T., Lee, R.D., and Coy, A.E. (2011). Registration of maize inbred line GT603. *J. Plant Reg.* 5, 211–214. https://doi.org/10.1007/s12298-018-0522-7

Han, S., Miedaner, T., Utz, H.F., Schipprak, W., Schrag, T.A., and Melchinger, A.E. (2018). Genomic prediction and GWAS of *Gibberella* ear rot resistance traits in dent and flint lines of a public maize breeding program. *Euphytica* 214, 1–20. https://doi.org/10.1007/s10681-017-2090-2

Han, S., Utz, H.F., Liu, W., Schrag, T.A., Stang, M., and Wurschum, T. (2016). Choice of models for QTL mapping with multiple families and design of the training set for prediction of Fusarium resistance traits in maize. *Theor. Appl. Genet.* 129, 431–444. https://doi.org/10.1007/s00122-015-2637-3

Hawkins, L.K., Mylroie, J.E., Oliveira, D.A., Smith, J.S., Ozkan, S., and Windham, G.L. (2015). Characterization of the maize chitinase genes and their effect on *Aspergillus flavus* and aflatoxin accumulation resistance. *PLoS ONE* 10, e0126185. https://doi.org/10.1371/journal.pone.0126185

Heussner, A.H., and Bingle, L.E. (2015). Comparative ochratoxin toxicity: A review of the available data. *Toxins* 7, 4253–4282. https://doi.org/10.3390/toxins7104253

Ismaiel, A.A., and Papenbrock, J. (2015). Mycotoxins: Producing fungi and mechanisms of phytotoxicity. *Agriculture* 5, 492–537. https://doi.org/10.3390/agriculture5030492

Janik, E., Niemcewicz, M., Ceremuga, M., Stela, M., Saluk-Bijak, J., and Siadkowski, A. (2020). Molecular aspects of mycotoxins—A serious problem for human health. *Int. J. Mol. Sci.* 21, 8187. https://doi.org/10.3390/ijms21218187

Janik, E., Niemcewicz, M., Podogrocki, M., Ceremuga, M., Gorniak, L., Stela, M., and Bijak, M. (2021). The existing methods and novel approaches in mycotoxins' detection. *Molecules* 26, 3981. https://doi.org/10.3390/molecules26133981

Jin, G., Liu, N., Yu, B., Jiang, Y., Luo, H., and Huang, L. (2023). Identification and pyramiding major QTL loci for simultaneously enhancing aflatoxin resistance and yield components in peanut. *Genes* 14, 625. https://doi.org/10.3390/genes14030625

Jose, A.M., Tejaswi, A., Kokiladevi, E., Sharma, N., and Yogendra, K. (2024). Genome-wide identification of the glycine-rich RNA-binding protein genes and their expression analysis upon *Aspergillus flavus* infection in groundnut (*Arachis hypogaea*). *Agronomy* 14, 165. https://doi.org/10.3390/agronomy14010165

Kamala, A., Shirima, C., Jani, B., Bakari, M., Sillo, H., and Rusibamayila, N. (2018). Outbreak of an acute aflatoxicosis in Tanzania during 2016. *World Mycotoxin J.* 11, 311–320. https://doi.org/10.3920/WMJ2018.2344

Kelley, R.Y., Williams, W.P., Mylroie, J.E., Boykin, D.L., Harper, J.W., and Windham, G.L. (2012). Identification of maize genes associated with host plant resistance or susceptibility to *Aspergillus flavus* infection and aflatoxin accumulation. *PLoS ONE* 7, e36892. https://doi.org/10.1371/journal.pone.0036892

Khaneghah, A.M., Fakhri, Y., Raeisi, S., Armoon, B., and Sant'Ana, A.S. (2018). Prevalence and concentration of ochratoxin A, zearalenone, deoxynivalenol and total aflatoxin in cereal-based products: A systematic review and meta-analysis. *Food Chem. Toxicol.* 118, 830–848. https://doi.org/10.1016/j.fct.2018.06.037

Khaneghah, A.M., Fakhri, Y., Sant'Ana, A.S., Oliveira, C.A.F., and Saeedi, R. (2017). Physico-chemical and sensory properties of Frankfurter sausage as affected by some new natural preservatives: Nisin, natamycin, and potassium sorbate. *J. Food Saf.* 37, e12328. https://doi.org/10.3390/foods10102418

Koch, A., and Kogel, K.H. (2014). New wind in the sails: Improving the agronomic value of crop plants through RNAi-mediated gene silencing. *Plant Biotechnol. J.* 12, 821–831. https://doi.org/10.1111/pbi.12226

Kuiper-Goodman, T., and Scott, P.M. (1989). Risk assessment of the mycotoxin ochratoxin A. *Biomed. Environ. Sci.* 2, 179–248. https://doi.org/10.1007/978-3-642-00725-5_18

Liu, L., Xie, M., and Wei, D. (2022). Biological detoxification of mycotoxins: Current status and future advances. *Int. J. Mol. Sci.* 23, 1064. https://doi.org/10.3390/ijms23031064

Liu, Z., Jian, Y., Chen, Y., Kistler, H.C., He, P., and Ma, Z. (2019). A phosphorylated transcription factor regulates sterol biosynthesis in *Fusarium graminearum*. *Nat. Commun.* 10, 1228. https://doi.org/10.1038/s41467-019-09145-6

Lye, M.S., Ghazali, A.A., Mohan, J., Alwin, N., and Nair, R.C. (1995). An outbreak of acute hepatic encephalopathy due to severe aflatoxicosis in Malaysia. *Am. J. Trop. Med. Hyg.* 53, 68–72.

Maier, F.J., Miedaner, T., Hadeler, B., Felk, A., Salomon, S., and Lemmens, M. (2006). Involvement of trichothecenes in fusarioses of wheat, barley and maize evaluated by gene disruption of the trichodiene synthase (Tri5) gene in three field isolates of different chemotype and virulence. *Mol. Plant Pathol.* 7, 449–461. https://doi.org/10.1111/j.1364-3703.2006.00351.x

Majumdar, R., Rajasekaran, K., and Cary, J.W. (2017). RNA interference (RNAi) as a potential tool for control of mycotoxin contamination in crop plants: Concepts and considerations. *Front. Plant Sci.* 8, 200. https://doi.org/10.3389/fpls.2017.00200

Marasas, W.F., Kellerman, T.S., Gelderblom, W.C., Coetzer, J.A., Thiel, P.G., and Lugt, V.D.J.J. (1988). Leukoencephalomalacia in a horse induced by fumonisin B1 isolated from *Fusarium moniliforme*. *Onderstepoort J. Vet. Res.* 55, 197–203.

Marc, R.A. (2022). Mycotoxins and food safety. Implications of mycotoxins in food safety. In: *Mycotoxins and Food Safety-Recent Advances*. IntechOpen. https://doi.org/10.5772/intechopen.102495

Maschietto, V., Colombi, C., Pirona, R., Pea, G., Strozzi, F., Marocco, A., and Lanubile, A. (2017). QTL mapping and candidate genes for resistance to Fusarium ear rot and fumonisin contamination in maize. *BMC Plant Biol.* 17, 1–21.

Mayfield, K., Betrán, F.J., Isakeit, T., Odvody, G., Murray, S.C., and Rooney, W.L. (2012). Registration of maize germplasm lines Tx736, Tx739, and Tx740 for reducing preharvest aflatoxin accumulation. *J. Plant Regist.* 6, 88–94. https://doi.org/10.3198/jpr2010.12.0675crg

Mayfield, K.L., Murray, S.C., Rooney, W.L., Isakeit, T., and Odvody, G.A. (2011). Confirmation Of QTL reducing aflatoxin in maize testcrosses. *Crop Sci.* 51, 2489–2498. https://doi.org/10.2135/cropsci2011.02.0112

Mazaheri, M., Maymand, M.M., Gilasgar, A., Akbarzadeh, A., and Manafi, M.H. (2021). Quantification of the zearalenone in maize oil with no clean-up. *Food Control.* 127108166. https://doi.org/10.1016/j.foodcont.2021.108166

Mesterházy, A., Oláh, J., and Popp, J. (2020). Losses in the grain supply chain: Causes and solutions. *Sustainability* 12, 2342. https://doi.org/10.3390/su12062342

Nathanail, A.V., Varga, E., Meng-Reiterer, J., Bueschl, C., Michlmayr, H., and Malachova, A. (2015). Metabolism of the *Fusarium* mycotoxins T-2 toxin and HT-2 toxin in wheat. *J. Agric. Food Chem.* 63 7862–7872. https://doi.org/10.1021/acs.jafc.5b02697

Nayak, S.N., Agarwal, G., Pandey, M.K., Sudini, H.K., Jayale, A.S., and Purohit, S. (2017). *Aspergillus flavus* infection triggered immune responses and host-pathogen cross-talks in groundnut during in-vitro seed colonization. *Sci. Rep.* 7, 9659. https://doi.org/10.1038/s41598-017-09260-8

Nunes, C.C., and Dean, R.A. (2012). Host-induced gene silencing: A tool for understanding fungal host interaction and for developing novel disease control strategies. *Mol. Plant Pathol.* 13, 519–529. https://doi.org/10.1111/j.1364-3703.2011.00766.x

Ogunola, O., Smith, J.S., Xu, W., Bhattramakki, D., Jeffers, D., and Williams, W.P. (2021). Characterization of a source of resistance to aflatoxin accumulation in maize. *Agrosyst. Geosci. Environ.* 4, e20203. https://doi.org/10.1002/agg2.20203

Omotayo, O.P., Omotayo, A.O., Mwanza, M., and Babalola, O.O. (2019). Prevalence of mycotoxins and their consequences on human health. *Toxicol. Res.* 35, 1–7. https://doi.org/10.5487/TR.2019.35.1.001

Oppong, A., Appiah Jnr, K., Appiah-Kubi, Z., Adu-Amoah, R., and Adu-Dapaah, H. (2022). Mitigating aflatoxin contamination in maize: Breeding for stable resistant cultivars. *Front. Hortic.* 1, 1029804. https://doi.org/10.3389/fhort.2022.1029804

Ostry, V., Malir, F., Toman, J., and Grosse, Y. (2017). Mycotoxins as human carcinogens—the IARC Monographs classification. *Mycotoxin Res.* 33, 65–73. https://doi.org/10.1007/s12550-016-0265-7

Ostry, V., Malir, F., Toman, J., and Grosse, Y. (2013). Mycotoxins as human carcinogens—the IARC Monographs classification. *Mycotoxin Res.* 29, 181–186. https://doi.org/10.1007/s12550-016-0265-7

Pandey, M.K., Kumar, R., Pandey, A.K., Soni, P., Gangurde, S.S., and Sudini, H.K. (2019). Mitigating aflatoxin contamination in groundnut through a combination of genetic resistance and post-harvest management practices. *Toxins* 11, 315. https://doi.org/10.3390/toxins11060315

Patriarca, A., and Pinto, V.F. (2017). Prevalence of mycotoxins in foods and decontamination. *Curr. Opin. Food Sci.* 14, 50–60. https://doi.org/10.1016/j.cofs.2017.01.011

Peng, W.X., Marchal, J.L.M., and van der Poel, A.F.B. (2018). Strategies to prevent and reduce mycotoxins for compound feed manufacturing. *Anim. Feed Sci. Technol.* 237, 129–153. https://doi.org/10.1016/j.anifeedsci.2018.01.017

Pinotti, L., Ottoboni, M., Giromini, C., Dellorto, V., and Cheli, F. (2016). Mycotoxin contamination in the EU feed supply chain: A focus on cereal byproducts. *Toxins* 8, 45. https://doi.org/10.3390/toxins8020045

Pitt, J.I. (1989). An introduction to mycotoxins. In: Semple, R.L., Frio, A.S., Hicks, P.A., and Lozare, J.V. (eds.) *Mycotoxin Prevention and Control in Food Grains*. Food and Agriculture Organization, Rome, Italy.

Pleadin, J., Frece, J., and Markov, K. (2019). Mycotoxins in food and feed. *Adv. Food Nutr. Res*. 89, 297–345. https://doi.org/10.1016/BS.AFNR.2019.02.007

Probst, C., Njapau, H., and Cotty, P.J. (2007). Outbreak of an acute aflatoxicosis in Kenya in 2004: Identification of the causal agent. *Appl. Environ. Microbiol.* 73, 2762–2764. https://doi.org/10.1128/AEM.02370-06

Reddy, B.N., and Raghavender, C.R. (2007). Outbreaks of aflatoxicoses in India. *Afr. J. Food Agric. Nutr. Dev.* 7, 1–15. https://doi.org/10.18697/ajfand.16.2750

Richard, J.L. (2007). Some major mycotoxins and their mycotoxicoses—an overview. *Int. J. Food Microbiol.* 119, 3–10. https://doi.org/10.1016/j.ijfoodmicro.2007.07.019

Rodrigues, I., and Naehrer, K. (2012). A three-year survey on the worldwide occurrence of mycotoxins in feedstuffs and feed. *Toxins* 4, 663–675. https://doi.org/10.3390/TOXINS4090663

Ropejko, K., and Twarużek, M. (2021). Zearalenone and its metabolites—General overview, occurrence, and toxicity. *Toxins* 13, 35. https://doi.org/10.3390/toxins13010035

Ruyck, D.K., De Boevre, M., Huybrechts, I., and Saeger, D.S. (2015). Dietary mycotoxins, co-exposure, and carcinogenesis in humans: Short review. *Mutat. Res. Rev. Mutat. Res.* 766, 32–41.

Schatzmayr, G., Zehner, F., Taubel, M., Schatzmayr, D., Klimitsch, A., and Loibner, A.P. (2006). Microbiologicals for deactivating mycotoxins. *Mol. Nutr. Food Res.* 50, 543–551. https://doi.org/10.1002/MNFR.200500181

Scott, G.E., and Zummo, N. (1990). Registration of Mp313E parental line of maize. *Crop Sci.* 30, 1378–1378. https://doi.org/10.2135/cropsci1990.0011183X003000060080x

Sipos, P., Peles, F., Brasso, D.L., Beri, B., Pusztahelyi, T., and Pocsi, I. (2021). Physical and chemical methods for reduction in aflatoxin content of feed and food. *Toxins* 13, 204. https://doi.org/10.3390/toxins13030204

Sun, L.H., Lei, M.Y., Zhang, N.Y., Gao, X., Li, C., and Krumm, C.S. (2015). Individual and combined cytotoxic effects of aflatoxin B1, zearalenone, deoxynivalenol and fumonisin B1 on BRL 3A rat liver cells. *Toxicon* 95, 6–12. https://doi.org/10.1016/J.TOXICON.2014.12.010

Taheur, F.B., Kouidhi, B., Al Qurashi, Y.M.A., Salah-Abbès, J.B., and Chaieb, K. (2019). Biotechnology of mycotoxins detoxification using microorganisms and enzymes. *Toxicon* 160, 12–22.

Takeshita, N., Higashitsuji, Y., Konzack, S., and Fischer, R. (2008). Apical sterol-rich membranes are essential for localizing cell end markers that determine growth directionality in the filamentous fungus *Aspergillus nidulans*. *Mol. Biol. Cell* 19, 339–351. https://doi.org/10.1091/mbc.e07-06-0523

Tao, Y., Xie, S., Xu, F., Liu, A., Wang, Y., and Chen, D. (2018). Ochratoxin, A. Toxicity, oxidative stress, and metabolism. *Food Chem. Toxicol.* 112, 320–331. https://doi.org/10.1016/j.fct.2018.01.002

Thakare, D., Zhang, J., Wing, R.A., Cotty, P.J., and Schmidt, M.A. (2017). Aflatoxin-free transgenic maize using host-induced gene silencing. *Sci. Adv.* 3, e1602382. https://doi.org/10.1126/sciadv.1602382

Theumer, M., Henneb, Y., Khoury, L., Snini, S., Tadrist, S., and Canlet, C. (2018). Genotoxicity of aflatoxins and their precursors in human cells. *Toxicol. Lett.* 287, 100–107. https://doi.org/10.1016/j.toxlet.2018.02.007

Umereweneza, D., Kamizikunze, T., and Muhizi, T. (2018). Assessment of mycotoxins types in some foodstuff consumed in Rwanda. *Food Control.* 85, 432–436. https://doi.org/10.1016/j.foodcont.2017.10.015

Urban, M., Mott, E., Farley, T., and Hammond-Kosack, K. (2003). The *Fusarium graminearum MAP1* gene is essential for pathogenicity and development of perithecia. *Mol. Plant Pathol.* 4, 347–359. https://doi.org/10.1046/j.1364-3703.2003.00183.x

Wagacha, J.M., and Muthomi, J. (2008). Mycotoxin problem in Africa: Current status, implications to food safety and health and possible management strategies. *Int. J. Food Microbiol.* 124, 1–12. https://doi.org/10.1016/j.ijfoodmicro.2008.01.008

Wang, Y., Liu, D., Yin, H., Wang, H., Cao, C., and Wang, J. (2023). Transcriptomic and metabolomic analyses of the response of resistant peanut seeds to *Aspergillus flavus* infection. *Toxins* 15, 414. https://doi.org/10.3390/toxins15070414

Wang, Y., Zhou, Z., Gao, J., Wu, Y., Xia, Z., and Zhang, H. (2016). The mechanisms of maize resistance to *Fusarium verticillioides* by comprehensive analysis of RNA-seq data. *Front. Plant Sci.* 7, 1–14. https://doi.org/10.3389/fpls.2016.01654

Warburton, M.L., Brooks, T.D., Windham, G.L., and Williams, W.P. (2011). Identification of novel QTL contributing resistance to aflatoxin accumulation in maize. *Mol. Breed.* 27, 491–499. https://doi.org/10.1007/s11032-010-9446-9

Warburton, M.L., Tang, J.D., Windham, G.L., Hawkins, L.K., Murray, S.C., and Xu, W. (2015). Genome-wide association mapping of *Aspergillus flavus* and afatoxin accumulation resistance in maize. *Crop Sci.* 55, 1857–1867. https://doi.org/10.2135/cropsci2014.06.0424

Wen, J., Mu, P., and Deng, Y. (2016). Mycotoxins: Cytotoxicity and biotransformation in animal cells. *Toxicol. Res.* 5, 377–387. https://doi.org/10.1039/c5tx00293a

Williams, W.P. (2006). Breeding for resistance to aflatoxin accumulation in maize. *Mycotoxin Res.* 22, 27–32. https://doi.org/10.1007/BF02954554/METRICS

Williams, W.P., and Windham, G.L. (2012). Registration of Mp718 and Mp719 germplasm lines of maize. *J. Plant Regist.* 6, 200–202. https://doi.org/10.3198/jpr2011.09.0489crg

Womack, E.D., Williams, W.P., Windham, G.L., and Xu, W. (2020). Mapping quantitative trait loci associated with resistance to aflatoxin accumulation in maize inbred Mp719. *Front. Microbiol.* 11, 45. https://doi.org/10.3389/fmicb.2020.00045

Wu, Q., Dohnal, V., Kuca, K., and Yuan, Z. (2013). Trichothecenes: Structure-toxic activity relationships. *Curr. Drug Metab.* 14, 641–660. https://doi.org/10.2174/1389200211314060002

Yin, Z., Wang, Y., Wu, F., and Xu, W. (2014). Quantitative trait locus mapping of resistance to *Aspergillus flavus* infection using a recombinant inbred line population in maize. *Mol. Breed.* 33, 39–49. https://doi.org/10.1007/s11032-013-9932-y

Yu, B., Huai, D., Huang, L., Kang, Y., Ren, X., and Chen, Y. (2019). Identification of genomic regions and diagnostic markers for resistance to aflatoxin contamination in peanut (*Arachis hypogaea* L.). *BMC Genet.* 20, 1–13. https://doi.org/10.1007/s00122-023-04306-0

Yu, B., Liu, N., Huang, L., Luo, H., Zhou, X., and Lei, Y. (2024). Identification and application of a candidate gene *AhAftr1* for aflatoxin production resistance in peanut seed (*Arachis hypogaea* L.). *J. Adv. Res.* 62, 15–26. https://doi.org/10.1016/j.jare.2023.09.014

Zhang, H., Wang, Q., Wang, J., Fan, S., Xu, W., and Jiao, X. (2019). Current situation and future trends of research on biological detoxification of mycotoxins in China. *Toxins* 11, 464. https://doi.org/10.1038/s41586-020-2008-3

Zhang, Y., Cui, M., Zhang, J., Zhang, L., Li, C., and Kan, X. (2016a). Confirmation and fine mapping of a major QTL for aflatoxin resistance in maize using a combination of linkage and association mapping. *Toxins (Basel)* 8, 1–15. https://doi.org/10.3390/toxins8090258

Zhang, Z., Hu, X., Zhang, Q., and Li, P. (2016b). Determination for multiple mycotoxins in agricultural products using HPLC–MS/MS via a multiple antibody immunoaffinity column. *J. Chromatogr. B*, 1021, 145–152.

Zhou, H., George, S., Hay, C., Lee, J., Qian, H., and Sun, X. (2017). Individual and combined effects of Aflatoxin B1, Deoxynivalenol and Zearalenone on HepG2 and RAW 264.7 cell lines. *Food Chem. Toxicol.* 103, 18–27. https://doi.org/10.1016/J.FCT.2017.02.017

Zhu, Y., Hassan, Y.I., Watts, C., and Zhou, T. (2017). Innovative technologies for the mitigation of mycotoxins in animal feed and ingredients—A review of recent patents. *Anim. Feed Sci. Technol.* 226, 77–90. https://doi.org/10.1016/j.anifeedsci.2016.03.030

11 Progress in Development of High-Oleic and/or Hypoallergenic Peanuts with Sensory and Textural Traits

U. Nikhil Sagar, Priya Shah, Deekshitha Bomireddy, H.V. Veerendrakumar, Sandip K. Bera, Chuanzhi Zhao, Xinping Wang, Sunil S. Gangurde, and Manish K. Pandey

11.1 INTRODUCTION

Groundnut or peanut (*Arachis hypogaea* L.), here onward will be referred as peanut, is a significant oil seed legume, cultivated globally for its oil, protein, and carbohydrate contents. Peanut is an allotetraploid (AABB, 4x = 40), with a genome size of 2.7 Gb, originated from crossing the two diploid progenitors, *A. duranensis* (AA) and *A. ipaensis* (BB) (Bertioli et al., 2019). Peanut is also rich in minerals, vitamins (vitamin B6, pantothenic acid, folate, thiamine, niacin, and other B complex vitamins), and antioxidants, such as resveratrol and p-coumaric acids. Peanuts and its products, because of their high nutritional value, can be marketed as functional food for underprivileged individuals deficient in protein, energy, and micronutrients (Arya et al., 2016). Peanut oil is composed of free fatty acids (0.1–0.4%), sterols (0.69–0.80%), glycolipids (0.10–0.14%), phospholipids (2.4–2.9%), and triglycerides (96.1–96.4%). Palmitic acid (8–11%), linoleic acid (24–43%), and oleic acid (36–52%) are the major fat in peanut oils (Parmar et al., 2022). Globally, 41% of the peanut yield is consumed as food, 49% for edible oil extraction, and 10% as seed and feed (Singh and Nigam, 2016). About 81% of the total peanuts produced in India are crushed for oil, 12% are used as seed, 6% are directly consumed, and 1% are exported. The high oleic/linoleic acid (O/L) ratio of peanut kernels results in extended shelf-life. High-oleate peanuts demand a higher price in international and domestic markets. High-oleate peanuts are desirable for extraction of oil, whereas those with high protein and low oil contents, are commercialized as confectionary and food-grade products (Gangurde et al., 2023). A range of physical, chemical, and

DOI: 10.1201/9781003467335-11

nutritional parameters influence the quality of peanuts, for which there is a significant genetic variation (Dwivedi and Nigam, 2005). An important objective of peanut improvement is breeding for increased oil, protein, and oleate content in the kernels.

Peanut oils with a high-oleate trait exhibit enhanced nutritional value and improved thermo-oxidative stability. O/L ratios in high-oleate peanut mutants ranged from 30 to 40, while those in regular-grown peanuts, from 1 to 4. Most high-oleate peanut cultivars that have been generated thus far are the result of conventional breeding efforts. However, during last decade, high-oleate peanut cultivars were developed using genomic-assisted breeding (Nawade et al., 2018; Pandey et al., 2024). High levels of linoleate in oil are generally not recommended for cooking because such oils can cause oxidative rancidity and thermodynamic instability when heated to high temperatures (Kamdar et al., 2017). The composition of fatty acids plays a major role in maintaining the shelf-life of peanut oil. Of the 12 fatty acids present in peanut oil, oleic acid (monounsaturated fat), and linoleic acid (polyunsaturated fat) constitute about 36–67% and 15–43% fat, respectively. Moreover, the remaining 10% of the oil is composed of up to nine different fatty acids, of which palmitic acid (saturated fatty acid), comprises about ~10% of the total oil (Janila et al., 2016b; Moore and Knauft, 1989). The shelf-life, flavor, and nutritional value of peanut products and seeds are all influenced by the different fatty acids in peanut oil, including monounsaturated, saturated, and polyunsaturated fatty acids (Derbyshire, 2014). Peanut is a healthy cooking oil due to its high ratio of unsaturated fatty acid (UFA) to saturated fatty acid (SFA), surpassing coconut and palm oil in health benefits (O'Byrne et al., 1997). Oleate is less likely to undergo oxidation, which increases the shelf-life of oil and the derived products. Reduced oil absorption and shorter cooking times are achieved when a larger amounts of unsaturated fatty acid oils are heated to high temperatures with no smoking (Miller et al., 1990). In terms of nutrition, oleic acid reduces the bad cholesterol like linoleic acid, while leaving good cholesterol levels unchanged, thereby balancing the cholesterol that is ideal for a healthy heart (Kris-Etherton et al., 2001).

Allergens limit the consumption of peanuts in Western and European countries. In Europe, 2.2% of people have a peanut allergy (Nwaru et al., 2014), which is responsible for roughly 59% of food allergic-related deaths (Bock et al., 2007). In the United States, 2.5% and 1.8%, of children and adults, respectively, suffer with peanut allergy (Gupta et al., 2019; Jiang et al., 2018). While peanut proteins offer substantial nutritional and physiological benefits, many who consume experience severe, fatal allergic reactions (Sicherer and Sampson, 2010). Globally, about 1–2% of the total population have peanut allergy, resulting in anaphylaxis or death (Pandey et al., 2019b). So far there is no readily available allergy vaccine or method to lessen the allergenicity of peanut-based foods. Peanut allergy is regarded as the most severe food allergy, because of the frequency, persistence, and degree of allergic reaction severity (Skolnick et al., 2001). Roasted or thermally processed peanuts could intensify the allergenic reaction because the Maillard reaction causes the protein alterations (Maleki et al., 2000). The Maillard reaction, named after the French physicist and chemist Louis Camille Maillard, refers to non-enzymatic browning. The peanuts processed at high temperature leads to a chemical reaction between reducing sugars and amino acids, resulting in the formation of Maillard reaction products (MRPs), either beneficial or toxic (Tamanna and Mahmood, 2015). It has been discovered that peanuts contain 17 allergenic proteins (Becker and Jappe, 2014; WHO, 2016). Of these, *Ara h 2* and *Ara h 6* were identified as the most prevalent peanut allergens in the United States (Kulis et al., 2012; Zhuang and Dreskin, 2012). Eighty-five percent of the protein composition of peanuts is constituted by all known peanut allergens, of which 75% is made up of *Ara h 1*, *Ara h 2*, and *Ara h 3* together (Koid et al., 2014).

The marketability and customer preference of peanuts are significantly influenced by their sensory and textural qualities. Textural quality in peanut refers to the sensory characteristics that include the seed hardness, crispness, and smoothness. Taste, smell, color, crunchiness, and mouthfeel are key sensory attributes that influence the whole eating experience (Choi et al., 2002). The textural attributes are influenced by the processing methods, such as roasting and grinding, and storage conditions. However, inherent variability in these traits across peanut cultivars can present challenges in meeting consumer expectations and maintaining market competitiveness. Inconsistencies in sensory profiles or undesirable textural properties may lead to reduced consumer satisfaction and market value.

A range of breeding methods and genetic engineering tools are explored for genetic improvement of these two traits in peanuts. Conventional breeding involves selecting and crossbreeding peanut varieties with desirable sensory and textural attributes (Holbrook, 2019). However, this process typically involves several generations of selective breeding to develop a variety with enhanced flavor. Advanced genomic methods like marker-assisted back-crossbreeding (MABB) or marker-assisted selection (MAS) accelerate the breeding procedure by identifying gene tags (DNA markers) associated with the target traits. Moreover, genetic engineering can be used for precise trait manipulation. Techniques like clustered regularly interspaced short palindromic repeats/cas9 (CRISPR/Cas9) gene editing cause specific gene modifications responsible for sensory and textural traits, potentially resulting in peanuts with enhanced flavor, aroma, color, crunchiness, and mouthfeel (Wang et al., 2018).

11.2 KEY CHARACTERISTICS OF PEANUT QUALITY

11.2.1 Sensory Proteins, Flavor Compounds, and Textural Traits

The high-oleate peanuts currently compete with olive oils having unsaturated fats (78%) and slightly high polyunsaturated fats (8%). The monounsaturated fatty acids are comparatively high in high-oleate peanuts when compared with other oil seed crops, including canola, sesame, palm, soybean, sunflower, safflower, and coconut (Figure 11.1). The important quality traits of peanut include physical characteristics such as dark discoloration and hardness during storage and chemical properties such as rancidity and off-flavor. These quality traits, along with visual and sensory features, in the food and oil processing industries are commonly referred as trader and consumer preferred traits. Roasted peanut flavor and color are crucial research areas as these traits significantly influence the consumer preference (Lee and Resurreccion, 2006; Young et al., 2005). Roasted peanuts' flavor and sensory attributes are linked to the carbohydrate components found in the kernel (Pattee et al., 2000). The roasted peanutty flavor is assessed using an organoleptic test conducted by trained human panel, called a taste panel. Coffee, fresh, nutty, popcorn, almond, smokey, sweet, and not off-flavor are among the desired flavors (Fletcher, 1987). The key ingredients that add flavor to the roasted peanuts are amino acids and sugars (Newell et al., 1967). Pyrazines play predominant role in determining the flavor of roasted peanuts. When an amino acid reacts with an α-dicarbonyl molecule, it undergoes decarboxylation and deamination, producing an α-amino ketone, and an aldehyde known as Strecker aldehyde—which has carbon atoms fewer than the original acid; this process is known as Strecker degradation (Mottram et al., 2006). The last stage of the Maillard reaction, known as Strecker degradation, yields a variety of heterocyclic chemicals, including pyrroles, pyrazines, and pyridines, and is responsible for the distinct flavors of thermally processed peanuts (Salehi, 2020).

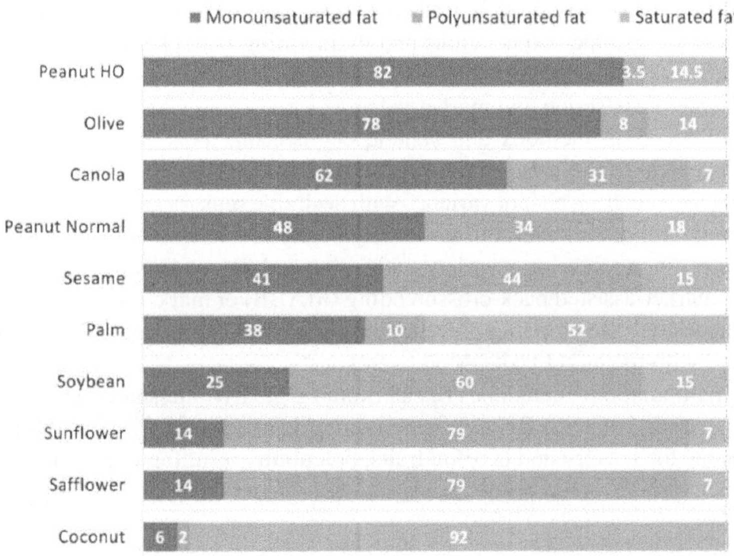

Pyrazine compounds, of which 2,5-dimethylpyrazine is the most common compound produce the aroma and roasted peanut flavor (Baker et al., 2003). Pyrazines and its derivative compounds like benzaldehyde, methylpyrazine, 2-ethyl-5-methylpyrazine, and 2,5-dimethylpyrazine are associated with typical nutty or roasted flavor and sweet aroma of peanut oil (Dun et al., 2019; Hu et al., 2021). Flavor content and sensory evaluation analysis demonstrate that the flavor of peanut oil is significantly affected by the pyrazines and the amino acids, like lysine, phenylalanine, arginine, glutamic acid, and isoleucine, show association with the pyrazines, thus improving the sensory qualities of peanut oil (Sun et al., 2024).

Peanuts that are used as an ingredient in foods must be preserved in a high-quality manner. High-oleate raw peanuts stored in their shells exhibit superior sensory and quality preservation over time (Martin et al., 2018). A distinctive flavor that can be produced by roasting peanuts is one of the key elements determining consumer acceptance and choice. Additionally, roasted peanuts contain volatile compounds such as pyrazines, but this does not always indicate that they significantly enhance the flavor of the food. When the concentrations of certain dietary components are below the sensory threshold, there is a chance that they have no effect on flavor (McGorrin, 2002).

The link between each fragrance constituent and its matching flavor can occasionally be impacted by associations with the other chemical components and the food matrix (Drake and Civille, 2003). Foods having a roast flavor can undergo the Maillard reaction when heated, which is a widely recognized explanation for how the flavor is altered when free amino acids and sugar polymerize (Nursten, 2005), and peanuts have been found to contain these amino acid compounds and sugars (Chiou et al., 1993; Liu et al., 2011; Pattee et al., 2001). When compared to frying, baking contributes to the development of color while reducing the smell of roasted peanuts and dark roast, as color and flavor develop through the similar pathway (Maillard reaction), thus increasing the aroma of raw beans (Lykomitros et al., 2016). The nutrients and flavor of peanuts can be altered by the raw ingredients as well as the processing techniques. Roasting peanuts usually improves their flavor, color, aroma, taste, and crunchy texture (Chang et al., 2016; Tamanna and Mahmood, 2015). It also inhibits the bacteria and fungi producing aflatoxin in the raw peanuts (Ostadrahimi et al., 2014).

Along with color and flavor, the texture of peanut butter is a crucial component, influencing eating quality and customer satisfaction (Shieh et al., 1996). In the United States, peanut butter constitutes about 50% of the total peanut consumption (Amba et al., 2019). Bitter taste and 'raw bean' aroma were less preferred, whereas 'dark roast', 'roasted peanut', and 'sweet' aromas, as well as the color grade, were more preferred attributes by the European consumers (Lykomitros, 2018). Consumers' preferences for peanut texture vary according to their gender and age in Japan, as the elder ones tend to prefer easily broken peanuts when compared to young individuals (Miyagi and Ogaki, 2014). Qualitative electron microscopy techniques are used to study the variations in texture during processing in relation to microstructure changes. The majority of these studies concluded that the disruption of the microstructure during roasting brought on by the generation of steam is the only reason for the texture to become crispier and crunchier during roasting (Young et al., 1993, 2004).

11.2.2 Allergens

According to National Institute of Allergy and Infectious Diseases (NIAID) standards, an *'adverse health effect resulting from a specific immune response that occurs reproducibly when exposed to a given food'* is the commonly accepted definition of the term food allergy (Panel, 2010). Asian nations have a lower incidence of peanut allergies but a higher incidence of fish allergies (Lee et al., 2013). A significant risk factor for potentially fatal anaphylaxis in almost 26% of the children is peanut allergy. Fatal allergy (FA) is an increasingly serious global public health concern, a form of food sensitivity (Loh and Tang, 2018). FA refers to the harmful immunological reaction that the human body may have to specific diets (Burks et al., 2012). Since FA has become more common in recent years, it has become more difficult for researchers, including nutritionists, to develop quick and precise screening procedures along with preventative and therapeutic strategies for those susceptible to allergic food ingredients. The absence of effective preventive methods, other than immunotherapy-induced desensitization, has made FA a global concern for food safety (Du Toit et al., 2015). Peanuts are the most well-characterized source of allergens among the common food allergens (wheat, peanut, soybean, tree nuts, shellfish, fish, milk, and eggs) and are linked to severe anaphylaxis in both adults and children. Compared to the other food allergies, these allergies tend to last for a lifetime (Lieberman et al., 2020; Sicherer and Sampson, 2010). According to reports, the frequency of peanut allergies has been doubled over the last ten years. There are two categories of peanut allergens: major allergens and mild allergens. Due to their fatal reactions that may be identified by IgE and result in anaphylaxis or death, *Ara h 1, Ara h 2, Ara h 3,* and *Ara h 6* are categorized as significant allergens. Tetraploid peanuts have two identical copies of the major allergen genes *Ara h 1* and *Ara h 2*, one in each subgenome (Ratnaparkhe et al., 2014).

The peanut allergens play a major role in controlling several biological activities like stabilization of cell membranes and structure, cell wall organization, resistance to various stresses, plant growth and development, and metabolic activities (Table 11.1). Six cysteine residues in the *Arah 1* gene stabilize its structure, preventing from digestive denaturation and inhibits trypsin activity (Viquez et al., 2003; Wichers et al., 2004). The seed storage proteins with 2S albumin are found in the *Ara h 2, Ara h 6*, and *Ara h 7* genes (Mylne et al., 2014) whereas Ara h 9, Ara h 16, and Ara h 17 genes control cell wall organization, signal transduction, stabilization of membranes, and resistance against biotic and abiotic stresses (Liu et al., 2015). *Ara h 10, Ara h 11, Ara h 14*, and *Ara h 15* are involved in various lipid and energy metabolism activities (Jolivet et al., 2013). Antimicrobial activity and the effects suppressing mold strains of the genera *Alternaria* and *Cladosporium* are regulated by the *Ara h 12* and *Ara h 13* genes (Vriens et al., 2014; Petersen et al., 2015).

TABLE 11.1
Peanut Allergens and Their Biological Functions

Peanut allergen	Biological function	Reference
Ara h 1	Trypsin inhibitory activity (defense role in plants against insects) and contains six cysteine residues to stabilize the cell membrane structure against digestive denaturation	Viquez et al. (2003), Wichers et al. (2004)
Ara h 2, Ara h 6, Ara h 7	Seed-storing proteins containing 2S albumin	Mylne et al. (2014)
Ara h 9, Ara h 17, Ara h 16	Involved in signal transduction, membrane stabilization, cell wall organization, growth and development of plants, and resistance against various stresses	Liu et al. (2015)
Ara h 10, Ara h 11, Ara h 14, Ara h 15	Involved in lipid and energy metabolism activities	Jolivet et al. (2013)
Ara h 12, Ara h 13	Antimicrobial and antifungal activity; show inhibitory action on the mold strains of the genera *Alternaria* and *Cladosporium*	Vriens et al. (2014) Petersen et al. (2015)

Peanuts contain several multiple allergen coding genes like *Ara h 1:3*, *Ara h 2:1*, *Ara h 3:8*, and *Ara h 6:2* (Ratnaparkhe et al., 2014). In seeds, *Ara h 1, 2, 3*, and *6* exhibit prominent expressions. Gene silencing results in a hypoallergenic peanut variety where the expression of the key allergen, *Ara h 2*, is reduced to 25% (Dodo et al., 2008). Additionally, the A genome contained 21 genes that code for allergens. Among them, nine have been already identified in peanuts, while the other ones were similar with other crops (Chen et al., 2016). The decrease of allergic potency in peanuts has been demonstrated by non-covalent changes that hinder proteolytic digestion by binding with substances including tannic acid, phytic acid, and phenolic chemicals to form insoluble complexes (Chung and Champagne, 2008; Chung and Reed, 2012). The allergenic proteins, *Ara h 1* and *Ara h 2,* from the roasted peanuts are fully hydrolyzed after three hours of treatment with chymotrypsin (0.1%) and trypsin (0.15%) (Yu et al., 2011). *Ara h 1, Ara h 2*, and *Ara h 3* genes are successfully removed through hydrolysis with flavorzyme and alcalase (Cabanillas et al., 2012). The allergens present in the raw peanut pulp are greatly reduced by fermenting and autoclaving it with *Bacillus natto* (Pi et al., 2021). Most of the people are allergenic to *Ara h 2* allergen protein (Shah et al., 2019; Wu et al., 2016a). Eradicating the *Ara h 2* genes may therefore reduce peanut allergy and may not have a harmful effect on the plant development.

11.3 DEVELOPING HIGH-OLEATE PEANUTS

11.3.1 Identification of DNA Markers Associated with High-Oleate Trait and Progress in Development of High-Oleate Peanuts

11.3.1.1 DNA Markers

Molecular markers like allele-specific polymerase chain reaction (AS-PCR) markers and cleaved amplified polymorphic sequences (CAPS) were successfully utilized for developing high-oleate peanuts (Bera et al., 2018; Janila et al., 2016b). Selections based on markers were also conducted in numerous germplasm collections (Mukri et al., 2012; Wang et al., 2015). The development of linked CAPS (Chu et al., 2007, 2009) and AS-PCR (Chen et al., 2010) markers for the two *Ah fatty acid desaturase* (*FAD2*) genes speed up the high-oleate peanut-breeding programs. Kompetitive allele-specific PCR (KASP) markers have been used for high-throughput discovery of *AhFAD2* mutants in the peanut, which is more appropriate than other techniques for identifying the *AhFAD2* genotype due to its high accuracy and ability to save time and resources (Zhao et al., 2017). KASP markers for the FAD genes governing the high oleic trait in peanuts have been developed and validated from the result of the sequencing of Girnar 4 and Girnar 5 (Pandey et al., 2024).

11.3.1.2 Breeding Methods

Several traditional and advanced breeding methods including DNA markers were used for the improvement of oleic acid (25–80%) and increased O/L ratio (5–7%) (Table 11.2). The oleic acid percentage was improved in TMV 7, an elite groundnut cultivar, by introducing a recessive mutation from ICG 15419 that is responsible for high oleic acid. The backcross-derived populations involving TMV7 (recurrent) and ICG 15419 (high-oleate donor line) improved oleate content by utilizing the allele-specific markers F435-F and F435SUB-R (Vargheese et al., 2024). Hybridization and selection identified 37 peanut cultivars with high oleate content of 77% to 81%

(Kamdar et al., 2020). The mutation breeding using gamma rays had improved 27.3% of oleate content, i.e., the O/L ratio nearly increased to 5.5 in TGM 192M (Mondal et al., 2011). Three lines (BC_1F_{3-76}, BC_1F_{3-278}, and BC_1F_{3-296}) with oleate content of >80% were identified by near-infrared reflectance (NIR) spectroscopy using simple sequence repeats (SSRs), allele-specific, and single-nucleotide polymorphisms (SNPs) (Deshmukh et al., 2020). Marker-assisted backcross (MABC) method was used to create 30 BC_2F_4 and 22 BC_3F_4 introgression lines (ILs) for FDR and 41 BC_2F_4 and 46 BC_3F_4 ILs for high oleate content using 58KAxiom_*Arachis* array (Shasidhar et al., 2020). Traditional breeding methods without DNA markers are simple and cost-effective and rely on natural selection and variation but they often take multiple generations to achieve the desirable traits and are less precise. MAS or MABC methods of breeding accelerates the breeding process by facilitating early identification of plants with the desirable traits and DNA markers improves the selection accuracy, but overdependence on specific markers can reduce the diversity of genes (Nawade et al., 2018; Tang et al., 2022).

TABLE 11.2
Traditional and Genomic-Assisted Breeding Methods Used for Developing High-Oleic Peanuts

Breeding method	Methodology	Outcome	Reference
Traditional breeding	Hybridization and selection	A total of 37 peanut breeding lines were identified, with high-oleate contents varying from 77% to 81%	Kamdar et al. (2020)
Mutation breeding	Gamma rays	The oleic acid content of TGM 192M had nearly improved by 27.3%, resulting in an increase in the O/L ratio to 5.5	Mondal et al. (2011)
MAS	Simple sequence repeats, single-nucleotide polymorphic markers and allele-specific markers	Three lines with >80% oleic acid (BC_1F_3-296, BC_1F_3-278, and BC_1F_3-76) were identified using near-infrared reflectance spectroscopy	Deshmukh et al. (2020)
MAS	Allele-specific markers, F435SUB-R and F435-F	Increased oleic acid containing introgressed lines were identified in the F_1, BC_1F_1, BC_2F_1 and BC_2F_2 populations	Vargheese et al. (2024)
MAS, MABC	Allele-specific markers governing *FAD2A* and *FAD2B* genes	Several advanced lines with high oleic acid (~80%) developed, of which, Girnar 4 (ICGV 15083) and Girnar 5 (ICGV 15090) identified as best performing lines	Janila et al. (2016a, 2016b)
MABC	58KAxiom_*Arachis* array	41 BC_2F_4 and 46 BC_3F_4 ILs for high oleate and 30 BC_2F_4 and 22 BC_3F_4 ILs foliar disease resistance and high-oleate trait	Shasidhar et al. (2020)
MAS, MABC	CAPS markers and AS-PCR markers	A total of 3 MABC and 22 MAS derived lines were produced with increased oleate content (78–80%)	Kamdar et al. (2022)

11.3.1.3 Progress

Biofortification offers a sustainable means to address malnutrition by increasing the nutritional value of crops using several breeding methods (Bouis and Saltzman, 2017). Through germplasm screening, plant breeders have discovered a high-oleate peanut germplasm, F435 containing around oleic acid content of 80%, and breeding for high-oleate peanuts began (Norden et al., 1987). SunOleic 95R (Runner botanical type) was the first peanut cultivar registered with oleic acid content of 82% (Gorbet and Knauft, 1997). F435 contains two different kinds of mutations. A 'G' to 'A' change occurs at base pair (bp) 448, in the *AhFAD2A* gene, leading to a missense mutation, while in the *AhFAD2B* gene, an 'A' insertion occurs between base pair (bp) 441 and 442 (441 442insA) (Jung et al., 2000). Since then, numerous cultivars with high O/L ratio were developed by utilizing the resultant mutant line as a parent in regular peanut breeding to improve peanut quality (Chu et al., 2009). Subsequently, in the United States, the Florida Agricultural Experiment Station successfully created a multiline variety, SunOleic 97R, which boasts an impressive oleic acid content of 81.8% (Gorbet and Knauft, 2000). Three lines that resulted from the cross between Sunrunner (F519-9), the recurrent parent, and F435-2-2-E-2-1-b4-E-b2-b3-1-E (high oleate) were used to develop this cultivar (Gorbet and Knauft, 2000; Norden et al., 1987). The most promising crosses to utilize genetic diversity for oleic acid were reported between Georgia-02C and SunOleic 97R and the line [(NC17090 × B1)-9-1 × KK 60-3] $F_{6\text{-}8\text{-}3}$ (Singkham et al., 2011). Digenic regulation of high oleate and the interaction of other allelic variations both among and within the Spanish-type peanut varieties was also reported (Lopez et al., 2001). The Spanish and Valencia types have high saturated fatty acid content than Virginia and Runner types; however, the Virginia and Runner types have high unsaturated fatty acid content compared to the other types (Bakal, 2020). Several Spanish bunch peanut varieties, such as Georgia-04S, Ole, and TAMrun OL06, with O/L ratios of 34:1, 20:1, and 21:1, and Schubert with oleic acid of 77.6%, have been developed in the USA using various traditional breeding methods (Baring et al., 2006; Branch, 2005; Burow et al., 2014; Chamberlin et al., 2015). Numerous elite germplasm lines and cultivars have been created through the strategic utilization of synthetic amphidiploids, which are the outcome of the crosses between wild *Arachis* species. These include Georgia-14N and TifNV, showing high-oleate trait along with resistance against root-knot nematodes and Tomato Spotted Wilt Virus (TSWV), all derived from wild introgressions (Branch and Brenneman, 2015; Holbrook et al., 2017). Puppala and Tallury (2014) were the first to introduce high-oleate trait in Valenica (*Arachis hypogea fastigiata* var. *fastigiata*) peanut market type. It was derived from crossing New Mexico Valencia A with Brantley and named as NuMex 01. It shows relatively high 100-seed weight than New Mexico Valenica A, O/L ratio of 23 as compared to 1 in New Mexico Valenica A, and adapted in New Mexico and western Texas, the major Valenica-peanut-growing region in the USA. Subsequently, TAMVal OL14 was released for cultivation in the same region, and is relatively less susceptible to sclerotinia blight (*Sclerotinia minor*) than New Mexico Valencia A (Burow et al., 2019).

Several high-oleate peanut varieties were developed by China and other countries that are producing high-oleate peanut foods for consumption by humans, including

Australia, Argentina, Brazil, Japan, Israel, and South Africa (Janila et al., 2016b). In China, the elite peanut cultivar huayu22 was crossed with the donor N176 containing high oleic acid and subjected to four successive generations of backcrossing as a recurrent parent utilizing fatty acid desaturase (*FAD2*) MABC. The advanced generations, obtained through selfing, were screened using KASP markers to assess oleic acid content. From these lines, YH61 was selected in BC_4F_6 for its high-oleate trait and yield stability across multiple locations (Tang et al., 2022). Twenty-four lines with high oleate content were obtained within five years, with a genome-wide recovery rate ranging between 79.49% and 92.31%, and the increase in the oleate percent of these lines is due to the inactivity of *AhFAD2A* and *AhFAD2B* genes resulting from mutations (Huang et al., 2019). Spontaneous mutations within the *AhFAD2A* and *AhFAD2B* genes resulted in most of the existing high-oleate peanut cultivars. However, gamma-ray and ethyl methyl sulphonate (EMS)-based mutagenesis of Minhua 8 and Minhua 6 generated three and four high-oleate lines, respectively with better agronomic performance (Zhuang et al., 2019).

In India, several backcross-derived lines with high-oleate trait have been developed, including ICGVs 05141, 06110, 06142, and 06420, GG 20, GJGHPS 1, GJG 9, and Kadiri 6 (Bera et al., 2018; Deshmukh et al., 2020; Janila et al., 2016b; Shasidhar et al., 2020). One of the developed high-oleic peanut lines, ICGV 15033 showed oleic acid content of 82%, improved from a normal oleic peanut (ICGV 13193) with oleic acid content of 44.65%, using *FAD2A* and *FAD2B* markers (Janila et al., 2016b) (Figure 11.2). The two high-oleate Virginia Bunch types, Girnar4 (ICGV 15083) and Girnar5 (ICGV 15090), have been created from the cross ICGV 06420 × (ICGV 06420 × Sun Oleic 95-R), whereas the three high-oleate Spanish Bunch varieties, GG 39 (ICGV 16697), ICRC-1 (ICGV 16690), and GG 40 (ICGV 16668), were developed by crossing ICGV 06110 × (ICGV 06110 × Sun Oleic 95-R) (Janila et al., 2016a, 2016b). Two *FAD2* mutant alleles were incorporated into a peanut breeding line (ICGV06100) with a high amount of oil (~55%) by MABC (Bera et al., 2019). The total amount of linoleic acid reduced by 92% while oleic acid increased by 97%, resulting in an O/L ratio of 25.

MABB method was utilized successfully to transfer the high-oleate trait in peanuts (Bera et al., 2018, 2019; Gangurde et al., 2019; Janila et al., 2016b). Both the AS-PCR and the CAPS markers were effective in tracking the *AhFAD2A* and *AhFAD2B* mutant alleles (Bera et al., 2018, 2019; Janila et al., 2016b). By using these techniques, the amount of time and breeding material required in subsequent generations was significantly reduced. The amount of oleic acid of two elite peanut varieties, G 2–52 and GPBD 4, was increased by using the MABC (Jadhav et al., 2021). An oleate content of 80.6% was obtained by transferring the *AhFAD2B* mutant allele from the donor SunOleic 95R (Jadhav et al., 2021). The well-known peanut varieties GG-7 and TKG19A were transformed using the *AhFAD2A* and *AhFAD2B* mutant alleles and AS-PCR, CAPS markers were used to confirm the increased oleate content (78–80%) in three MABC and 22 MAS derived lines (Kamdar et al., 2022). Three popular Indian peanut cultivars have been improved using MABC (GJG 9, GG 20, and GJGHPS 1) to increase their resistance to foliar diseases and high-oleate trait (Shasidhar et al., 2020). In a similar way, MAS was used to create a high-oleic variant of the well-known cultivar

ICGV 05141 (Bera et al., 2018). Advances in speed-breeding technology and seed-chipping technique are facilitating the marker-based early generation selection to accelerate cultivar development in peanut (Parmar et al., 2021). This method would reliably advance 3.5 generations in one year by reducing the operational requirement of labor and saves 6–8 months in peanut breeding programs.

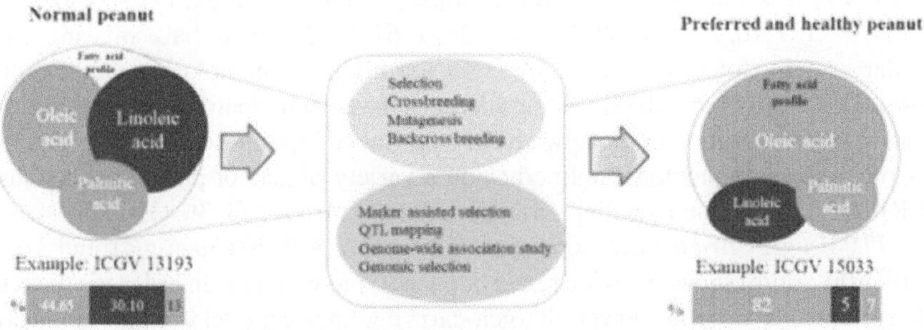

11.3.2 Identification of DNA Markers Associated with Peanut Allergens and Progress in the Development of Hypoallergenic Peanuts

11.3.2.1 DNA Markers

Peanut is one of the most examined and clinically significant allergens worldwide. A new DNA-based marker technique was developed to analyze the variations between 21 peanut accessions, revealing polymorphic fingerprints. This technique was most successful in identifying the analyzed accessions, with vicilin-based amplicon polymorphism (VBAP) being the most effective. The marker potentiality to identify polymorphisms was identical, with a mean polymorphism information content (PIC) of 0.47 (Ziarovska et al., 2023). Peanut allergens in the processed food products can be determined using real-time PCR, with optimal targets showing interspecific variation and high copy numbers. Chloroplast markers can be used for specific detection, and DNA isolation protocols were evaluated. The method's specificity and sensitivity were assessed using primers specific for the three chloroplast markers, namely, mat k, rpl16 and trnH-psbA, and TaqMan probes (Sanchiz et al., 2021). To specifically determine peanut allergies in a variety of foods, a multiplex real-time PCR assay utilizing the chloroplast markers (*matK, rpl16,* and *trnH-psbA*) was used. This technique is regarded as a useful tool to detect peanut allergies in various kinds of food items (Lelievre and Eischeid, 2018).

11.3.2.2 Breeding Methods

Traditional breeding efforts targeting hypoallergenic variations in peanuts have been made, despite the time-consuming nature of screening germplasm to discover segregants with decreased allergen levels (Pandey et al., 2019a; Perkins et al., 2006). Previously, patient sera were used to screen 35 US cultivars of peanuts, but no significant variations in the allergen levels were found (Dodo et al., 2002), and monoclonal antibodies have also been utilized to evaluate peanut-based products (Filep et al.,

2018). Similarly, when human sera were used to determine the allergen level of 53 Chinese peanut cultivars, no variance was found (Wu et al., 2016b).

Using a large-scale phenotyping method, several peanut lines at ICRISAT were shown to have reduced levels of the main allergenic proteins, namely, *Ara h 1, Ara h 2, Ara h 3, Ara h 6*, and *Ara h 8* (Pandey et al., 2019a). Several genotypes were found to have either lower levels of one or more than one of the four main allergens—*Ara 1, Ara h 2, Ara h 3,* and *Ara h* 6 (for example, P-189, P-11, P-64, P-194, P-60, P-29, P-120, P-21, P-16, P-55, P-69, P-143, P-26, P-67, P-22)—or to have missing a particular protein (for example, P-84A lacks the highly immunologic peanut protein *Ara h 1*) (Rustgi et al., 2022). Major allergens have been identified through the successful standardization of a sandwich enzyme-linked immunosorbent assay (ELISA) approach using monoclonal antibodies on a variety of sets of peanut accessions at ICRISAT. Each of the five allergens examined *i.e., Ara h 1* (4–36,833 μg/g), *Ara h 2* (41–77,041 μg/g), *Ara h 3* (22–106,765 μg/g), *Ara h 6* (829–103,892 μg/g), and *Ara h 8* (0.01–70.12 μg/g) showed a wide range of phenotypic variance. In order to create low allergen peanut varieties, lower-allergen-carrying lines were selected for application in future breeding programs (Pandey et al., 2019b). The identified peanut lines with low allergen content holds an immense potential to create vaccines or medicinal goods and to lower the incidence of peanut allergies. These hypoallergenic peanut genotypes (ICG 311, ICG 532, ICG 3992) with low allergen levels for *Ara h 1* (4 μg/g), *Ara h 2* (41 μg/g), *Ara h 3* (22 μg/g), *Ara h 6* (829 μg/g), and *Ara h 8* (0.01 μg/g) have enabled for their efficient use in various breeding and genomic studies (Pandey et al., 2019a).

11.3.3 Non-Genetic Options to Reduce Peanut Allergens

Several methods are assessed and contrasted to lessen the allergenicity of peanuts: biological (enzyme catalysis, microbial fermentation, and genetic engineering), chemical (magnetic-bead adsorption, acid treatment, and glycosylation), and physical (radiation, heat, and high-pressure treatments). Acid treatment, irradiation, the other physical and chemical techniques may be useful situations, but heat treatment and high-pressure treatments seem to be the most feasible and successful ways to lessen peanut allergenicity overall (Pi et al., 2019). For instance, roasting peanuts increases IgE binding activity by 90% and, due to Maillard reactions, renders the allergens *Ara h 1* and *Ara h 2* immune to proteolysis by enzymes used for digestion (Maleki et al., 2000).

Additional research has also attempted to reduce peanut allergy using a particular RNA interference (RNAi) gene silencing technique to suppress *Ara h 2* protein (Dodo et al., 2008). Transgenic peanut cultivars with reduced activity of the *Ara h 2* and *Ara h 6* genes were stable for numerous generations and minimal collateral changes were identified in the allergenic proteins, *Ara h 1* and *Ara h 3,* throughout three generations, namely, T_3, T_4, and T_5. The impact of varying soil sulfur levels on sulfur-rich proteins (*Ara h 2, Ara h 3*, and *Ara h 6*) and sulfur-poor proteins (*Ara h 1*) in non-transgenic and transgenic peanuts was examined and the results showed that soil sulfur content had no influence on *Ara h 1*, while low sulfur significantly reduced *Ara h 3* in both transgenic and non-transgenic seeds, as well as *Ara h* 2 and *Ara h 6* in non-transgenic seeds (Chandran et al., 2015).

11.4 BREEDING AND GENOMIC TOOLS FOR SENSORY AND TEXTURAL TRAITS IN PEANUTS

Limited efforts were made in the past to select peanuts with good sensory traits as there is lack of sufficient genetic variance for sensory characteristics and high-throughput phenomics to screen breeding population for sensory traits. Advanced breeding lines and released cultivars are normally subjected to sensory evaluation to identify lines with best sensory characteristics for commercialization. For example, sensory characterization of Malawian peanuts (Baka, CG7, Chitala, Chalimbana, Nsinjiro, and Kakoma) and volatiles associated with good roasted peanut flavor revealed that Nsinjiro and Baka have best sensory score of roasted peanut flavor and aroma. The predominant volatile compounds in these cultivars were pyrazines and furans. 2,5 dimethyl pyrazine and 2-ethyl-3,5-dimethyl pyrazine ($r = 0.916$), among pyrazines, were strongly correlated (r between 0.916 and 0.927) with roasted peanutty flavor, meaning thereby that pyrazine synthesis pathways may provide more details about the sources of roasted peanutty flavor (Gama and Adhikari, 2019). The key aroma compounds in peanut oil are alcohols, aldehydes, and heterocyclic compounds among different peanut cultivars, with odor activity value (OAV) of 25 key aroma compounds exceeded a value of 1. Pyrazines had a major influence on the peanut oil's flavor profile. Many amino acids, for example lysine, phenylalanine, arginine, glutamic acid, and isoleucine, were significantly correlated with flavors of both pyrazine and nuts (Sun et al., 2024).

Comparison of sensory characteristics and volatiles between roasted normal and high-oleate peanut showed that high-oleate peanuts contain highly stable pyrazines and stable peanutty flavor with low peroxide values and low hexanol content compared to normal peanuts, doubling the shelf-life of high-oleate peanuts (Braddock et al., 1995). The distribution of volatile substances varies significantly in large and small peanut seeds wherein the pyrazine levels were high in smaller peanuts compared to larger ones; 2,5-dimethylpyrazine was the most common aromatic compound, followed by 2,3-dihydrobenzofuran and 2-ethyl-3,5-dimethylpyrazine, the vital component influencing the roasted peanut flavor in the two types of seeds and had the highest OAV (Eker et al., 2023). The progenitor of large-seeded Virginia market type, Jenkins Jumbo, negatively affect the flavor of peanuts. When tested, nine large-seeded selections together with NC 7 and Florunner as controls, NC 7 had the least intense roasted peanut flavor, while UF714021 had the best flavor profile. Japan Jumbo derived X90053 and X90037, had intermediate and very low flavor scores for roasted peanut and sweet, respectively. Sensory characteristics were not significantly correlated with agronomic traits, except for the relationship between astringent attribute and percentage of extra-large kernels (Pattee et al., 2002).

Sugar content was positively correlated with sweetness, while bitterness and astringency were negatively correlated, suggesting assaying for total sugars could improve selection efficiency for sweetness and roasted peanut quality within the peanut market-types (Pattee et al., 2000). In addition to quantifying free amino acids, fatty acids, and tocopherols, the metabolites of raw Runner and Virginia peanuts were analyzed using various techniques. The results showed that among a total of 365 metabolites detected, 52 were significantly different between the two market types. The eicosenoate and ethylmalonate content of runner-type peanuts distinguished them from Virginia-type peanuts, which contained higher quantities of oxylipins, purine metabolites, amino acids, and gamma-glutamylalanine, produced from alpha-ketoglutarate (Klevorn et al., 2019). The comparison of a novel high-oleate peanut cultivar (K-Ol) with two traditional varieties, Daekwang and Jopyung (O/L ratio 2.69 and 2.16), revealed that K-Ol had a high O/L ratio, a lower total tocopherol content, increased oxidative stability, and a lower antioxidant capacity. Overall scores for flavor were high in K-Ol in the sensory characteristics analysis, except for strength (Lim et al., 2017).

Sucrose content is the major component that influences the sweet flavor of fresh peanuts and contributes to the roasted peanut flavor. In a study involving the F_2 population from a hybrid between Zhonghuatian 1 and Nanyangbaipi, a genomic region associated with oil and sucrose content was detected on chromosome A06, which represents 47.5% of the phenotypic variance. A candidate gene that controls the gene expression related to oil content and sucrose was identified as *Arahy.42CAD1*. Additionally, this study suggested that disturbances in TAG transit or storage processes could be the source of sucrose accumulation, offering new information about sucrose accumulation in peanuts (Huai et al., 2024). A comparative transcriptome analysis comprising a low-sucrose variety (Zhonghua 10) and a high-sucrose variety (ICG 12625) on samples taken at seven seed development stages identified 8334 genes involved in sucrose metabolic process in peanut seed. A strong association was found between the sucrose level and many differentially expressed genes (DEGs), genes encoding enzymes, 90 TF, and three SWEET genes (Li et al., 2021). More recently, two major QTLs (*qSCA06.2* and *qSCB06.2*), explained 31.41% and 24.13% of the phenotypic variance, respectively, associated with sucrose contents, were mapped in homologous genomic regions between the A and B subgenomes. *qSCA06.2* was exclusively present in Valencia-type peanut, while *qSCB06.2* in other peanut types, with both alleles in elite genetic background resulting in significantly higher sucrose content. The diagnostic markers were validated, which may facilitate selection of increased sucrose content via marker-assisted breeding in peanut (Wang et al., 2024b).

The assessment of the sensory and chemical stability of raw in-shell high-oleate and normal-oleate peanuts during prolonged storage (about 22 months) showed that high-oleate samples have increased O/L ratio than normal-oleic samples. All the stored samples showed increase in conjugated dienes, free fatty acids, p-anisidine value, peroxide value, and oxidized flavors, particularly in normal-oleic peanuts, whereas the roasted peanutty flavor was lowered in high-oleic samples. Unshelled samples of normal- or high-oleic peanuts do not exhibit apparent sensory or chemical alterations, which could imply that the peanut shell prevents the kernels from the damage (Martin et al., 2018). The impact of the process and raw material technologies on the organoleptic properties of roasted peanuts was examined. One hundred and thirty-four distinct samples from the twelve different raw peanuts have undergone eleven different treatments. Selection process had a major impact on the sensory attributes; dark roast aroma and roasted peanuts were reduced by baking and increased by maceration (Lykomitros, 2018). Two genomic regions (qSUCA08a and qSUCB06a) that regulate sugar content were discovered on chromosomes A08 and B06, with a major QTL, qSUCA08.2, showing PVE of 5.43–17.84%. A natural population including 172 peanut varieties was used to develop and validate an InDel marker. Of these, 93 germplasm with the ICG 12625 allele had a mean sucrose content that was considerably higher than 79 varieties with the Zhonghua 10 allele (Li et al., 2023).

A linkage group B09 with a major stable QTL (qSCB09/qSSCB09) linked to sucrose and soluble sugar content, was discovered to examine the genetic basis of sugar accumulation in peanut kernels. This region accounts for 21.51–33.58% of the phenotypic variance in the peanut kernels, where six potential candidate genes were identified (Wang et al., 2024a). Significant variations in the lipid and protein content of 10 local peanut types were identified, and these changes were positively correlated with the Taihua 6 variety. Taihua 9725 and 9922 exhibited low volatile components, while Taihua 4 and 6 had the largest amounts of bitter and sweet amino acids (Ding et al., 2023). A GWAS analysis involving 102 US peanut mini-core accessions identified 12,526 SNP markers linked to three characteristics (sucrose, total sugars, and total tocopherols). Seven and twenty-two significant quantitative trait loci were discovered for sucrose and total sugars, respectively. QTLs for total tocopherols were not significant (Zhang et al., 2023).

11.5 PROSPECTS OF GENOME-EDITING TECHNOLOGY IN DEVELOPING HIGH-OLEATE AND/OR ALLERGEN-FREE PEANUTS

Sustainable quality enhancement in peanuts is achieved through genetic transformation (trans-, intra-, and cisgenesis) and gene editing (CRISPR/CAS9 system, homologous recombination, transcription activator-like effector nucleases [TALENs], and zinc finger nucleases [ZFNs]) (Rajyaguru and Maheshala, 2024). CRISPR/Cas9- or RNAi-mediated gene editing can be utilized in genetic engineering and/or molecular breeding to change the expression levels of allergens (Jouanin et al., 2018; Saurabh et al., 2014). Additionally, a great potential for successful genetic changes arises from the application of technology advancements such as the identification and characterization of allergenic proteins (Brackett et al., 2022).

Genetic transformation is essential for studying the plant gene functions, even though its low transformation energy limits its applications in some systems (Taylor et al., 2006). Legume gene function was studied using *Agrobacterium tumefaciens*–mediated transformation (Somers et al., 2003). Moreover, this gene transfer approach has the advantages of cost-effectiveness and enhanced availability (Rafiq, 2014). To change genes related to nutritional quality, yield, and stress resilience, genome-editing technologies—which are relatively new are frequently applied in plant research (Kaniganti et al., 2022). Improved agricultural yields can be achieved more quickly using gene-editing techniques that include site-directed endonucleases to break chromosomal double strands (Kim et al., 1996; Christian et al., 2010; Jinek et al., 2012). By utilizing the normal cellular DNA repair process, double-stranded breaks (DSBs) can provide disease resistance and specific genomic modifications to increase agronomic traits like quality, yield, and nutrition contents (Yin et al., 2017; Borrelli et al., 2018).

Genome-editing technique has been proved to be the most promising method to create a better genotype. Techniques for modifying the genome of peanuts were used to target allergens and high oleate content (Yuan et al., 2019; Biswas et al., 2022). To date, natural mutations occurring in the *AhFAD2A* and *AhFAD2B* genes that encode fatty acid desaturase resulted in the high-oleate peanut varieties. Genome-editing technique utilizing CRISPR/Cas9 technology was applied to generate novel mutations in the *AhFAD2* genes using models, such as hairy root cultures and peanut protoplasts (Yuan et al., 2019). This study has resulted in G448A mutation in *AhFAD2A*, 441_442insA mutation and a unique G451T mutation in the peanut's *AhFAD2B* gene's coding region. In these genes, the target region served a hotspot for spontaneous mutations. Appropriate single guide RNAs (sgRNAs) were created and inserted into a plasmid for CRISPR/Cas9 expression. CRISPR/Cas9 activity led to the discovery of three mutations: G451T in *AhFAD2B*, 441 442insA, and G448A in *AhFAD2A*. The G451T mutation is novel, although the G448A and 441 442insA variants are present in high-oleate peanuts. After the oleic acid content of transformed lines has been validated, the mutations created in *AhFAD2B* through gene editing may be helpful in creating high-oleate lines with a range of genetic origins as the occurrence of natural spontaneous mutations in the *AhFAD2A* gene are more prevalent than in the *AhFAD2B* gene of *A. hypogaea* var. *hypogaea*. An additional benefit of high-oleate peanut is the G448A mutation in *AhFAD2A*.

TALENs were used to create targeted mutations in the conserved coding regions of *AhFAD2* genes (Wen et al., 2018). The frequencies of mutations in the *AhFAD2* mutant lines were closely associated with the accumulation of oleic acid, and these mutant lines showed a 0.5-to-2-fold rise in oleic acid content. A popular peanut cultivar, Huayu 23, with normal oleic acid content having defective *FAD2A* and functional *FAD2B* genes, was altered utilizing the node injection approach using a CRISPR/Cas9 construct that targets *FAD2B* gene (Han et al., 2023). The results demonstrated that T_0 seeds containing oleic acid content of 80% were obtained. Rajyaguru and Maheshala (2024) introduced CRISPR_GG20_*AhFAD2B* construct into the de-embryonated cotyledon of the GG20 plant, where the *AhFAD2B* gene lacks natural mutations. This CRISPR/CAS gene-editing technique has been used to alter the *AhFAD2B* gene and the modified GG20 plants were examined to determine the O/L ratio, which showed an increase from 3.1 to 7.3 in the edited plants. Similarly, using an endogenous tRNA-processing system, multiplex CRISPR/Cas9 gene editing was carried out in the protoplast culture to modify the main allergy gene *Ara h2* (Chu et al., 2009). During this process, two sgRNAs were created targeting the allergenic gene *Ara h 2*, protoplast separation and transformation using the green fluorescent protein (GFP) plasmid was effectively optimized, and their effectiveness was evaluated using Cas9 in vitro digestion (Biswas et al., 2022). The high-oleate concentration and allergens in peanuts were the targets of genome-editing techniques (Parmar et al., 2022).

Ultimately, deep-sequencing analysis revealed several changes in the desired gene following polyethylene glycol (PEG)-mediated transformation in protoplasts bearing a Cas9 and sgRNA-containing vector. This approach offers a productive pathway for creating gene-editing structures for different genes and peanut modification. *Agrobacterium*-mediated transformation or other delivery methods can be used to create stable transformants once they have been optimized. To help basic research for understanding molecular pathways and gene functions, an effective gene-editing framework in peanuts must be developed. In fact, additional editing methods, like allele substitution, can be investigated to further optimize the CRISPR/Cas9 system in peanuts to expand the target features and expedite the breeding process (Biswas et al., 2022).

11.6 CONCLUSIONS

Peanut products with high oleate contents have enhanced nutritional value and fetch a greater price in domestic and international markets. DNA markers associated with the high-oleate trait were successfully deployed to develop high-oleate peanut germplasm in all botanical or market types. The quality factors like visual, sensory, and textural traits are desirable in the oil processing and food industries. Peanut breeding based on quality and sensory traits is carried out more effectively and efficiently using genomic-assisted breeding approaches. High oleate levels and/or reduced allergenic protein in peanuts are the targets of genome-editing techniques. The gene *AhFAD2*, which codes for fatty acid and transforms oleic acid into linoleic acid, was modified following CRISPR/Cas9 system. Other options are also being used to reduce peanut allergens. An endogenous tRNA-processing mechanism, multiplex CRISPR/Cas9 genome editing, was carried out in the peanut protoplasts to disable a key allergy gene *Ara h 2*. As the genome-editing technology advance in peanuts, it may be feasible to produce peanuts with high quality attributes introducing foreign genes. Speed-breeding technology and marker-based rapid generation advancement can be utilized to increase the oleate content and reduce the allergenic proteins. High-throughput genotyping methods can be used to detect mutations governing low allergen contents and developing gene specific markers associated with high-oleate trait in peanut.

REFERENCES

Amba, V., Murphy, G., Etemadi, A., Wang, S., Abnet, C.C., and Hashemian, M. (2019). Nut and peanut butter consumption and mortality in the National Institutes of Health-AARP Diet and Health Study. *Nutrients* 11, 1508. https://doi.org/10.3390/nu11071508

Arya, S.S., Salve, A.R., and Chauhan, S. (2016). Peanuts as functional food: A review. *J. Food Sci. Technol.* 53, 31–41. https://doi.org/10.1007/s13197-015-2007-9

Bakal, H. (2020). Comparison of fatty acids composition and oil quality factors of different market type peanut (*Arachis hypogaea* L.) varieties. *Authorea* 51, 877–884. https://doi.org/10.22541/au.160684332.20470764/v1

Baker, G.L., Cornell, J.A., Gorbet, D.W., O'Keefe, S.F., Sims, C.A., and Talcott, S.T. (2003). Determination of pyrazine and flavor variations in peanut genotypes during roasting. *J. Food Sci.* 68, 394–400. https://doi.org/10.1111/j.1365–2621.2003.tb14171.x

Baring, M.R., López, Y., Simpson, C.E., Cason, J.M., Ayers, J., and Burow, M.D. (2006). Registration of 'Tamnut OL06' peanut. *Crop Sci.* 46, 2720a–2721a. http://doi.org/10.2135/cropsci2006.06.0412

Becker, W.M., and Jappe, U. (2014). Peanut allergens. *Chem. Immunol. Allergy* 100, 256–267. https://doi.org/10.1159/000359916

Bera, S.K., Kamdar, J.H., Kasundra, S.V., Maurya, A.K., Jasani, M.D., and Chandrashekar, A.B. (2018). Improving oil quality by altering levels of fatty acids through marker-assisted selection of AhFAD2 alleles in peanut (*Arachis hypogaea* L.). *Euphytica* 214, 162. https://doi.org/10.1007/s10681-018-2241-0

Bera, S.K., Kamdar, J.H., Kasundra, S.V., Patel, S.V., Jasani, M.D., and Maurya, A.K. (2019). Steady expression of high oleic acid in peanut bred by marker assisted backcrossing for fatty acid desaturase mutant alleles and its effect on seed germination along with other seedling traits. *PLoS ONE*. 14, e0226252. https://doi.org/10.1371/journal.pone.0226252

Bertioli, D.J., Jenkins, J., Clevenger, J., Dudchenko, O., Gao, D., Seijo, G., et al. (2019). The genome sequence of segmental allotetraploid peanut *Arachis hypogaea*. *Nat. Genet.* 51, 877–884. https://doi.org/10.1038/s41588-019-0405-z

Biswas, S., Wahl, N.J., Thomson, M.J., Cason, J.M., McCutchen, B.F., and Septiningsih, E.M. (2022). Optimization of protoplast isolation and transformation for a pilot study of genome editing in peanut by targeting the allergen gene Ara h 2. *Int. J. Mol. Sci.* 23, 837. https://doi.org/10.3390/ijms23020837

Bock, S.A., Munoz-Furlong, A., and Sampson, H.A. (2007). Further fatalities caused by anaphylactic reactions to food, 2001–2006. *J. Allergy. Clin. Immunol.* 119, 1016–1018. https://doi.org/10.1016/j.jaci.2006.12.622

Borrelli, V.M., Brambilla, V., Rogowsky, P., Marocco, A., and Lanubile, A. (2018). The enhancement of plant disease resistance using CRISPR/Cas9 technology. *Front. Plant Sci.* 9, 1245. https://doi.org/10.3389/fpls.2018.01245

Bouis, H.E., and Saltzman, A. (2017). Improving nutrition through biofortification: A review of evidence from HarvestPlus, 2003 through 2016. *Glob. Food Sec.* 12, 49–58. https://doi.org/10.1016/j.gfs.2017.01.009

Brackett, N.F., Pomés, A., and Chapman, M.D. (2022). New frontiers: Precise editing of allergen genes using CRISPR. *Front. Allergy*. 2, 109. https://doi.org/10.3389/falgy.2021.821107

Braddock, J.C., Sims, C.A., and O'Keefe, S.F. (1995). Flavor and oxidative stability of roasted high oleic acid peanuts. *J. Food Sci.* 60, 489–493. https://doi.org/10.1111/j.1365-2621.1995.tb09809.x

Branch, W., and Brenneman, T.B. (2015). Registration of 'Georgia-14N' peanut. *J. Plant Reg.* 9, 159–161. https://doi.org/10.3198/jpr2014.11.0082crc

Branch, W.D. (2005). Registration of 'Georgia-04S' peanut. *Crop Sci.* 45, 1653a–1654a. http://doi.org/10.2135/cropsci2004-059

Burks, A.W., Tang, M., Sicherer, S., Muraro, A., Eigenmann, P.A., Ebisawa, M., et al. (2012). ICON: Food allergy. *J. Allergy Clin. Immunol.* 129, 906–920. https://doi.org/10.1016/j.jaci. 2012.02.001

Burow, M.D., Baring, M.R., Chagoya, J., Trostle, C., Puppala, N., Simpson, C.E., et al. (2019). Registration of 'Tamval Ol14'peanut. *J. Plant Reg.* 13, 134–138. https://doi.org/10.3198/jpr2017.10.0073crc

Burow, M.D., Baring, M.R., Puppala, N., Simpson, C.E., Ayers, J.L., and Cason, J. (2014). Registration of 'Schubert' peanut. *J. Plant Reg.* 8, 122–126. http://doi.org/10.3198/jpr2013.07.0042crc

Cabanillas, B., Pedrosa, M.M., Rodriguez, J., Muzquiz, M., Maleki, S.J., Cuadrado, C., et al. (2012). Influence of enzymatic hydrolysis on the allergenicity of roasted peanut protein extract. *Int. Arch. Allergy Immunol.* 157, 41–50. https://doi.org/10.1159/000324681

Chamberlin, K.D., Bennett, R.S., Damicone, J.P., Godsey, C.B., Melouk, H.A., and Keim, K. (2015). Registration of 'OLé' peanut. *J. Plant Reg.* 9, 154–158. http://doi.org/10.3198/jpr2014.10.0072crc

Chandran, M., Chu, Y., Maleki, S.J., and Ozias-Akins, P. (2015). Stability of transgene expression in reduced allergen peanut (*Arachis hypogaea* L.) across multiple generations and at different soil sulfur levels. *J. Agric. Food Chem.* 63, 1788–1797. https://doi.org/10.1021/jf504892

Chang, S.K., Alasalvar, C., and Bolling, B.W. (2016). Nuts and their co-products: The impact of processing (roasting) on phenolics, bioavailability, and health benefits—a comprehensive review. *J. Funct. Foods* 26, 88–122. https://doi.org/10.1016/j.jff.2016.06.029

Chen, X., Li, H., Pandey, M.K., Yang, Q., Wang, X., Garg, V., et al. (2016). Draft genome of the peanut A-genome progenitor (*Arachis duranensis*) provides insights into geocarpy, oil biosynthesis, and allergens. *Proc. Nat. Acad. Sci. U. S. A.* 113, 6785–6790. https://doi.org/10.1073/pnas.1600899113

Chen, Z., Wang, M.L., Barkley, N.A., and Pittman, R.N. (2010). A simple allele specific PCR assay for detecting *FAD2* alleles in both A and B genomes of the cultivated peanut for high-oleate trait selection. *Plant Mol. Biol. Rep.* 28, 542–548. https://doi.org/10.1007/s11105-010-0181-5

Chiou, R.Y.Y., Cheng, S.L., Tseng, C.Y., and Lin, T.C. (1993). Flavor fortification and characterization of raw peanut oils subjected to roasting with partially defatted peanut meal under various atmospheric conditions. *J. Agric. Food Chem.* 41, 1110–1113. https://doi.org/10.1021/jf00031a019

Choi, I.D., Resurreccion, A.V.A., and Phillips, R.D. (2002). Sensory profiles of an extruded snack product based on peanut flour and rice flour. In: *Development, Physical, and Sensory Characterization of Extruded, Indirectly Puffed Peanut-Based Snack Products*, p. 93. http://getd.libs.uga.edu/pdfs/choi_in-duck_200205_phd.pdf

Christian, M., Cermak, T., Doyle, E.L., Schmidt, C., Zhang, F., and Hummel, A. (2010). Targeting DNA double-strand breaks with TAL effector nucleases. *Genetics* 186, 757–761. https://doi.org/10.1534/genetics.110.120717

Chu, Y., Holbrook, C.C., and Ozias-Akins, P. (2009). Two alleles of *AhFAD2B* control the high oleic acid trait in cultivated peanut. *Crop Sci.* 49, 2029–2036. https://doi.org/10.2135/cropsci2009.01.0021

Chu, Y., Ramos, L., Holbrook, C.C., and Ozias-Akins, P. (2007). Frequency of a loss-of-function mutation in oleoyl-PC desaturase in the mini-core of the US peanut germplasm collection. *Crop Sci.* 47, 2372–2378. https://doi.org/10.2135/cropsci2007.02.0117

Chung, S., and Champagne, E.T. (2008). Using phenolic compounds to reduce the allergenic properties of peanut extracts and peanut butter slurries. *J. Allergy Clin. Immunol.* 121, 249. https://doi.org/10.1016/j.jaci.2007.12.985

Chung, S.Y., and Reed, S. (2012). Removing peanut allergens by tannic acid. *Food Chem.* 134, 1468–1473. https://doi.org/10.1016/j. foodchem.2012.03.057

Derbyshire, E.J. (2014). A review of the nutritional composition, organoleptic characteristics and biological effects of the high oleic peanut. *Int. J. Food Sci. Nutr.* 65, 781–790. https://doi.org/10.3109/09637486.2014.937799

Deshmukh, D., Marathi, B., Sudini, H.K., Variath, M.T., Chaudhari, S., Manohar, S.S., et al. (2020). Combining high oleic acid trait and resistance to late leaf spot and rust diseases in groundnut (*Arachis hypogaea* L.). *Front. Genet.* 11, 514. https://doi.org/10.3389/fgene.2020.00514

Ding, B., Wang, F., Zhang, B., Feng, M., Chang, L., Shao, Y., et al. (2023). Flavor characteristics of ten peanut varieties from China. *Foods.* 12, 4380. https://doi.org/10.3390/foods12244380

Dodo, H., Marsic, D., Callender, M., Cebert, E., and Viquez, O. (2002). Screening 34 peanut introductions for allergen content using ELISA. *Food Agric. Immunol.* 14, 147–154. https://doi.org/10.1080/09540100220145179

Dodo, H.W., Konan, K.N., Chen, F.C., Egnin, M., and Viquez, O.M. (2008). Alleviating peanut allergy using genetic engineering: The silencing of the immunodominant allergen Ara h 2 leads to its significant reduction and a decrease in peanut allergenicity. *Plant Biotechnol. J.* 6, 135–145. https://doi.org/10.1111/j.1467-7652.2007.00292.x

Drake, M.A., and Civille, G.V. (2003). Flavor lexicons. *Compr. Rev. Food Sci. Food Saf.* 2, 33–40. https://doi.org/10.1111/j.1541-4337.2003.tb00013.x

Du Toit, G., Roberts, G., Sayre, P.H., Bahnson, H.T., Radulovic, S., Santos, A.F., et al. (2015). Randomized trial of peanut consumption in infants at risk for peanut allergy. *N. Engl. J. Med.* 372, 803–813. https://doi.org/10.1056/NEJMoa1414850

Dun, Q., Yao, L., Deng, Z., Li, H., Li, J., Fan, Y., et al. (2019). Effects of hot and cold-pressed processes on volatile compounds of peanut oil and corresponding analysis of characteristic flavor components. *LWT-Food Sci. Tech.* 112, 107648. https://doi.org/10.1016/j.lwt.2018.11.084

Dwivedi, S.L., and Nigam, S.N. (2005). Confectionery groundnuts: Issues and opportunities to promote export and food uses in India. *J. Oilseeds Res.* 22, 1–4. http://oar.icrisat.org/id/eprint/792

Eker, T., Darici, M., and Cabaroglu, T. (2023). The impact of seed size on volatile composition in roasted peanuts extracted by the Purge and Trap method. *J. Food Res.* 17, 211–223. https://doi.org/10.1007/s11694-022-01583-z

Filep, S., Block, D.S., Smith, B.R., King, E.M., Commins, S., Kulis, M., et al. (2018). Specific allergen profiles of peanut foods and diagnostic or therapeutic allergenic products. *J. Allergy Clin. Immunol.* 141, 626–631. https://doi.org/10.1016/j.jaci.2017.05.049

Fletcher, M.M. (1987). Evaluation of peanut flavor quality. In: *Peanut Quality: Its Assurance and Maintenance from the Farm to End Product*. Bharti Publications, New Delhi, p. 94.

Gama, A.P., and Adhikari, K. (2019). Sensory characterization of dominant Malawi peanut varieties after roasting. *J. Food Sci.* 84, 1554–1562. https://doi.org/10.1111/1750-3841.14641

Gangurde, S.S., Khan, A.W., Janila, P., Variath, M.T., Manohar, S.S., Singam, P., et al. (2023). Whole genome sequencing based discovery of candidate genes and diagnostic markers for seed weight in groundnut. *Plant Genome* 16, e20265. https://doi.org/10.1002/tpg2.20265

Gangurde, S.S., Kumar, R., Pandey, A.K., Burow, M., Laza, H.E., Nayak, S.N., et al. (2019). Climate-smart groundnuts for achieving high productivity and improved quality: Current status, challenges, and opportunities. In: *Genomic Designing of Climate-smart Oilseed Crops*. Springer, Cham, pp. 133–172. https://doi.org/10.1007/978-3-319-93536-2_3

Gorbet, D.W., and Knauft, D.A. (1997). Registration of 'SunOleic 95R' peanut. *Crop Sci.* 37, 1392. https://doi.org/10.2135/cropsci1997.0011183X003700040081x

Gorbet, D.W., and Knauft, D.A. (2000). Registration of 'SunOleic 97R' peanut. *Crop Sci.* 40, 1190.

Gupta, R.S., Jiang, J., Bozen, A., Johnson, J., Arbes, S.J., Cooper, S., et al. (2019). Implementation, practices, and barriers to the 2017 Peanut Allergy Prevention Guidelines among pediatricians. *J. Allergy Clin. Immunol.* 143, 84. https://doi.org/10.1016/j.jaci.2018.12.260

Han, H.W., Yu, S.T., Wang, Z.W., Yang, Z., Jiang, C.J., Wang, X.Z., et al. (2023). In planta genetic transformation to produce CRISPRed high-oleic peanut. *Plant Growth Regul.* 101, 443–451. https://doi.org/10.1007/s10725-023-01031-y

Holbrook, C.C. (2019). *Peanut Breeding and Genetics*. Peanut Academic Press, pp. 51–77.

Holbrook, C.C., Ozias-Akins, P., Chu, Y., Culbreath, A.K., Kvien, C.K., and Brenneman, T.B. (2017). Registration of 'TifNV-High O/L' peanut. *J. Plant Reg.* 11, 228–230. https://doi.org/10.3198/jpr2016.10.0059crc

Hu, H., Shi, A., Liu, H., Liu, L., Fauconnier, M.L., and Wang, Q. (2021). Study on key aroma compounds and its precursors of peanut oil prepared with normal-and high-oleic peanuts. *Foods*. 10, 3036. https://doi.org/10.3390/foods10123036

Huai, D., Zhi, C., Wu, J., Xue, X., Hu, M., Zhang, J., et al. (2024). Unveiling the molecular regulatory mechanisms underlying sucrose accumulation and oil reduction in peanut kernels through genetic mapping and transcriptome analysis. *Plant Physiol. Biochem.* 208, 108448. https://doi.org/10.1016/j.plaphy.2024.108448

Huang, B., Qi, F., Sun, Z., Miao, L., Zhang, Z., Liu, H., et al. (2019). Marker-assisted backcrossing to improve seed oleic acid content in four elite and popular peanut (*Arachis hypogaea* L.) cultivars with high oil content. *Breed. Sci.* 69, 234–243. https://doi.org/10.1270/jsbbs.18107

Jadhav, M.P., Patil, M.D., Hampannavar, M., Venkatesh Dattatreya, P., Shirasawa, K., Pasupuleti, J., et al. (2021). Enhancing oleic acid content in two commercially released peanut varieties through marker-assisted backcross breeding. *Crop Sci.* 61, 2435–2443. https://doi.org/10.1002/csc2.20512

Janila, P., Pandey, M.K., Manohar, S.S., Variath, M.T., Nallathambi, P., Nadaf, H.L., et al. (2016a). Foliar fungal disease-resistant introgression lines of groundnut (*Arachis hypogaea* L.) record higher pod and haulm yield in multilocation testing. *Plant Breed.* 135, 355–366. https://doi.org/10.1111/pbr.12358

Janila, P., Pandey, M.K., Shasidhar, Y., Variath, M.T., Sriswathi, M., Khera, P., et al. (2016b). Molecular breeding for introgression of fatty acid desaturase mutant alleles (AhFAD2A and AhFAD2B) enhances oil quality in high and low oil containing peanut genotypes. *Plant Sci.* 242, 203–213. https://doi.org/10.1016/j.plantsci.2015.08.013

Jiang, J., Bushara, O., Ponczek, J., Warren, C., Blumenstock, J., Smith, B., et al. (2018). Updated pediatric peanut allergy prevalence in the United States. *Ann. Allergy Asthma Immunol.* 121, 14. https://doi.org/10.1016/j.anai.2018.09.042

Jinek, M., Chylinski, K., Fonfara, I., Hauer, M., Doudna, J.A., and Charpentier, E. (2012). A programmable dual-RNA–guided DNA endonuclease in adaptive bacterial immunity. *Science* 337, 816–821. https://doi.org/10.1126/science.1225829

Jolivet, P., Acevedo, F., Boulard, C., d'Andréa, S., Faure, J.D., Kohli, A., et al. (2013). Crop seed oil bodies: From challenges in protein identification to an emerging picture of the oil body proteome. *Proteomics* 13, 1836–1849. https://doi.org/10.1002/pmic.201200431

Jouanin, A., Boyd, L., Visser, R.G., and Smulders, M.J. (2018). Development of wheat with hypo immunogenic gluten obstructed by the gene editing policy in Europe. *Front. Plant Sci.* 9, 1523. https://doi.org/10.3389/fpls.2018.01523

Jung, S., Swift, D., Sengoku, E., Patel, M., Teule, F., Powell, G., et al. (2000). The high oleate trait in the cultivated peanut [*Arachis hypogaea* L.]. I. Isolation and characterization of two genes encoding microsomal oleoyl-PC desaturases. *Mol. Genet. Genomic.* 263, 796–805. https://doi.org/10.1007/s004380000244

Kamdar, J.H., Jasani, M.D., Chandrashekar, A.B., Janila, P., Pandey, M.K., Georrge, J.J., et al. (2022). Does improved oleic acid content due to marker-assisted introgression of AhFAD2 mutant alleles in peanuts alter its mineral and vitamin composition? *Front. Plant Sci.* 13, 942617. https://doi.org/10.3389/fpls.2022.942617

Kamdar, J.H., Jasani, M.D., and Georrge, J.J. (2020). Development of peanut breeding lines with high oleic acid content. In: *Proceedings of the National Conference on Innovations in Biological Sciences (NCIBS)*. https://doi.org/10.2139/ssrn.3573337

Kamdar, J.H., Jasani, M.D., Kasundra, S.V., Bera, S.K., and Georrge, J.J. (2017). Recent advances on development of groundnut with high oleate traits. In: *Proceedings of International Science Symposium on Recent Trends in Science and Technology*, Vol. 493. Bharti Publications, New Delhi, pp. 12–34.

Kaniganti, S., Chaitanya, A.K., Shah, P., Magar, N.D., Jamedar, H.V.R., and Sajjad, M. (2022). Status and prospects of genetic improvement of grain legumes using in vitro techniques: A review. *SSRN*, 4100545. https://doi.org/10.2139/ssrn.4100545

Kim, Y.G., Cha, J., and Chandrasegaran, S. (1996). Hybrid restriction enzymes: Zinc finger fusions to Fok I cleavage domain. *Proc. Natl. Acad. Sci. U. S. A.* 93, 1156–1160. https://doi.org/10.1073/pnas.93.3.1156

Klevorn, C.M., Dean, L.L., and Johanningsmeier, S.D. (2019). Metabolite profiles of raw peanut seeds reveal differences between market-types. *J. Food Sci.* 84, 397–405. https://doi.org/10.1111/1750-3841.14450

Koid, A., Chapman, M., and Hamilton, R. (2014). Ara h 6 complements Ara h 2 as an important marker for IgE reactivity to peanut. *J. Agric. Food Chem.* 62, 206–213. https://doi.org/10.1021/jf4022509

Kris-Etherton, P.M., Zhao, G., Binkoski, A.E., Coval, S.M., and Etherton, T.D. (2001). The effect of nuts on coronary heart disease risk. *Nutr. Rev.* 59, 103–111. https://doi.org/10.1111/j.1753-4887.2001.tb06996.x

Kulis, M., Chen, X., and Lew, J. (2012). The 2S albumin allergens of *Arachis hypogaea*, Ara h 2 and Ara h 6, are the major elicitors of anaphylaxis and can effectively desensitize peanut-allergic mice. *Clin. Exp. Allergy* 42, 326–336. https://doi.org/10.1111/j.1365-2222.2011.03934.x

Lee, A.J., Thalayasingam, M., and Lee, B.W. (2013). Food allergy in Asia: How does it compare? *Asia Pac. Allergy* 3, 3–14. https://doi.org/10.5415/apallergy.2013.3.1.3

Lee, C.M., and Resurreccion, A.V.A. (2006). Consumer acceptance of roasted peanuts affected by storage temperature and humidity conditions. *LWT-Food Sci. Tech.* 39, 872–882. https://doi.org/10.1016/j.lwt.2005.06.011

Lelievre, C.P., and Eischeid, A.C. (2018). Development and evaluation of a real-time PCR multiplex assay for the detection of allergenic peanut using chloroplast DNA markers. *J. Agric. Food Chem.* 66, 8623–8629. https://pubs.acs.org/doi/abs/10.1021/acs.jafc.8b02053

Li, W., Huang, L., Liu, N., Chen, Y., Guo, J., Yu, B., et al. (2023). Identification of a stable major sucrose-related QTL and diagnostic marker for flavor improvement in peanut. *Theor. Appl. Genet.* 136, 78. https://doi.org/10.1007/s00122-023-04306-0

Li, W., Huang, L., Liu, N., Pandey, M.K., Chen, Y., Cheng, L., et al. (2021). Key regulators of sucrose metabolism identified through comprehensive comparative transcriptome analysis in peanuts. *Int. J. Mol. Sci.* 22, 7266. https://doi.org/10.3390/ijms22147266

Lieberman, J.A., Bingemann, T.A., and Wang, J. (2020). Diagnostic challenges in anaphylaxis. *J. Allergy Clin. Immunol.* 8, 1177–1184.

Lim, H.J., Kim, D.S., Pan, J.H., Pae, S.B., Kim, H.S., Shin, E.C., et al. (2017). Characterization of physicochemical and sensory attributes of a novel high-oleic peanut oil cultivar (*Arachis hypogaea* ssp. *Fastigiata* L.). *Appl. Biol. Chem.* 60, 653–657. https://doi.org/10.1007/s13765-017-0324-6

Liu, F., Zhang, X., Lu, C., Zeng, X., Li, Y., Fu, D., et al. (2015). Non-specific lipid transfer proteins in plants: Presenting new advances and an integrated functional analysis. *J. Exp. Bot.* 66, 5663–5681. https://doi.org/10.1093/jxb/erv313

Liu, R., Jin, Q., Huang, J., Liu, Y., Wang, X., Mao, W., et al. (2011). Photodegradation of Aflatoxin B1 in peanut oil. *Eur. Food Res. Technol.* 232, 843–849. https://doi.org/10.1007/s00217-011-1452-6

Loh, W., and Tang, M. (2018). Adjuvant therapies in food immunotherapy. *Immunol. Allergy Clin. North Am.* 38, 89–101. https://doi.org/10.1016/j.iac.2017.09.008.

Lopez, Y., Smith, O.D., Senseman, S.A., and Rooney, W.L. (2001). Genetic factors influencing high oleic acid content in Spanish market-type peanut cultivars. *Crop Sci.* 41, 51–56. http://doi.org/10.2135/cropsci2001.41151x

Lykomitros, D. (2018). *Multivariate Relationships Between Instrumental Attributes, Microstructure, Sensory Profiles and Consumer Preference in Roasted Peanuts (Arachis spp.)* (Doctoral dissertation, Wageningen University and Research). https://research.wur.nl/en/publications/multivariate-relationships-between-instrumental-attributes-micros

Lykomitros, D., Fogliano, V., and Capuano, E. (2016). Flavor of roasted peanuts (*Arachis hypogaea*)—Part II: Correlation of volatile compounds to sensory characteristics. *Food Res. Int.* 89, 870–881. https://doi.org/10.1016/j.foodres.2016.08.01

Maleki, S.J., Kopper, R.A., Shin, D.S., Park, C.W., Compadre, C.M., Sampson, H., et al. (2000). Structure of the major peanut allergen Ara h 1 may protect IgE-binding epitopes from degradation. *J. Immunol.* 164, 5844–5849. https://doi.org/10.4049/jimmunol.164.11.5844

Martin, M.P., Grosso, A.L., Nepote, V., and Grosso, N.R. (2018). Sensory and chemical stabilities of high-oleic and normal-oleic peanuts in shell during long-term storage. *J. Food Sci.* 83, 2362–2368. https://doi.org/10.1111/1750-3841.14295

McGorrin, R.J. (2002). Character impact compounds: Flavors and off-flavors in foods. *Food Sci. Tech.* 375–413.

Miller, M.F., Shackelford, S.D., Hayden, K.D., and Reagan, J.O. (1990). Determination of the alteration in fatty acid profiles, sensory characteristics and carcass traits of swine fed elevated levels of monounsaturated fats in the diet. *J. Animal Sci.* 68, 1624–1631. https://doi.org/10.2527/1990.6861624x

Miyagi, A., and Ogaki, Y. (2014). Instrumental evaluation and influence of gender and/or age among consumers on textural preference of deep-fried peanuts. *Food Res. Int.* 64, 227–233. https://doi.org/10.1016/j.foodres.2014.06.023

Mondal, S., Badigannavar, A.M., and d'Souza, S.F. (2011). Induced variability for fatty acid profile and molecular characterization of high oleate mutant in cultivated groundnut (*Arachis hypogaea* L.). *Plant Breed.* 130, 242–247. https://doi.org/10.1111/j.1439-0523.2010.01787.x

Moore, K.M., and Knauft, D.A. (1989). The inheritance of high oleic acid in peanut. *J. Hered.* 80, 252–253.

Mottram, D.S., Low, M.Y., and Elmore, J.S. (2006). The Maillard reaction and its role in the formation of acrylamide and other potentially hazardous compounds in foods. In: *Acrylamide and Other Hazardous Compounds in Heat-treated Foods*, pp. 3–22. https://doi.org/10.1533/9781845692018.1.3

Mukri, G., Nadaf, H.L., Bhat, R.S., Gowda, M.V.C., Upadhyaya, H.D., and Sujay, V. (2012). Phenotypic and molecular dissection of ICRISAT mini core collection of peanut (*Arachis hypogaea* L.) for high oleic acid. *Plant Breed.* 131, 418–422. https://doi.org/10.1111/j.1439-0523.2012.01970.x

Mylne, J.S., Hara-Nishimura, I., and Rosengren, K.J. (2014). Seed storage albumins: Biosynthesis, trafficking and structures. *Funct. Plant Biol.* 41, 671–677. https://doi.org/10.1071/FP14035

Nawade, B., Mishra, G.P., Radhakrishnan, T., Dodia, S.M., Ahmad, S., Kumar, A., et al. (2018). High oleic peanut breeding: Achievements, perspectives, and prospects. *Trends Food Sci. Technol.* 78, 107–119. https://doi.org/10.1016/j.tifs.2018.05.022

Newell, J.A., Mason, M.E., and Matlock, R.S. (1967). Precursor of typical and atypical roasted peanut flavor. *J. Agric. Food Chem.* 15, 767–772. https://doi.org/10.1021/jf60153a010

Norden, A.J., Gorbet, D.W., Knauft, D.A., and Young, C.T. (1987). Variability in oil quality among peanut genotypes in the Florida breeding program. *Peanut Sci.* 14, 7–11. https://doi.org/10.3146/i0095-3679-14-1-3

Nursten, H.E. (2005). The Maillard reaction: Chemistry, biochemistry, and implications. *R. Soc. Chem.* 37–91. https://doi.org/10.1039/9781847552570

Nwaru, B.I., Hickstein, L., Panesar, S.S., Roberts, G., Muraro, A., and Sheikh, A. (2014). EAACI Food Allergy and Anaphylaxis Guidelines Group. Prevalence of common food allergies in Europe: A systematic review and meta-analysis. *Allergy* 69, 992–1007. https://doi.org/10.1111/all.12423

O'Byrne, D.J., Knauft, D.A., and Shireman, R.B. (1997). Low fat, mono-saturated rich diets containing high oleic peanuts improve the serum lipoprotein profiles. *Lipids* 32, 687–695.

Ostadrahimi, A., Ashrafnejad, F., and Kazemi, A. (2014). Aflatoxin in raw and salt-roasted nuts (pistachios, peanuts and walnuts) Sold in Markets of Tabriz, Iran, Jundishapur. *J. Microbiol.* 7, 1. https://doi.org/10.5812%2Fjjm.8674

Pandey, A.K., Sudini, H.K., Upadhyaya, H.D., Varshney, R.K., and Pandey, M.K. (2019b). Hypoallergen peanut lines identified through large-scale phenotyping of global diversity panel: Providing hope toward addressing one of the major global food safety concerns. *Front. Genet.* 10, 1177. https://doi.org/10.3389/fgene.2019.01177

Pandey, A.K., Varshney, R.K., Sudini, H.K., and Pandey, M.K. (2019a). An improved enzyme-linked immunosorbent assay (ELISA) based protocol using seeds for detection of five major peanut allergens Ara h 1, Ara h 2, Ara h 3, Ara h 6, and Ara h 8. *Front. Nutr.* 6, 68. https://doi.org/10.3389/fnut.2019.00068

Pandey, M.K., Gangurde, S.S., Shasidhar, Y., Sharma, V., Kale, S.M., Khan, A.W., et al. (2024). High-throughput diagnostic markers for foliar fungal disease resistance and high oleic acid content in groundnut. *BMC Plant Biol.* 24, 262. https://doi.org/10.1186/s12870-024-04987-9

Panel, N.S.E. (2010). Guidelines for the diagnosis and management of food allergy in the United States: Report of the NIAID-sponsored expert panel. *J. Allergy Clin. Immunol.* 126, 1–58. https://doi.org/10.1016/j.jaci.2010.10.007

Parmar, S., Deshmukh, D.B., Kumar, R., Manohar, S.S., Joshi, P., Sharma, V., et al. (2021). Single seed-based high-throughput genotyping and rapid generation advancement for accelerated groundnut genetics and breeding research. *Agronomy* 11, 1226. https://doi.org/10.3390/agronomy11061226

Parmar, S., Sharma, V., Bomireddy, D., Soni, P., Joshi, P., Gangurde, S.S., et al. (2022). Recent advances in genetics, genomics, and breeding for nutritional quality in groundnut. *Accel. Plant Breed.* 4, 111–137. https://doi.org/10.1007/978-3-030-81107-5_4

Pattee, H.E., Isleib, T.G., Giesbrecht, F.G., and McFeeters, R.F. (2000). Relationship of sweet, bitter and roasted peanuts sensory attributes with carbohydrate components in peanuts. *J. Agric. Food Chem.* 48, 757–763. https://doi.org/10.1021/jf9910741

Pattee, H.E., Isleib, T.G., Gorbet, D.W., and Giesbrecht, F.G. (2002). Selection of alternative genetic sources of large-seed size in Virginia-type peanut: Evaluation of sensory, composition, and agronomic characteristics. *J. Agric. Food Chem.* 50, 4885–4889. https://doi.org/10.1021/jf025601j

Pattee, H.E., Isleib, T.G., Gorbet, D.W., Giesbrecht, F.G., and Cui, Z. (2001). Parent selection in breeding for roasted peanut flavor quality. *Peanut Sci.* 28, 51–58. https://doi.org/10.3146/i0095-3679-28-2-3

Perkins, T., Schmitt, D.A., Isleib, T.G., Cheng, H., and Maleki, S.J. (2006). Breeding a hypoallergenic peanut. *J. Allergy Clin. Immunol.* 117, 328. https://doi.org/10.1016/j.jaci.2005.12.1293

Petersen, A., Kull, S., Rennert, S., Becker, W.M., Krause, S., Ernst, M., et al. (2015). Peanut defensins: Novel allergens isolated from lipophilic peanut extract. *J. Allergy Clin. Immunol.* 136, 1295–1301. https://doi.org/10.1016/j.jaci.2015.04.010

Pi, X., Fu, G., Dong, B., Yang, Y., Wan, Y., and Xie, M. (2021). Effects of fermentation with Bacillus natto on the allergenicity of peanut. *Lwt.* 141, 110862. https://doi.org/10.1016/j.lwt.2021.110862

Pi, X., Wan, Y., Yang, Y., Li, R., Wu, X., Xie, M., et al. (2019). Research progress in peanut allergens and their allergenicity reduction. *Trends Food Sci. Technol.* 93, 212–220. https://doi.org/10.1016/j.tifs.2019.09.014

Puppala, N., and Tallury, S.P. (2014). Registration of 'NuMex 01'high oleic Valencia peanut. *J. Plant Reg.* 8, 127–130. https://doi.org/10.3198/jpr2013.11.0070crc

Rafiq, S. (2014). *Analysis of Factors Influencing Successful Agrobacterium-Mediated Genetic Transformation in Two Different Explants of Peanut (Arachis hypogaea L.) Variety BINA Chinabadam–2.* BRAC University. http://hdl.handle.net/10361/3038

Rajyaguru, R., and Maheshala, N.G.K. (2024). Genetic improvement in peanut: Role of genetic engineering. In: *Genetic Engineering of Crop Plants for Food and Health Security Singapore*, Vol. 1. Springer Nature, pp. 271–288. https://doi.org/10.1007/978-981-99-5034-8_13

Ratnaparkhe, M.B., Lee, T.H., Tan, X., Wang, X., Li, J., Kim, C., et al. (2014). Comparative and evolutionary analysis of major peanut allergen gene families. *Genome Biol. Evol.* 6, 2468–2488. https://doi.org/10.1093/gbe/evu189

Rustgi, S., Alam, T., Jones, Z.T., Brar, A.K., and Kashyap, S. (2022). Reduced-immunogenicity wheat and peanut lines for people with foodborne disorders. *Chem. Proc.* 10, 67. https://doi.org/10.3390/IOCAG2022-12221

Salehi, F. (2020). Physico-chemical properties of fruit and vegetable juices as affected by pulsed electric field: A review. *Int. J. Food Prop.* 23, 1036–1050. https://doi.org/10.1080/10942912.2020.1775250

Sanchiz, A., Sánchez-Enciso, P., Cuadrado, C., and Linacero, R. (2021). Detection of peanut allergen by real-time PCR: Looking for a suitable detection marker as affected by processing. *Foods* 10, 1421.

Saurabh, S., Vidyarthi, A.S., and Prasad, D. (2014). RNA interference: Concept to reality in crop improvement. *Planta* 239, 543–564. https://doi.org/10.1007/s00425-013-2019-5

Shah, F., Shi, A., Ashley, J., Kronfel, C., Wang, Q., Maleki, S.J., et al. (2019). Peanut allergy: Characteristics and approaches for mitigation. *Compr. Rev. Food Sci. Food Saf.* 18, 1361–1387. https://doi.org/10.1111/1541-4337.12472

Shasidhar, Y., Variath, M.T., Vishwakarma, M.K., Manohar, S.S., Gangurde, S.S., Sriswathi, M., et al. (2020). Improvement of three popular Indian groundnut varieties for foliar disease resistance and high oleic acid using SSR markers and SNP array in marker-assisted backcrossing. *Crop J.* 8, 1– 15. https://doi.org/10.1016/j.cj.2019.07.001

Shieh, C.J., Akoh, C.C., and Koehler, P.E. (1996). Optimizing low fat peanut spread containing sucrose polyester. *J. Food Sci.* 61, 1227–1229. https://doi.org/10.1111/j.1365-2621.1996.tb10966.x

Sicherer, S.H., and Sampson, H.A. (2010). Food allergy. *J. Allergy Clin. Immunol.* 125, 116–125. https://doi.org/10.1016/j.jaci.2009.08.028

Singh, A.K., and Nigam, S.N. (2016). Arachis gene pools and genetic improvement in groundnut. In: Rajpal, V., Rao, S., and Raina, S. (eds.) *Gene Pool Diversity and Crop Improvement. Sustainable Development and Biodiversity*, Vol. 10. Springer, Cham. https://doi.org/10.1007/978-3-319-27096-8_2

Singkham, N., Jogloy, S., Kesmala, T., Swatsitang, P., Jaisil, P., Puppala, N., et al. (2011). Combining ability for oleic acid in peanut (*Arachis hypogaea* L.). *Oil Crop Sci.* 6, 175–179. https://doi.org/10.1016/j.ocsci.2021.10.005

Skolnick, H.S., Conover-Walker, M.K., and Koerner, C.B. (2001). The natural history of peanut allergy. *J. Allergy Clin. Immunol.* 107, 367–374. https://doi.org/10.1067/mai.2001.112129

Somers, D.A., Samac, D.A., and Olhoft, P.M. (2003). Recent advances in legume transformation. *Plant Physiol.* 131, 892–899. https://doi.org/10.1104/pp.102.017681

Sun, J., Zhang, C., Song, Y., Chu, B., Wang, M., Zhang, Z., et al. (2024). Characterization of key aroma compounds and main contributing amino acids in hot-pressed oil prepared from various peanut varieties. *Molecules* 29, 1947. https://doi.org/10.3390/molecules29091947

Tamanna, N., and Mahmood, N. (2015). Food processing and maillard reaction products: Effect on human health and nutrition. *Int. J. Food Sci.* 2015, 526762. https://doi.org/10.1155/2015/526762

Tang, Y., Qiu, X., Hu, C., Li, J., Wu, L., Wang, W., et al. (2022). Breeding of a new variety of peanut with high-oleic-acid content and high-yield by marker-assisted backcrossing. *Mol. Breed.* 42, 42. https://doi.org/10.1007/s11032-022-01313-9

Taylor, C.G., Fuchs, B., Collier, R., and Lutke, W.K. (2006). Generation of composite plants using *Agrobacterium rhizogenes*. *Agrobact. Prot.* 155–168. https://doi.org/10.1385/1-59745-130-4:155

Vargheese, R.L., Saravanan, S., Hepziba, S.J., Kumari, S.M.P., Pushpam, A.K., Kanagarajan, S., et al. (2024). Marker-assisted introgression to improve the oleic acid content in the TMV 7 groundnut (*Arachis hypogaea* L.) variety suitable for the oil industry. *BMC Plant Biol.* 24, 733. https://doi.org/10.1186/s12870-024-05387-9

Viquez, O.M., Konan, K.N., and Dodo, H.W. (2003). Structure and organization of the genomic clone of a major peanut allergen gene, Ara h 1. *Mol. Immunol.* 40, 565–571. https://doi.org/10.1016/j.molimm.2003.09.002

Vriens, K., Cammue, B.P., and Thevissen, K. (2014). Antifungal plant defensins: Mechanisms of action and production. *Molecules* 19, 12280–12303. https://doi.org/10.3390/molecules190812280

Wang, F., Miao, H., Zhang, S., Hu, X., Li, C., Chu, Y., et al. (2024a). Identification of a major QTL underlying sugar content in peanut kernels based on the RIL mapping population. *Front. Plant Sci.* 15, 1423586. https://doi.org/10.3389/fpls.2024.1423586

Wang, M.L., Tonnis, B., An, Y.Q.C., Pinnow, D., Tishchenko, V., and Pederson, G.A. (2015). Newly identified natural high-oleate mutant from *Arachis hypogaea* L. subsp. *hypogaea*. *Mol. Breed.* 35, 1–9. https://doi.org/10.1007/s11032-015-0377-3

Wang, Q., Wang, H., and Huang. (2018). *Peanut Processing Characteristics and Quality Evaluation*. Springer, pp. 97–98. https://doi.org/10.1007/978-981-10-6175-2

Wang, Z., Zhang, Y., Huai, D., Chen, Y., Wang, X., Kang, Y., et al. (2024b). Detection of two homologous major QTLs and development of diagnostic molecular markers for sucrose content in peanut. *Theor. Appl. Genet.* 137, 61. https://doi.org/10.1007/s00122-024-04549-5

Wen, S., Liu, H., Li, X., Chen, X., Hong, Y., Li, H., et al. (2018). TALEN-mediated targeted mutagenesis of fatty acid desaturase 2 (FAD2) in peanut (*Arachis hypogaea* L.) promotes the accumulation of oleic acid. *Plant Mol. Biol.* 97, 177–185. https://doi.org/10.1007/s11103-018-0731-z

Wichers, H.J., De Beijer, T., Savelkoul, H.F., and Van Amerongen, A. (2004). The major peanut allergen Ara h 1 and its cleaved-off N-terminal peptide; possible implications for peanut allergen detection. *J. Agric. Food Chem.* 52, 4903–4907. https://doi.org/10.1021/jf049697o

World Health Organization (2016). [www.who.org] WHO/IUIS: Allergen Nomenclature Subcommittee. http://www.allergen.org

Wu, Z., Lian, J., Han, Y., Zhou, N., Li, X., Yang, A., et al. (2016a). Crosslinking of peanut allergen Ara h 2 by polyphenol oxidase: Digestibility and potential allergenicity assessment. *J. Sci. Food Agric.* 96, 3567–3574. https://doi.org/10.1002/jsfa.7542

Wu, Z., Zhou, N., Xiong, F., Li, X., Yang, A., Tong, P., et al. (2016b). Allergen composition analysis and allergenicity assessment of Chinese peanut cultivars. *Food Chem.* 196, 459–465. https://doi.org/10.1016/j.foodchem.2015.09.070

Yin, K., Gao, C., and Qiu, J.L. (2017). Progress and prospects in plant genome editing. *Nat. Plants* 3, 1–6. https://doi.org/10.1038/nplants.2017.107

Young, C.T., Pattee, H.E., Schadel, W.E., and Sanders, T.H. (2004). Microstructure of peanut (*Arachis hypogaea* L. cv. "NC 7") cotyledons during development. *LWT—Food Sci. Technol.* 37, 439–445. https://doi.org/10.1016/j.lwt.2003.10.016

Young, C.T., Schadel, W.E., and Heertje, I. (1993). A comparison of the effects of oven roasting and oil cooking on the microstructure of peanut (*Arachis hypogaea* L. cv. *Florigiant*) cotyledon. *Food Struct.* 12, 59–66.

Young, N.D., Sanders, T.H., Drake, M.A., Osborne, J., and Civille, G.V. (2005). Descriptive analysis and US consumer acceptability of peanuts from different origins. *Food Qual. Prefer.* 16, 37–43. https://doi.org/10.1016/j.foodqual.2003.12.006

Yu, J., Ahmedna, M., Goktepe, I., Cheng, H., and Maleki, S. (2011). Enzymatic treatment of peanut kernels to reduce allergen levels. *Food Chem.* 127, 1014–1022. https://doi.org/10.1016/j.foodchem. 2011.01.074

Yuan, M., Zhu, J., Gong, L., He, L., Lee, C., and Han, S. (2019). Mutagenesis of FAD2 genes in peanut with CRISPR/Cas9 based gene editing. *BMC Biotechnol.* 19, 1–7. https://doi.org/10.1186/s12896-019-0516-8

Zhang, H., Dean, L., Wang, M.L., Dang, P., Lamb, M., and Chen, C. (2023). GWAS with principal component analysis identify QTLs associated with main peanut flavor-related traits. *Front. Plant Sci.* 14, 1204415. https://doi.org/10.3389/fpls.2023.1204415

Zhao, S., Li, A., Li, C., Xia, H., Zhao, C., Zhang, Y., et al. (2017). Development and application of KASP marker for high throughput detection of AhFAD2 mutation in peanut. *Electron. J. Biotechnol.* 25, 9–12. https://doi.org/10.1016/j.ejbt.2016.10.010

Zhuang, W., Chen, H., Yang, M., Wang, J., Pandey, M.K., Zhang, C., et al. (2019). The genome of cultivated peanut provides insight into legume karyotypes, polyploid evolution and crop domestication. *Nat. Genet.* 51, 865–876. https://doi.org/10.1038/s41588-019-0402-2

Zhuang, Y., and Dreskin, S.C. (2012). Redefining the major peanut allergens. *Immunol. Res.* 55, 125–134. https://doi.org/10.1007/s12026-012-8355-x

Ziarovska, J., Urbanova, L., Montero-Torres, J., Kovacik, A., Klongova, L., Bharati, R., et al. (2023). Polymorphism of Bolivian accessions of *Arachis hypogaea* L. revealed by allergen coding DNA markers. *Plant Soil Environ.* 69, 615–627. https://doi.org/10.17221/306/2023-PSE

12 Reducing Bitterness in *Cucurbitaceae* and Glucosinolates in *Brassicaceae*

Sangam L. Dwivedi and Autar K. Mattoo

12.1 INTRODUCTION

Cucurbitaceae is the second largest fruit and vegetable family, next only to Solanaceae, widely distributed worldwide, and contain a variety of fruit and vegetable crops. Cucumber (*Cucumis sativus*), zucchini (*C. pepo*), pumpkin (*C. maxima*, *C. moschata*, and *C. argyrosperma*), wax gourd (*Benincasa hispida*), bottle gourd (*Lagenaria siceraria*), bitter gourd (*Momordica charantia*), ridge gourd (*Luffa acutangula*), sponge gourd (*Luffa cylindrica*), chayote (*Sechium edule*), and snake gourd (*Trichosanthes anguina*), among vegetables, and melon (*Cucumis melo*), horned cucumber (*Cucumis metuliferus*), watermelon (*Citrullus lanatus*), and luohanguo (*Siraitia grosvenorii*), among fruits, are major fruit and vegetable crops in Cucurbitaceae, with many of the Cucurbitaceae genomes sequenced (Ma et al., 2022, and references therein).

Brassica genus includes important vegetables, oilseed crops, and forage species, which are grouped into six species—*Brassica nigra*, *B. oleracea*, and *B. rapa,* the diploid species, and *B. carinata*, *B. juncea*, and *B. napus* L., the amphidiploids. The dominant vegetable species in *B. oleracea* includes a wide range of morphological variants, such as broccoli (var. *italica*), cabbage (var. *capitata*), kohlrabi (var. *gongylodes*), cauliflower (var. *botrytis*), brussels sprouts (var. *gemmifera*), and kale (var. *acephala*). Rapeseed or canola (*B. napus*) and mustard (*B. nigra*) are important oilseed crops (Chalhoub et al., 2014; Salehi et al., 2021; Šamec et al., 2017). They are a rich source of GLSs, both in leaves and seeds. The concentration and composition of GLSs may vary between species as well between plant organs within a plant of a given species (roots, leaves, seeds). The levels of GLS may vary greatly during plant growth and development and under abiotic stresses.

Cucurbitacins are structurally diverse triterpenes mainly found in Cucurbitaceae and several other plant families. They cause bitterness in cucumber, melon, pumpkin, squash, and watermelon while possessing immense pharmacological potential. They provide health benefits such as antitumor, anti-inflammatory, atherosclerosis, blood circulation promoter, immunosuppressant, and antidiabetic to humans (Varela

DOI: 10.1201/9781003467335-12

et al., 2022). Cucurbitacin level varies between tissues but are mostly concentrated in fruits and roots of mature plants. The seeds generally contain very low concentration of cucurbitacins. Three forms of cucurbitacins are known—cucurbitacin B causes bitterness in melon, cucurbitacin C in cucumber, and cucurbitacin E in watermelon (Ma et al., 2022). Cucurbitacins are also highly toxic and when consumed as food plants can be fatal sometimes, for example, consumption of bitter bottle gourd (Kaushik et al., 2015; Qing et al., 2022; Silvestre et al., 2022; Zieniuk and Pawełkowicz, 2023). Cucurbitacins discourage insect pests to cause damage to the plant (Bruno et al., 2023; Jaccard et al., 2023).

Leafy vegetables belonging to *Brassica* are the source of sulfur-containing compounds known as glucosinolates (GLSs) and their secondary metabolites (Verkerk et al., 2009). The regular consumption of *Brassica* leafy vegetables provides health benefits to humans, particularly in fighting several kinds of cancers and other diseases such as human intestinal inflammation (Li et al., 2022; Costa-Pérez et al., 2024). GLSs are involved in plant growth and development and stress tolerance (del Carmen Martinez-Ballists et al., 2013; Hasanuzzaman et al., 2017) and confer health benefits to humans (Miękus et al., 2020) while providing unique flavors, bitter taste, and pungent odor (Connolly et al., 2021). However, it is also considered as antinutrient, limiting the bioavailability and absorption of nutrients in animals and humans (Salim et al., 2023).

Wild cucurbits invariably have bitter fruits. Cucurbitaceae fruit crops, except for bitter gourd (*Momordica charantia*), lost their bitterness through domestication (Chomicki et al., 2020; Shang et al., 2014).

High-performance liquid chromatography/quadrupole-time-of-flight mass spectrometry combined with in-source fragmentation and alkali adduct ions (HPLC-Q-TOF-MS) is recommended for isolating cucurbitacins and their glycosides (Chen et al., 2022), while a visible-near infrared spectroscopy (Vis-NIRS) and UHPLC-Q-Orbitrap-MS or by combining UHPLC-HRMS/MS and UHPLC-QqQ-MS/MS may be used as an optimal extraction methods for detection and quantitation limits of GLSs and their hydrolyzed products in *Brassica* (Gahain et al., 2021; Hwang et al., 2019; Yan et al., 2023).

12.2 *BRASSICA* AND CUCUMBER GERMPLASM HOLDINGS ACROSS GENE BANKS

Brassica remains the largest oil-seed crop, after soybean, and is cultivated across 124 countries. Genesys data suggests a total of over 85,474 *Brassica* accessions conserved in gene banks at the global level. Another source details at least 81,752 accessions of *Brassica* conserved in 148 institutions across 81 countries, with major holdings in India, Australia, Korea, the UK, Russia, Germany, Pakistan, the USA, and Japan (Subramanian et al., 2023).

A global strategy was developed for the conservation of *Brassica* genetic resources. Brassicaceae contains 348 genera and approximately 3,700 species. The species in this family are known to produce secondary metabolites and GLSs, and a characteristic flower morphology wherein petals are arranged in the shape of a cross, leading to the original family name of Crucifereae. *Brassica* contains between

36 and 41 species, with majority of the species not cultivated. This conservation strategy focuses on six cultivated species: *Brassica rapa*, *B. oleracea*, *B. nigra*, *B. napus*, *B. carinata*, and *B. juncea*. Within each species, different crop types have been selected, including vegetable, oilseed, condiment, and fodder types. *Brassica* are diverse groups of crops, including broccoli, cabbage, bok choy, turnip, kale, rapeseed, and mustard. *B. oleracea* and *B. rapa* show immense degree of intraspecific variation, with former containing 14 and later 15 sub-specific taxa, each corresponding to a different crop type with its own unique morphology and characteristics (Allender and Giovannini, 2023).

Genesys database lists close to 7000 cucumber (*Cucumis sativus*) germplasm preserved, with major holding in the USA, Bulgaria, the Netherlands, Czechoslovakia, Germany, Poland, Spain, Taiwan, Ukraine, and Hungary. Other countries and germplasm repositories with significant cucurbit holdings not currently documented in Genesys include those in China, Japan, Russia (Vavilov Institute of Plant Industry), India, Korea, Uzbekistan, and Taiwan (World Vegetable Center) (Grumet et al., 2021). Crop Trust developed a global conservation strategy for preservation of Cucurbitaceae germplasm, including cucumber. This strategy promotes the rationalization of conservation efforts at national, regional, and global scales and identify priority actions to strengthen the conservation of cucurbit genetic resources. It encourages partnerships and the sharing of responsibilities, facilities, and tasks for the conservation and sustainable use of Cucurbitaceae crop genetic resources (Ebert et al., 2023). The genus included are *Citrullus* (watermelon), *Cucumis* (cucumbers, melons), and *Cucurbita* (pumpkins, squash); all three rank among the top 10 fruit and vegetables crops globally. The genus *Cucumis* contains two major crops, cucumber (*C. sativus*) and muskmelon (*C. melo*).

Most wild cucurbits produce highly bitter fruits due to high concentrations of cucurbitacins. This toxic compound is also present in their leaves and roots, which protects plants from herbivory damage and enhance abiotic stress adaptation. Cucurbitacins have immense pharmacological and pharmaceutical values (Maja et al., 2022). The bitterness of most Cucurbitaceae fruit crops, except bitter gourd (*M. charantia*), has decreased through domestication (Ebert et al., 2023). Cucurbitaceae family germplasms are challenged by multiple fungal, oomycete, bacterial, and viral diseases and insects that cause vector disease and feeding damage. They are also constrained by market demands, climate vulnerability, domestication bottlenecks, and limited capacity for interspecific hybridization, creating limited opportunity for crop improvement (Grumet et al., 2021).

12.3 GENETICS AND MOLECULAR BASIS OF BITTERNESS IN CUCURBITACEAE

The bHLH TF genes, *Bl* (bitter leaf) and *Bt* (bitter fruit), regulate cucurbitacin C biosynthesis (Shang et al., 2014; Xu et al., 2022). *Bt* confers bitterness to the whole plant (Huang et al., 2009). One hundred and twelve putative domestication sweeps were noted in cucurbit genomes, of which one of these had a gene involved in the loss of bitterness (*Bt*) in fruits through domestication (Qi et al., 213). Cucumber, melon, and watermelon homologues revealed a convergent domestication sweep at the *Bt* loci and the loss of bitterness in Cucurbitaceae (Shang et al., 2014; Qi et al., 2013; Zhou et al., 2016). The genes associated with bitterness have conserved micro-synteny for cucurbitacin biosynthesis in Cucurbitaceae (Ding et al., 2024). *CsMATE1* is physically clustered with cucurbitacin C (CuC) biosynthetic genes (Ma et al., 2023). KASP-based markers, *BiHRM1* and *Bi-KASP*, accurately differentiated bitter-free segregants in cucumber breeding programs (Venkatesh et al., 2018).

Two QTLs, *QBt.1* mapped on chromosome 6 and *QBt.2* on chromosome 7, regulate bitterness in bottle gourd. The causal genes underlying *QBt.1* and QBt.2 are not orthologs to the known bitterness genes in cucumber (Wu et al., 2019). Metabolome analysis differentiated bitter and non-bitter bottle gourd accessions, while transcriptome analysis revealed significantly higher expression of *Bi*, *CYP450s*, and *ACT* of the cucurbitacin biosynthesis pathway genes (Zhao et al., 2022).

12.4 GENETIC AND MOLECULAR BASIS OF GLUCOSINOLATES IN BRASSICACEAE

Assessment of *Brassica* gene pools revealed wide variation in GLSs in roots, leaves, and seeds, with some compounds significantly greater than others (Assefa et al., 2023; Kim et al., 2023a, 2023b; Kittipol et al., 2019; Klopsch et al., 2018; Missinou et al., 2022; Rhee et al., 2020). The middle layer leaves have significantly greater concentrations of GLSs compared to older (outer) leaves and younger inner leaves. Green pigmented leaves were superior in total GLS content than those with red-pigmented accessions (Rhee et al., 2020).

There has been an increased interest in using rapeseed's flowering stalk tissues as green vegetable, especially in China. The flowering stalk tissues are harvested as vegetables during the bud formation stage, which are excellent source of GLSs, offering unique flavor and taste. Continuous and wide variation in GLS content in flowering stalk tissues were reported among rapeseed germplasms, with some compounds such as progoitrin, glucobrassicanapin, and gluconapin greater than others. Five QTLs were significantly associated with three GLS compounds in flowering stalk tissues. The QTL region *GTL_A01* and *4OH_A06* contain four candidate genes (*BnaA01. SOT16*, *BnaA06. SOT17*, *Bna*A06. *MYB51a*, and *Bna*A06. *MYB51b*) providing new insights into GLS biosynthesis in flowering stalk tissues, which after functional validation, may be deployed in improving the GLS content in flowering stalk tissues (Gao et al., 2023).

The nutritional value of *Brassica* meal is reduced due to the presence of GLSs, the antinutrient. GLS transporters (GTRs), *GTR1* and *GTR2*, involved in relocation of GLS from source to sink translocation, regulate tissue-specific accumulation of GLSs in *B. juncea*. The single knockout lines display different phenotypes though both show ubiquitous expression patterns and broad substrate specificity. *GTR2*-knockdown lines show a significant reduction of GLSs in seeds but higher accumulation in leaves and pods, thereby have higher resistance toward the generalist pest, *Spodoptera litura*. *GTR1*-knockdown plants display a smaller reduction of GLSs in seeds and significantly lower GLS levels in leaves. Clearly, *GTR2* has great potential in breeding for enhanced nutrition and plant defense (Nambiar et al., 2021). In rapeseed (*Brassica napus*), *Bna.HAG1.A9* and *Bna.HAG1.C2* regulate GLS (aliphatic type) content in roots, leaves, and seeds, with *Bna.HAG3.A3* aromatic GLS present in the root (Kittipol et al., 2019). A major QTL controls aliphatic GLS accumulation in rapeseed leaves. This QTL region contains five candidate genes, of which, *BrMAM-3* expression is significantly correlated with the accumulation of aliphatic GLSs. A naturally occurring insertion within exon 1 of *BrMAM-3*, predicted to be a loss-of-function mutation, is associated with the low GLS content in rapeseed accessions. Transgenic *Arabidopsis* overexpressing *BrMAM-3* enhances total aliphatic GLSs in leaves (Zhang et al., 2018). The three major form of GLSs include aliphatic, aromatic, and indole, depending on precursor amino acids (Sønderby et al., 2010).

12.5 PRODUCT (CULTIVAR) DEVELOPMENT

GLSs are the potential target for genetic manipulation and application in crop improvement programs. Plant breeders often face dilemma of how much GLSs to reduce via breeding to ensure that new cultivars resist pest damage as well the edible produce (seed or leaves) is nutritionally beneficial to animal/human health. One approach could be to maintain GLS levels in leaves at levels beneficial to plants to resists pest damage but reducing it to the extent that seed or feed is beneficial to animal and human health. The blockage of GLS transport into seeds (achieving low GLS content beneficial to animal and human health) while maintaining high aliphatic GLS content in vegetative tissues (protecting plants from pest damage) could be a viable strategy. Thus, identifying such genetic resources from the gene pool represents an excellent opportunity to be explored.

The mutation of genes encoding two *GTRs* reduces GLSs in *Brassica* species. The seed GLS content in *B. rapa* and *B. juncea* is reduced by 60–70%, which is stably inherited over multiple generations across field locations. Thus, loss-of-function phenotype via transport engineering could be applied to reduce seed GLS content in other oilseed crops (Nour-Eldin et al., 2017).

Many germplasm lines with acceptable GLS levels (<30 μmoles g^{-1} seed dry weight, SDW) are available in rapeseed (*B. napus*) and mustard (*B. juncea*). However, such lines have become more susceptible to pest and diseases. CASPR/CAS9-based genome editing of *GTR* family genes resulted in mustard lines with the desired low seed GLS content (reducing from 146.09 μmoles g^{-1} to as low as 6.21 μmoles g^{-1} SDW) while maintaining higher GLS levels in leaves to defend plants against biotic stresses. The seed yield and seed quality of such lines were comparable with wild type (Mann et al., 2023). The knockout of two independent GLS biosynthesis genes (*BnMYB28* or *BnCYP79F1*) in rapeseed (*B. napus*) resulted in mutants with significantly lower seed GLSs. The GLS content as compared to WT (36.3 μmol g^{-1} SDW) reduced by 55.3% and 32.4% in *BnMYB28* (16.20 μmol g^{-1} SDW) and *BnCYP79F1* (24.5 μmol g^{-1} SDW) mutants, respectively (Jhingan et al., 2023). Clearly, targeted editing of genes in GLS biosynthesis pathway is a feasible approach to reduce the seed GLSs globally accepted levels without disrupting high leaf GLS or seed yield and seed quality.

RNAi-based targeted suppression of *BjMYB28* in transgenic mustard resulted in significant reduction in aliphatic GLS fractions, without altering the desirable non-aliphatic GLS pool, both in leaves and seeds. Transgenic plants stably expressed the low GLS levels in subsequent generations with no adverse effects on plant growth and development. The performance of transgenic lines under containment facility field evaluation were comparable with wild types for fatty acid composition, oil and protein contents, and seed weight (Augustine et al., 2013).

Divergent mass selection (DMS) in plant breeding refers to generating groups of individuals that share the same genetic background but with extreme values for a given trait practiced the selection. DMS has been effectively used to modify total GLSs as well individual GLS contents in *Brassica* species (Sotelo et al., 2016; Stowe and Marquis, 2011). The DMS method has been recently used to modify GSLs content in the leaves of *B. napus* and *B. rapa*. The response to selection was linear with no adverse effect on loss of variability for GLS modification in either direction. DMS was more effective in reducing the selected GLS than increasing it. It also modified GLSs in other plant parts, suggesting GLS translocation in plant or a modification in their synthesis pathway is not organ-specific. Moreover, the derived populations, as a result of MDS, had no adverse effects on yield or excelled better than original populations (Coves et al., 2023).

12.6 CONCLUSION

The Cucurbitaceae family of fruits and vegetables, except for bottle gourd, lost bitter genes through domestication. Genes associated with cucurbitacin biosynthesis pathway show conserved micro-synteny for cucurbitacin biosynthesis in Cucurbitaceae. Manipulating seed GLS levels at globally acceptable levels, without disrupting GLSs in vegetative tissues, is a significant breeding challenge, which could be achieved by targeted editing of GTR genes or by adopting divergent mass selection. The identification of novel sources of variations and advances in biosynthesis pathways and high-throughput assays provide greater opportunity to genetically manipulate the genomes of these two families toward the development of non-bitter cucumbers and their relatives as well as *Brassica*, with desired levels of seed and vegetative tissue GLSs, beneficial to plants as well as the animals and humans that consume them.

REFERENCES

Allender, C., and Giovannini, P. (2023). *Global Strategy for the Conservation of Brassica Genetic Resources*. Global Crop Diversity Trust, Bonn, Germany. https://doi.org/10.5281/zenodo.7544810

Assefa, A.D., Kim, S.-H., Ko, H.C., Ro, N., Subramanian, P., Chung, Y.J., et al. (2023). Leaf mustard (*Brassica juncea*) germplasm resources showed diverse characteristics in agromorphological traits and glucosinolate levels. *Foods* 12, 4374. https://doi.org/10.3390/foods12234374

Augustine, R., Mukhopadhyay, A., and Bisht, N.C. (2013). Targeted silencing of BjMYB28 transcription factor gene directs development of low glucosinolate lines in oilseed *Brassica juncea*. *Plant Biotechnol. J.* 11, 855–866. https://doi.org/10.1111/pbi.12078

Bruno, P., Arce, C.C.M., Machado, R.A.R., Besomi, G., Spescha, A., Glauser, G., et al. (2023). Sequestration of cucurbitacins from cucumber plants by *Diabrotica balteata* larvae provides little protection against biological control agents. *J. Pest Sci.* 96, 1061–1075. https://doi.org/10.1007/s10340-022-01568-3

Chalhoub, B., Denoeud, F., Liu, S., Parkin, I.A.P., Tang, H., Wang, X., et al. (2014). Early allopolyploid evolution in the post-neolithic Brassica napus oilseed genome. *Science* 345, 950–953. https://doi.org/10.1126/science.1253435

Chen, S.-Y., Zhou, Q.Y.-J., Chen, L., Li, J.-Y., Xie, T., and Zhang, S.-H. (2022). Screening and identifying cucurbitacins and cucurbitacin glycosides in *Cucumis sativus* using high-performance liquid chromatography/quadrupole-time-of-flight mass spectrometry combined with in-source fragmentation and alkali adduct ions. *Rapid Commun. Mass Spectrom.* 36, e9323. https://doi.org/10.1002/rcm.9323

Chomicki, G., Schaefer, H., and Renner, S.S. (2020). Origin and domestication of Cucurbitaceae crops: Insights from phylogenies, genomics and archaeology. *New Phytol.* 226, 1240–1255. https://doi.org/10.1111/nph.16015

Connolly, E.L., Sim, M., Travica, N., Marx, W., Beasy, G., Lynch, G.S., et al. (2021). Glucosinolates from cruciferous vegetables and their potential role in chronic disease: Investigating the preclinical and clinical evidence. *Front. Pharmacol.* 12, 767975. https://doi.org/10.3389/fphar.2021.767975

Costa-Pérez, A., Sánchez-Bravo, P., Medina, S., Domínguez-Perles, R., and García-Viguera, C. (2024). Bioaccessible organosulfur compounds in broccoli stalks modulate the inflammatory mediators involved in inflammatory bowel disease. *Int. J. Mol. Sci.* 25, 800. https://doi.org/10.3390/ijms25020800

Coves, S., Soengas, P., Velasco, P., Fernández, J.C., and Cartea, M.E. (2023). New vegetable varieties of *Brassica rapa* and *Brassica napus* with modified glucosinolate content obtained by mass selection approach. *Front. Nutr.* 10, 1198121. https://doi.org/10.3389/fnut.2023.1198121

del Carmen Martínez-Ballesta, M., Moreno, D.A., and Carvajal, M. (2013). The physiological importance of glucosinolates on plant response to abiotic stress in *Brassica*. *Int. J. Mol. Sci.* 14, 11607–11625. https://doi.org/10.3390/ijms140611607

Ding, X., Yang, Z., Wang, H., Zeng, J., Dai, H., and Mei, W. (2024). *Aquilaria sinensis*: An upstart resource for cucurbitacin production offers insights into the origin of plant bitter (*Bi*) gene clusters. *Plants (Basel)* 13, 260. https://doi.org/10.3390/plants13020260

Ebert, A.W., Drummond, E.B.M., Giovannini, P., and van Zonneveld, M. (2023). *A Global Conservation Strategy for Crops in the Cucurbitaceae Family.* Global Crop Diversity Trust, Bonn, Germany. Version 2 (minor corrections). https://doi.org/10.5281/zenodo.7696528

Gahain, B., Kumar, P., Malhotra, B., Augustine, R., Prdhan, A.K., and Bisht, N.C. (2021). A comprehensive Vis-NIRS equation for rapid quantification of seed glucosinolate content and composition across diverse *Brassica* oilseed chemotypes. *Food Chem.* 354, 129527. https://doi.org/10.1016/j.foodchem.2021.129527

Gao, C., Zhang, F., Hu, Y., Song, L., Tang, L., Zhang, X., et al. (2023). Dissecting the genetic architecture of glucosinolate compounds for quality improvement in flowering stalk tissues of *Brassica napus*. *Hortic. Plant J.* 9, 553e562. https://doi.org/10.1016/j.hpj.2022.09.001

Grumet, R., McCreight, J.D., McGregor, C., Weng, Y., Mazourek, M., Reitsma, K., et al. (2021). Genetic resources and vulnerabilities of major *Cucurbit* crops. *Genes (Basel)* 12, 1222. https://doi.org/10.3390/genes12081222

Hasanuzzaman, M., Nahar, K., Anee, T.I., and Fujita, M. (2017). Glutathione in plants: Biosynthesis and physiological role in environmental stress tolerance. *Physiol. Mol. Biol. Plants* 23, 249–268. https://doi.org/10.1007/s12298-017-0422-2

Huang, S., Li, R., Zhang, Z., Li, L., Gu, X., Fan, W., et al. (2009). The genome of the cucumber, *Cucumis sativus* L. *Nat. Genet.* 41, 1275–1281. https://doi.org/10.1038/ng.475

Hwang, I.M., Park, B., Dang, Y.M., Kim, S.-Y., and Seo, H.-Y. (2019). Simultaneous direct determination of 15 glucosinolates in eight *Brassica* species by UHPLC-Q-Orbitrap-MS. *Food Chem.* 282, 127–133. https://doi.org/10.1016/j.foodchem.2018.12.036

Jaccard, C., Ye, W., Bustos-Segura, C., Glauser, G., Kaplan, I., and Benrey, B. (2023). Consequences of squash (*Cucurbita argyrosperma*) domestication for plant defence and herbivore interactions. *Planta* 257, 106. https://doi.org/10.1007/s00425-023-04139-7

Jhingan, S., Harloff, H.-J., Abbadi, A., Welsch, C., Blümel, M., Tesdemir, D., et al. (2023). Reduced glucosinolate content in oilseed rape (*Brassica napus* L.) by random mutagenesis of *BnMYB28* and *BnCYP79F1* genes. *Sci. Rep.* 13, 2344. https://doi.org/10.1038/s41598-023-28661-6

Kaushik, U., Aeri, V., and Mir, S.R. (2015). Cucurbitacins—An insight into medicinal leads from nature. *Phcog. Rev.* 9, 12–18. https://doi.org/10.4103/0973-7847.156314

Kim, S.-H., Ochar, K., Hwang, A., Lee, Y.J., and Kang, H.J. (2023a). Variability of glucosinolates in pak choy (*Brassica rapa* subsp. *chinensis*) germplasm. *Plants (Basel)* 13, 9. https://doi.org/10.3390/plants13010009

Kim, S.H., Subramanian, P., and Hahn, B.-S. (2023b). Glucosinolate diversity analysis in choy sum (*Brassica rapa* subsp. *chinensis* var. *parachinensis*) germplasms for functional food breeding. *Foods* 12, 2400. https://doi.org/10.3390/foods12122400

Kittipol, V., He, Z., Wang, L., Doheny-Adams, T., Langer, S., and Bancroft, I. (2019). Genetic architecture of glucosinolate variation in *Brassica napus*. *J. Plant Physiol.* 240, 152988. https://doi.org/10.1016/j.jplph.2019.06.001

Klopsch, R., Witzel, K., Artemyeva, A., Ruppel, S., and Hanschen, F.S. (2018). Genotypic variation of glucosinolates and their breakdown products in leaves of *Brassica rapa*. *J. Agric. Food Chem.* 66, 5481–5490. https://doi.org/10.1021/acs.jafc.8b01038

Li, Y.-Z., Yang, Z.-Y., Gong, T.-T., Liu, Y.-S., Liu, F.H., Wen, Z.Y., et al. (2022). Cruciferous vegetable consumption and multiple health outcomes: An umbrella review of 41 systematic reviews and meta-analyses of 303 observational studies. *Food Funct.* 13, 4247–4259. https://doi.org/10.1039/d1fo03094a

Ma, L., Wang, Q., Zheng, Y., Guo, J., Yuan, S., Fu, A., et al. (2022). Cucurbitaceae genome evolution, gene function, and molecular breeding. *Hortic. Res.* 9, uhab057. https://doi.org/10.1093/hr/uhab057

Ma, Y., Li, D., Zhong, Y., Wang, X., Li, L., Osbourn, A., et al. (2023). Vacuolar MATE/DTX protein-mediated cucurbitacin C transport is co-regulated with bitterness biosynthesis in cucumber. *New Phytol.* 238, 995–1003. https://doi.org/10.1111/nph.18786

Maja, D., Mavengahama, S., and Mashilo, J. (2022). Cucurbitacin biosynthesis in cucurbit crops, their pharmaceutical value and agricultural application for management of biotic and abiotic stress: A review. *S. Afr. J. Bot.* 145, 3–12. https://doi.org/10.1016/j.sajb.2021.08.044

Mann, A., Kumari, J., Kumar, R., Kumar, P., Pradhan, A.K., Pental, D., et al. (2023). Targeted editing of multiple homologues of *GTR1* and *GTR2* genes provides the ideal low-seed, high-leaf glucosinolate oilseed mustard with uncompromised defence and yield. *Plant Biotechnol. J.* 21, 2182–2195. https://doi.org/10.1111/pbi.14121

Miękus, N., Marszałek, K., Podlacha, M., Iqbal, A., Puchalski, C., and Świergiel, A.H. (2020). Health benefits of plant-derived sulfur compounds, glucosinolates, and organosulfur compounds. *Molecules* 25, 3804. https://doi.org/10.3390/molecules25173804

Missinou, A.A., de Carvalho, J.F., Marnet, N., Delhaye, T., Hamzaoui, O., Sayed, D.A., et al. (2022). Identification and quantification of glucosinolates and phenolics in a large panel of *Brassica napus* highlight valuable genetic resources for chemical ecology and breeding. *J. Agric. Food Chem.* 70, 5245–5261. https://doi.org/10.1021/acs.jafc.1c08118

Nambiar, D.M., Kumari, J., Augustine, R., Kumar, P., Bajpai, P.K., and Bisht, N.C. (2021). GTR1 and GTR2 transporters differentially regulate tissue-specific glucosinolate contents and defence responses in the oilseed crop Brassica juncea. *Plant Cell Environ.* 44, 2729–2743. https://doi.org/10.1111/pce.14072

Nour-Eldin, H.H., Madsen, S.R., Engelen, S., Jøgensen, M.E., Olsen, C.E., Andersen, J.S., et al. (2017). Reduction of antinutritional glucosinolates in Brassica oilseeds by mutation of genes encoding transporters. *Nat. Biotechnol.* 35, 377–382. https://doi.org/10.1038/nbt.3823

Qi, J., Liu, X., Shen, D., Miao, H., Xie, B., Li, X., et al. (2013). A genomic variation map provides insights into the genetic basis of cucumber domestication and diversity. *Nat. Genet.* 45, 1510–1515. https://doi.org/10.1038/ng.2801

Qing, Z., Shi, Y., Han, L., Li, P., Zha, Z., Liu, C., et al. (2022). Identification of seven undescribed cucurbitacins in *Cucumis sativus* (cucumber) and their cytotoxic activity. *Phytochemistry* 197, 113123. https://doi.org/10.1016/j.phytochem.2022.113123

Rhee, J.-H., Choi, S., Lee, J.-E., Hur, O.-S., Ro, N.-Y., Hwang, A.-J., et al. (2020). Glucosinolate Content in Brassica genetic resources and their distribution pattern within and between inner, middle, and outer leaves. *Plants (Basel)* 9, 1421. https://doi.org/10.3390/plants9111421

Salehi, B., Quispe, C., Butnariu, M., Sarac, I., Marmouzi, I., Kamle, M., et al. (2021). Phytotherapy and food applications from *Brassica* Genus. *Phytother. Res.* 35, 3590–3609. https://doi.org/10.1002/ptr.7048

Salim, R., Nehvi, I.B., Mir, R.A., Tyagi, A., Ali, S., and Bhat, O.M. (2023). A review on antinutritional factors: Unraveling the natural gateways to human health. *Front. Nutr.* 10, 1215873. https://doi.org/10.3389/fnut.2023.1215873

Šamec, D., Pavlović, I., and Salopek-Sondi, B. (2017). White cabbage (Brassica oleracea Var. *capitata* f. *alba*): Botanical, phytochemical and pharmacological overview. *Phytochem. Rev.* 16, 117–135. https://doi.org/10.1007/s11101-016-9454-4

Shang, Y., Ma, Y., Zhou, Y., Zhang, H., Duan, L., Chen, H., et al. (2014). Biosynthesis, regulation, and domestication of bitterness in cucumber. *Science* 346, 1084–1088. https://doi.org/10.1126/science.1259215

Silvestre, G.F.G., de Lucena, R.P., and da Sivla Alves, H. (2022). Cucurbitacins and the immune system: Update in research on anti-inflammatory, antioxidant, and immunomodulatory mechanisms. *Curr. Med. Chem.* 29, 3774–3789. https://doi.org/10.2174/0929867329666220107153253

Sønderby, I.E., Geu-Flores, F., and Halkier, B.A. (2010). Biosynthesis of glucosinolates—gene discovery and beyond. *Trends Plant Sci.* 15, 283–290. https://doi.org/10.1016/j.tplants.2010.02.005

Sotelo, T., Velasco, P., Soengas, P., Rodríguez, V.M., and Cartea, M.E. (2016). Modification of leaf glucosinolate contents in *Brassica oleracea* by divergent selection and effect on expression of genes controlling glucosinolate pathway. *Front. Plant Sci.* 7, 1012. https://doi.org/10.3389/fpls.2016.01012

Stowe, K., and Marquis, R. (2011). Costs of defense: Correlated responses to divergent selection for foliar glucosinolate content in *Brassica rapa*. *Evol. Ecol.* 25, 763–775. https://doi.org/10.1007/s10682-010-9443-9

Subramanian, P., Kim, S.-H., and Hhn, B.-S. (2023). *Brassica* biodiversity conservation: Prevailing constraints and future avenues for sustainable distribution of plant genetic resources. *Front. Plant Sci.* 14, 1220134. https://doi.org/10.3389/fpls.2023.1220134

Varela, C., Melim, C., Neves, B.G., Sharifi-Rad, J., Calina, D., Mamurova, A., et al. (2022). Cucurbitacins as potential anticancer agents: New insights on molecular mechanisms. *J. Transl. Med.* 20, 630. https://doi.org/10.1186/s12967-022-03828-3

Venkatesh, J., Song, K., Lee, J.-H., and Kwon, J.-K. (2018). Development of *Bi* gene-based SNP markers for genotyping for bitter-free cucumber lines. *Hort. Environ. Biote.* 59, https://doi.org/10.1007/s13580-018-0029-8

Verkerk, R., Schreiner, M., Krumbein, A., Ciska, E., Holst, B., and Rowland, I. (2009). Glucosinolates in *Brassica* vegetables: The influence of the food supply chain on intake, bioavailability and human health. *Mol. Nutr. Food Res.* 53. https://doi.org/10.1002/mnfr.200800065

Wu, X., Wu, X., Wang, Y., Wang, B., Lu, Z., Xu, P., et al. (2019). Molecular genetic mapping of two complementary genes underpinning fruit bitterness in the bottle gourd (*Lagenaria siceraria* [Mol.] Standl.). *Front. Plant Sci.* 10, 1493. https://doi.org/10.3389/fpls.2019.01493

Xu, Y., Zhang, H., Zhong, Y., Jiang, N., Zhong, X., Zhang, Q., et al. (2022). Comparative genomics analysis of bHLH genes in cucurbits identifies a novel gene regulating cucurbitacin biosynthesis. *Hortic. Res.* 9, uhac038. https://doi.org/10.1093/hr/uhac038

Yan, C., Huang, Y., Zhang, S., Cui, L., Jiao, Z., Peng, Z., et al. (2023). Dynamic profiling of intact glucosinolates in radish by combining UHPLC-HRMS/MS and UHPLC-QqQ-MS/MS. *Front. Plant Sci.* 14, 1216682. https://doi.org/10.3389/fpls.2023.1216682

Zhang, J., Wang, H., Liu, Z., Liang, J., Wu, J., Cheng, F., et al. (2018). A naturally occurring variation in the *BrMAM-3* gene is associated with aliphatic glucosinolate accumulation in *Brassica rapa* leaves. *Hortic. Res.* 5, 69. https://doi.org/10.1038/s41438-018-0074-6

Zhao, G., Wang, M., Luo, C., Li, J., Gong, H., Zheng, X., et al. (2022). Metabolome and transcriptome analyses of cucurbitacin biosynthesis in *Luffa* (*Luffa acutangula*). *Front. Plant Sci.* 13, 886870. https://doi.org/10.3389/fpls.2022.886870

Zhou, Y., Ma, Y., Zeng, J., Duan, L., Xue, X., Wang, H., et al. (2016). Convergence and divergence of bitterness biosynthesis and regulation in Cucurbitaceae. *Nat. Plants* 2, 1251–1263. https://doi.org/10.1038/nplants.2016.183

Zieniuk, B., and Pawełkowicz, M. (2023). Recent advances in the application of cucurbitacins as anticancer agents. *Molecules* 13, 1081. https://doi.org/10.3390/metabo13101081

13 Profiling Germplasm and Identifying Genes and Markers Associated with Horticultural and Sensory Traits, and Developing Improved Germplasm of Cucurbitaceae

Sangam L. Dwivedi and Autar K. Mattoo

13.1 INTRODUCTION

Bitter gourd (*Momordica charantia*), bottle gourd (*Lagenaria siceraria*), chayote (*Sechium edule*), cucumber (*Cucumis sativus*), pumpkin and squash (*Cucurbita maxima*, *C. moschata*, and *C. argyrosperma*), ridge gourd (*Luffa acutangula*), snake gourd (*Trichosanthes anguina*), sponge guard (*Luffa cylindrica*), wax gourd (*Benincasa hispida*), and zucchini (*Cucurbita pepo*) are the major vegetables that are widely consumed globally. The quality (irrespective of fruits or vegetables) refers to variation in physical appearance (size and shape; spine color, size, and density; skin color and presence of ridges or speckles; presence or absence of wax on the surface), the flesh proximate composition (bitterness, soluble solids, vitamins, and minerals), flesh color (white to green or yellow to orange), and flavor (volatile and nonvolatile compounds) (Zhang et al., 2021).

Quality attributes can play a significant role in consumers' perception of likes or dislikes. The preference is highly correlated with color, total soluble solid content (TSSC), sweetness, texture, flavor, freshness, and willingness to pay, which the producers and breeders should factor in as indicators of overall liking when developing new cultivars or producers organizing mass production (Ayres et al., 2019; Moreno et al., 2023; Montero et al., 2024; Oliver et al., 2018). Integrating consumer preference mapping and consumer preference data (for determining the flavors contributing to liking) together with use of trained sensory panels (humans) for quantitative descriptive analysis are ideal tools to assess overall liking or disliking.

 DOI: 10.1201/9781003467335-13

13.2 NATURAL VARIATION ASSOCIATED WITH HORTICULTURAL AND SENSORY TRAITS

13.2.1 Bitter Gourd (*Momordica charantia*)

Assessment of bitter gourd gene pool unlocked wide variation for fruit color, fruit length, fruit diameter, average fruit weight, and yield of the plant^{-1} (Alhariri et al., 2021; Mallikarjuna et al., 2023). Nearly 20 market types of bitter gourd have been reported from Asia, involving fruit color, shape, skin pattern, and size that vary between and within countries. The fruit colors range from white or cream to light green to dark-green, while shapes include cylindrical, elliptical, spindle or conical. Fruits develop irregular longitudinal ridges and warty skin, depending on genotype (Dhillon et al., 2017).

Mutagenesis induced by ethyl methanesulfonate (EMS) is a powerful tool to generate new variability for use in crop improvement programs and functional genomics research. EMS exhibits high mutation rate but causes less harm to the plants (Chen et al., 2022). EMS concentration of 0.2% and mutagen time of 10 h is recommended as an optimum dose for bitter gourd. However, cultivar-specific mutagen dose may be needed to achieve the desired effects. The mutant resource contains 199 M_2 mutants with visible phenotypes for variation for dwarfness, leaf color and shape, tubercles, apical meristem defect, and fruit size and shape (Yu et al., 2022). Variability for degree of bitterness among bitter gourd genetic resources have not been investigated, but it must be done to identify lines with relatively less bitterness and without losing its nutritional benefits.

13.2.2 Bottle Gourd (*Lagenaria siceraria*)

The two cultivated subspecies are *L. siceraria siceraria* (African) and *L. siceraria asiatica* (Asian). Substantial genetic variability exists in bottle gourd gene pool for variation in fruit shape, size, length, color, and texture. The fruit shapes are various: club-shaped, globular, bottle-shaped, flat, pear-shaped cylindrical, elongated straight, pyriform, round, elongated curved, oval, etc. (Mahapatra et al., 2022; Mashilo et al., 2015; Mkhize et al., 2021, and references therein). Accessions with characteristic flavor, for example, strong flavor in J120 and weak flavor in G32 (Wang et al., 2022) or USVL10 enriched with distinct volatile organic compounds (Zia et al., 2023), have been documented.

13.2.3 Chayote (*Sechium edule*)

Fruits differ in shape, size, surface texture, and color. They are a rich source of complex carbohydrates including dietary fibers, starch, minerals, vitamins, and bioactive compounds (Pu et al., 2021; Joshi et al., 2020).

13.2.4 Cucumber (*Cucumis sativus*)

The cucumber gene pool has been systematically characterized for fruit size and shape, bitterness caused by variation in cucurbitacin C, aroma, volatile organic compounds, crispiness, and juiciness plus texture (Jo et al., 2022; Min et al., 2023; Sun et al., 2022; Zhang et al., 2022a; Williams et al., 2022). The key sensory attributes in cucumber include crispy and juiciness texture, fresh and green aromas, and sweet taste. Accessions with significant variation in fresh aroma, crispiness and juiciness texture, a pickling cucumber with bitter and salty taste, or accessions with highest astringency scores are useful genetic resources to breed cucumber lines with specific end use preference (Williams et al., 2022). The differences in cell number and shape in longitudinal and cross-section, driven in turn by differences in orientation, timing, and duration of cell division and expansion, both at pre- and post-anthesis phase, caused variation in fruit size and shape (Colle et al., 2017; Liu et al., 2020). Modeling variations in temperature and light conditions together with differences in cell number and shape in longitudinal and cross-section may provide opportunity for precision management for monitoring fruit growth and forecast the harvest time (Zhang et al., 2022b).

13.2.5 Pumpkin and Squash (*Cucurbita* spp.)

Cucurbita maxima, *C. moschata*, and *C. pepo* are widely cultivated and offer tremendous diversity in growth habit, fruit morphotypes and market class, surface color and texture, flesh color and thickness, flesh aroma (including taro-like aroma), and bioactive compounds with substantial antioxidant and pharmacological functions (Chen et al., 2023; Ferriol and Picó, 2008; Chomicki et al., 2020; Dhillon et al., 2020; Hernandez et al., 2023; Junxing et al., 2022; Paris and Brown, 2005; Paris et al., 2012). The 2-acetyl-1-pyrroline is the key contributor to the taro-like aroma (Junxing et al., 2022). The world's largest fruit, the giant pumpkin (also termed as Atlantic Giant), is *C. maxima* (Savage et al., 2015). *C. moschata* is cross-compatible to various degrees with *C. pepo* and *C. maxima*, which can be used as a bridge to shuffle genes across species (Sun et al., 2017).

13.2.6 Ridge Gourd (*Luffa acutangula*) and Sponge Gourd (*Luffa aegyptiaca*)

Luffa germplasm shows sufficient variability for fruit color (green, light green, dark green), fruit shape (elliptical, elongated, elongated tapered length), fruit ribs (absent, deep, intermediate, superficial), and fruit skin texture (smooth, grainy), fruit weight and fruits plant^{-1} (Tyagi et al., 2020; Dhillon et al., 2020), sensory and textural traits (Chaubey et al., 2022; Kumari et al., 2019), as well the oil and protein content and quality (Tyagi et al., 2023).

13.2.7 Wax Gourd (*Benincasa hispida*)

Mature fruits of wax gourd (also known as ash gourd or white gourd) have thick waxy cuticle. Four major cultivar groups and sixteen morphotypes based on fruit size, color, shape, and waxy bloom intensity are reported in wax gourd (Dhillon et al., 2017).

13.2.8 Zucchini (*Cucurbita pepo*)

The species contains two cultivated subspecies, *pepo* and *olivera*, and each with four fruit-shape morphotypes (or cultivar groups). The former consists of pumpkin, vegetable marrow, cocozelle, and zucchini, while the latter acorn, crookneck, scallop, and straightneck group of cultivars (Paris, 2001). Eight fruit morphotypes have been described (Xanthopoulou et al., 2019).

13.3 GENES AND MARKERS ASSOCIATED WITH HORTICULTURAL AND SENSORY TRAITS

Cucumbers and gourds in Cucurbitaceae are most extensively studied to unlock functional genes and DNA markers associated with fruit characteristics (Table 13.1). The presence of tubercules, warts, spines, ridges, and grooves on fruit epidermis determines the market value as well as influences consumer preferences. The peel color also influences the quality of fruits. Cucumbers with dense tubercules and spines on the fruit peel are popular in East Asia, while those with smooth peel are most liked present throughout Europe (Yang et al., 2014). The presence or absence of wart in fruit determines/differentiates warty (pearl) and non-warty (smooth) bitter gourds and end use. Bitter gourd fruits without warts and with longitudinally smooth ridges on the surface are popular in southern China, while those from other parts of China, Taiwan, and India prefer fruits with granular or spiny warts on the fruit surface (Cui et al., 2022). Thus, external quality of fruits determines the edible use of fruits and vegetables.

13.3.1 Bitter Gourd

A single dominant locus, *Fwa*, was fine-mapped into a 22.70 kb region and was found to harbor four protein-coding genes, and *MC04g1399* encoding an epidermal patterning factor 2-like protein is a candidate gene for *Fwa*. Further, a 1-bp insertion and deletion (InDel) polymorphism within *MC04g1399*, when converted to a CAPS marker, accurately differentiated warty and non-warty fruit types (Liu et al., 2024). *McAPRR2* determines fruit pericarp color in bitter gourd. It has three haplotypes, *Hap1* with a green pericarp in wild accession, *Hap2* with a green haplotype from South and Southeast Asia region, and *Hap 3* with a light green pericarp from Southeast Asia region. The latter two class belongs to cultigen gene pool. The variation in *McAPRR2* haplotype is significantly correlated with both pericarp color and ecological type (Tian et al., 2023b).

13.3.2 Bottle Gourd

A single dominant gene, *LsAPRR2*, regulates the green color of peels in bottle gourd. *LsAPRR2* expression is higher in all-green-skinned bottle gourds at various stages of fruit development than in the white-skinned bottle gourds (Huo et al., 2023). Fruit bitterness is a serious biological constraint in the consumption of the bottle gourd worldwide. Two QTLs, *QBt.1* mapped in LG2 on chromosome 6 and *QBt.2* in LG9 on chromosome 7, regulate fruit bitterness. The causal genes underlying *QBt.1* and *QBt.2* are not orthologs of the known cucurbit bitterness genes (Wu et al., 2019). *Qfs.zaas-6*, involved in transition between round and slender/straight fruit (Xu et al., 2021), and *LsOVATE1*, homologous to *OVATE*, are potential candidate genes affecting the fruit shape (Feng et al., 2022). Cell number and size significantly influence bottle gourd fruit size. The genomic region associated with fruit size was delimited to 17,747,353 bp-18,185,825 bp on chromosome 9, encompassing 0.44 Mb and containing 44 genes. The differentially expressed genes (DEGs) were predominantly enriched with cell wall biosynthesis, sugar metabolism, and hormone signaling that detected six genes within the localized region as potential regulators of fruit size (Fang et al., 2024).

13.3.3 Cucumber

Multiple genes control fruit size and shape (*Csa4G337340*, *CsSF4*, *CsCRC*^{G}, *CsKTN1*, *CsTRM5*, *FS5.2*, *fa4.1*, *fa6.1*, *CsYUC10b*), fruit color (*CsARC6*, *CsPC1*, *Csa3G904140*, *w*), fruit appearance (*CsDULL*, *CsGLF1*), fruit cuticular wax and glossiness (*CsCER6*, *CsCER7*), fruit skin netting (*H* and *Rs*), fruit hollowness (*CsALMT2*), fruit aroma (*qol8–2.1*, *fgr*), orange endocarp (*CsaV3_6G040750*), green flesh fruit (*Csa5G021320*), and fruit spine density (*Csa6G421750*). Hollow heart in cucumber is a physiological defect, seriously impacting yield, quality, and economic value. The low auxin levels in the carpel affects enzymes related to cell wall biosynthesis at the early stage of fruit development, resulting in incomplete development of carpel cells, thus forming a hollow heart in cucumber. The differentially expressed TFs, such as MYB, NAC, ERF, and bHLH as observed in this study, may have role in regulating hollow hearts in cucumber (Li et al., 2024).

13.3.4 Pumpkin and Squash

Fruit size is a quantitatively inherited trait controlled by multiple genes. Thirteen QTLs, including phytohormone-related genes (*CmaCh17G011340* and *CmaCh04G029660*), were associated with fruit size in F_2 population of a cross between Atlantic Giant and its small-fruit ancestor Hubbard. The differentially accumulated genes were significantly enriched in assimilate accumulation in fruits. This population showed abundant transgressive segregants. One InDel and three SNPs were successfully transformed to KASP markers tightly linked with fruit formation may be deployed to effect marker-assisted breeding to modify fruit size in pumpkins (Pan et al., 2022b). Several SNPs related to 289 genes involved in plastid biogenesis, protein ubiquitination, and transcriptional regulation at the carotenoid biosynthesis and transport, or fruit ripening, may further facilitate breeding for enhanced carotenoids in winter squashes (Reddy et al., 2023). A single dominant locus, *CmaMg*, fine-mapped to a region of 32.34 kb on chromosome 11, regulates mature green peel color, and *CmaKNAT6* as a probable candidate gene in this region. *CmaKNAT6* expression increases sharply during the early fruit development and decreases gradually at late fruit development. Allelic diversity among pumpkin germplasm resources revealed 580 bp deletion in the coding region of *CmaKNAT6* is associated with peel color, an important genomic resource to optimize peel color-based breeding strategies in pumpkin (Wang et al., 2024).

13.3.5 Ridge and Sponge Gourds

Consumers prefer to eat aromatic than non-aromatic sponge gourds as vegetable. The volatile compound 2-acetyl-1-pyrroline (2AP) produces popcorn-like or pandan-like aroma in plants (Arikit et al., 2011; Ruangnam et al., 2017; Wakte et al., 2017). *LcBADH* is associated with aroma in the sponge gourd. A SNP in exon 5 of *LcBADH* causes an amino acid change (from tyrosine to cysteine) at position 163, promoting substrate formation that bind pocket of the BADH enzyme. Further, a TaqMan marker *AroLuff* perfectly co-segregated with aroma phenotypes, supporting Mendelian's simple recessive inheritance. *AroLuff* is equally effective in improving ridge gourd aroma through interspecific crosses, thus it may be useful to breed aromatic fruit in both species (Saensuk et al., 2022). Monogenic recessive and a digenic recessive genes, respectively, regulate cluster fruit-bearing habit and hermaphrodite sex form in *Luffa* (Bhutia et al., 2023).

13.3.6 Wax Gourd (or Winter Melon)

It is commonly consumed vegetable in Asian countries due to its nutritional and medicinal significance. Also, 2-acetyl-1-pyrroline (2AP), like in other crops, causes a pandan-like aroma in winter melon. *BhAMADH* regulates the pandan-like aroma in winter melon. A sequence comparison of *BhAMADH* between the aromatic and non-aromatic accessions revealed an 804-bp deletion encompassing exons 11–13 in the aromatic accession. A functional marker based on the 804-bp deletion perfectly co-segregate, in biparental population, with aroma phenotypes, suggesting that the *BhAMADH* a major gene and functional marker could be efficiently used in breeding programs to select for the pandan-like aroma in winter melon (Ruangnam et al., 2017). Fruit and seed size are strongly correlated. A single gene, *BhLS*, regulates fruit/seed size, with large size being dominant over small. *BhLS* is delimited within a 47.098-kb physical interval on chromosome 10, containing only one gene, *Bhi10M000649*, and showing homology to *Arabidopsis HOOKLESS1*. A nonsynonymous mutation (G to C) in the second exon of *Bhi10M000649* is significantly associated with both fruit and seed size variation in wax gourd (Yan et al., 2024).

The mature fruit cuticular wax (MFCW) is an important trait that affects the commercial value of wax gourd fruit. A single dominant gene, *BhWAX*, on chromosome 9 regulates MFCW in the wax gourd fruit. Fine mapping delimited *BhWAX* to a 0.5 Mb region containing 12 genes, with *Bhi09G001428* encoding a membrane bound O-acyltransferase as the putative gene. CAPS marker discriminated waxy and nonwaxy fruit with 80% accuracy (Yan et al., 2022). Contrarily, *BhSAUR60*, a *SAUR* family gene, is highly expressed in mature fruit cuticular wax (Luo et al., 2022). *BhiPRR6* interacts with multiple genes to regulate the absorption of light signals and thereby changes the pericarp color and synthesis of flavonoids in wax gourd fruit (Xie et al., 2022). *BhAPRR2* expression at 20 days after pollination in the black peel accession was significantly higher than in the green-, light-green-, and white-peel lines. This suggests that the color difference between green-, black-, light-green-, and white-peeled wax gourds is caused by its allelic variation, and the color difference between black and light-green peels is caused by the difference in *BhAPRR2* expression. The accuracy of two InDel markers in detecting wax gourd peel color were close to 98% (Nong et al., 2023).

Fruit shape and peel color influence quality of wax gourd fruit. The InDel markers (GX1-GX9) linked with the fruit-shape gene (*Bch02G016830*) had 84–91% accuracy, while those with the peel-color gene (*Bch05G003950*) show 56–90% and 100% accuracy with 27 (PS1-PS27) and 6 pairs (PS28-PS33) of peel-color InDel markers, respectively. Furthermore, the purity of hybrid cultivars as determined by using GX1, GX2, PS13, and PS14 was consistent with the field evaluation (Huang et al., 2022a).

13.3.7 Zucchini

Peel color is one of the most important quality traits, affecting nutritional quality and consumer preference, controlled by the *Y* locus. Genetic analysis revealed that yellow peel is controlled by a single dominant genetic factor, *Y* mapped in a ~170 kb region on chromosome 10, with *CpCHLH* regarded as a promising candidate gene. *CpCHLH* encodes a magnesium chelatase H subunit involved in chlorophyll biosynthesis. Mutation in it resulted in a reduction in chlorophyll content and yellow phenotype (Niu et al., 2023). Fruit rind color is another important agronomic trait affecting commodity quality and consumer choice in zucchini. *CpW*, a dominant locus, regulates white immature fruit rind color. *CpW* locus was mapped to a 100.4 kb region on chromosome 5 and further narrowed down the candidate region to 37.5 kb, with *CpAPRR2*, a promising candidate gene. Significantly higher expression level of *CpAPRR2* was noted in dark green rind than that in white rind, induced by light. Allelic diversity analysis in zucchini germplasm resources revealed that rind color is associated with the deletion of *CpAPRR2*. NILs with dark green and white rind revealed that genes involved in photosynthesis and porphyrin metabolism pathways were enriched in dark green than while rind lines (Ding et al., 2024).

TABLE 13.1
Genes and Markers Associated with Aroma and Fruit Morphotypes in the Cucurbitaceae Family

Gene	Phenotype	Reference
Bitter gourd (*Momordica charantia*)		
Fwa (*MC04g1399*)	Fruit wart	Liu et al. (2024)
McAPRR2	Fruit pericarp color	Tian et al. (2023b)
McEPFL2	Continuous ridges on fruit skin	Yang et al. (2022b)
Bottle gourd (*Lagenaria siceraria*)		
LsAPRR2	Green color fruit pericarp	Huo et al. (2023)
LsOVATE1	Fruit shape	Feng et al. (2022)

(Continued)

Table 13.1 ***(Continued)***

Genes and Markers Associated with Aroma and Fruit Morphotypes in the Cucurbitaceae Family

Gene	Phenotype	Reference
Qfs.zaas-6	Fruit shape	Xu et al. (2021)
QBt.1, QBt.1	Fruitt bitterness	Wu et al. (2019)
Cucumber (*Cucumus sativus*)		
Csa4G337340	Fruit length	Xing et al. (2023)
CsSF4	Short fruit length	Zhang et al. (2023)
*CsCRC^G^*	Fruit elongation	Che et al. (2023)
CsKTN1	Fruit elongation	Wang et al. (2021)
CsTRM5	Fruit shape	Xie et al. (2023)
fa4.1, fa6.1	Fruit apex shape	Zhu et al. (2022)
FS5.2	Fruit size and shape	Pan et al. (2022a)
CsARC6	Fruit color (white)	Sun et al. (2023)
CsGATA1 (CsPC1)	Pericarp color (fruit)	Huang et al. (2022b)
CsMYB60	Mature fruit skin color	Xu et al. (2024)
Chr3CG52930; Chr3CG53640	Light-green and white fruit skin color	Kishore et al. (2021a)
Csa3G904140	Immature fruit color	Tang et al. (2018)
w	Immature white fruit color	Liu et al. (2015)
CsDULL (D)	Dull fruit appearance	Zhai et al. (2022)
CsGLF1	Fruit glossiness	Yang et al. (2022a)
CsCER6; CsCER7	Fruit cuticular wax and fruit glossiness	Liu et al. (2023)
H and *Rs*	Fruit skin netting	Zhang et al. (2022)
CsALMT2	Fruit hollowness	Zhou et al. (2022)
qol8–2.1	Fruit aroma	Sun et al. (2022)
CsaV3_6G040750	Orange endocarp	Kishor et al. (2021b)
CsYUC10b	Straight fruits	Li et al. (2020)
Csa5G021320 (qgf5.1)	Green flesh fruit	Bo et al. (2019a)
Csa6G421750	Fruit spine density	Bo et al. (2019b)
fgr	Fruit fragrance	
Pumpkin (*Cucurbita maxima*)		
CmaKNAT6	Regulates dark green peel color	Wang et al. (2024)
SNPs	Carotenoid biosynthesis and transport, fruit ripening	Reddy et al. (2023)
PAL, C4H, 4CL, CSE, HCT, CAD, CCoAOMT	Flavonoids and coumarins accumulation	Chen et al. (2023)
KASPs	Tightly linked with the Atlantic Giant phenotype	Pan et al. (2022b)
Ridge gourd (*Luffa acutangula*)		
LcBADH	Fruit aroma	Saensuk et al. (2022)
Sponge gourd (*Luffa cylindrica*)		
Loss of function mutation (recessive)	Fruit aroma	Chaubey et al. (2022)

(Continued)

Table 13.1 ***(Continued)***
Genes and Markers Associated with Aroma and Fruit Morphotypes in the Cucurbitaceae Family

Gene	Phenotype	Reference
Wax gourd (*Benincasa hispida*)		
BhAPRR2	Fruit peel color	Nong et al. (2023)
InDels	Fruit-shape and peel-color genes	Huang et al. (2022a)
BhSAUR60	Fruit development	Luo et al. (2022)
Bhi12M000742	Flavonoid and light-color pericarp	Xie et al. (2022)
BhWAX	Mature fruit cuticular wax	Yan et al. (2022)
BFS	Fruit shape	Cheng et al. (2021)
BhSSC2.1	Soluble solid content	Wu et al. (2022)
Core genes	Taste or nutrition	Xue et al. (2022)
Zucchini (*Cucurbita pepo*)		
Cpgp	Controls green fruit color	Gebretsadik et al. (2024)
CpAPRR2	Regulates immature fruit rind color	Ding et al. (2024)
CpCHLH	Yello fruit color	Niu et al. (2023)

13.4 GERMPLASM DEVELOPMENT WITH HORTICULTURAL AND SENSORY TRAITS

The size and shape of edible plant organs are the primary determinants of consumer preference for many vegetable crops. The shape is also associated with many aspects of marketing, ease in mechanical harvest, and postharvest processing. Both genetics and human intervention have played a significant role in creating phenotypic diversity in edible plant organ shape. Selecting for visually pleasing shapes in vegetable crops has been achieved by modifying supernumerary cambia, allelic variation at loci that control cell division, cell elongation, transposon-mediated variation, and photosynthate partitioning. Genes regulating cell division patterning are probably universal shape regulators in horticultural crops. Assessing and exploiting such variation in gene pools, including wild and weedy relatives, and newly emergent information about key genes and proteins regulating variation in edible organ size and shape may provide additional opportunities to modify the form and contour of vegetable crops (Goldman et al., 2023).

To date, several Cucurbitaceae genomes, including bitter gourd, bottle gourd, cucumber, pumpkin, and various gourds (ridge, sponge, and wax gourd) have been assembled (Ma et al., 2022), which has provided new impetus to studying Cucurbitaceae genome, understanding the biological mechanisms, and improving the nutritional quality of Cucurbitaceae crops at the molecular level. Advances in next-generation genome sequencing and breeding techniques, together with bioinformatics algorithms, have greatly facilitated unlocking crop diversity, and associated differences at molecular levels for rapid genetic gains. Further, the integration of multi-omics data with genetic and phenotypic data will accelerate discovery and functional characterization of genes associated with agriculturally beneficial traits, which may be deployed in the improvement of Cucurbitaceae crops. Furthermore, advances in genetic transformation and genome editing of Cucurbitaceae plants may provide new avenues with significant development potential to obtain genetic variants with ideal traits. The resequencing of 56 artificially selected cucumber inbred lines with distinct phenotypes, using highly informative SNPs, grouped these lines into two classes with distinct genetic structure, significantly different from 115 previously reported cucumber lines. Thus, an excellent genetic (56 inbred lines) and genomic (highly informative SNPs) resource for use in mapping and breeding of cucumber germplasm with specific characteristics is possible (Liu et al., 2019).

The establishment of mutant resource database (via reverse genetics) may provide additional source of variation to study gene function, for which either exists little or no variability in each crop gene pool. The *C. pepo* EMS mutant platform consists of 3,751 independent M_2 families differing in plant vigor, leaf size and shape, sex expression and determination, and fruit set and development. The predominant EMS-induced mutations were C-to-T and G-to-A transitions, randomly distributed along the 20 chromosomes of *C. pepo* (García et al., 2018). Thus, establishing such mutant database (platform) are ideal resource to discover novel alleles for functional genomics and breeding of *Cucurbita* via direct- or reverse-genetic approaches.

Improving fruit quality of squash is resource-intensive, especially the space occupied by breeding population, and the biology of squash makes it difficult to practice selection on both parents. These restrictions translate to opting for smaller breeding populations and limited use of greenhouse generations, which, in turn, limit genetic gain per breeding cycle and increases cycle length. Genomic selection (GS) is a promising approach to enhance crop-breeding efficiency, including horticultural crops. The predictive ability of whole-genome models for fruit quality traits is low to moderate, but sufficient for implementation. Three rounds of genomic recurrent selection in a butternut squash population based on a fruit quality index, derived from a multi-trait genomic selection model, revealed significant improvement in fruit quality index value and changes in correlated traits (Hernandez et al., 2020). GS uses genome-wide markers to predict genomic estimated breeding value (GEBV). The key steps involved in GS include training population, predicting GEBV, and selection of favorable genotypes based on predicted GEBV. Marker diversity and density, the size and composition of training population, number of QTL, and heritability affect GS accuracies (Gidamo, 2022). Thus, a promising strategy to realize rapid genetic gain per unit time for complex traits with low heritability.

In addition, conventional crossing and selection alone or in combination with genomic assisted breeding led to the development and release of advanced breeding lines and cultivars in the Cucurbitaceae family (Mkhize et al., 2021). Interspecific hybridization between sponge gourd (*Luffa cylindrica*) and satputia (*L. hermaphrodita*) resulted in isolation of segregants combining high productivity, bisexual flowers, and cluster-bearing fruit (Sidhu and Kaur, 2021). Thus, satputia is a potential resource for genetics and mapping of agronomic traits in *Luffa*. The World Vegetable Center (WorldVeg; https://avrdc.org/) in partnerships with public-private program has developed several improved germplasm lines in bitter gourd, tropical pumpkin, ridge gourd, and sponge gourd. Concerted efforts are on toward broadening the genetic base of these crops by bringing unexploited landraces to diversify gene pools for enhanced yield, quality, and stress tolerance. The institute then organizes crop-specific field days for selection by partners, both private and public sectors, for subsequent use in local institutions breeding programs (Dhillon et al., 2020).

Traditional breeding has been the predominant approach to develop improved germplasm/cultivars. Mutation in *DMP* homologs, a highly conserved gene in monocots and dicots, has been used to induce *in vivo* haploid and produce double haploids (DHs) in several crops (Li et al., 2022; Tian et al., 2023b; Zhong et al., 2022). In vivo maternal haploid production technology using *csdmp* mutant has been recently perfected to produce haploids and double haploids in major cucurbit crops (Yin et al., 2024), raising the prospects for cucurbit breeding. More recently, parthenogenic embryo technique is perfected to produce haploids via pollination with gamma-irradiated (400 to 500 Gy doses) pollen with optimum timeframe for embryo rescue 25 to 30 days after pollination in cucumber. This technique will streamline the breeding process, enhance genetic gain, and produce superior cultivars (Pradeepkumara et al., 2024).

13.5 CONCLUSIONS

Both genetics and human intervention have played a significant role in shaping phenotypic diversity in edible plant organ shape. The advances in genome sequencing, bioinformatics algorithms, genetic transformation, and genome editing of Cucurbitaceae plants may provide a new impetus to assess and exploit functional diversity and unlock the biological mechanisms to improve nutritional quality at the molecular level. Allelic variation at loci that control cell division, cell elongation, and photosynthate partitioning and modeling the effect of genomic selection may provide an additional opportunity to modify significant improvement in quality of edible plant organs. In vivo maternal production of haploids and doubled haploids has been perfected to use in cucurbit breeding programs.

REFERENCES

Alhariri, A., Behera, T.K., Jat, G.S., Devi, M.B., Boopalakrishnan, G., Hemeda, N.F., et al. (2021). Analysis of genetic diversity and population structure in bitter gourd (*Momordica charantia* L.) using morphological and SSR markers. *Plants (Basel)* 10, 1860. https://doi.org/10.3390/plants10091860

Arikit, S., Yoshihashi, T., Wanchana, S., Tanya, P., Juwattanasomran, T., Srinives, P., et al. (2011). A PCR-based marker for a locus conferring aroma in vegetable soybean (*Glycine max* L.). *Theor. Appl. Genet.* 122, 311–316. https://doi.org/10.1007/s00122-010-1446-y

Ayres, E.M.M., Lee, S.M., Boyden, L., and Guinard, J.-X. (2019). Sensory properties and consumer acceptance of cantaloupe melon cultivars. *J. Food Sci.* 84, 2278–2288. https://doi.org/10.1111/1750–3841.14724

Bhutia, N.D., Sureja, A.K., Verma, M., Krishnan, G., Arya, L., Bhardwaj, R., et al. (2023). Inheritance and molecular mapping of solitary/cluster fruit-bearing habit in Luffa. *Mol. Genet. Genom.* 298, 995–1006. https://doi.org/10.1007/s00438-023-02034-0

Bo, K., Miao, H., Wang, M., Xie, X., Song, Z., Xie, Q., et al. (2019b). Novel loci fsd6.1 and Csgl3 regulate ultra-high fruit spine density in cucumber. *Theor. Appl. Genet.* 132, 27–40. https://doi.org/10.1007/s00122-018-3191-6

Bo, K., Wei, S., Wang, W., Miao, H., Dong, S., Zhang, S., et al. (2019a). QTL mapping and genome-wide association study reveal two novel loci associated with green flesh color in cucumber. *BMC Plant Biol.* 19, 243. https://doi.org/10.1186/s12870-019-1835-6

Chaubey, T., Sagar, V., Singh, R.K., Chanotiya, C.S., Pandey, S., Singh, P.M., et al. (2022). Volatile compounds governed by single recessive gene impart aroma in sponge gourd (*Luffa cylindrica* L. Roem). *Plants (Basel)* 11, 2881. https://doi.org/10.3390/plants11212881

Che, G., Pan, Y., Liu, X., Li, M., Zhao, J., Yan, S., et al. (2023). Natural variation in *CRABS CLAW* contributes to fruit length divergence in cucumber. *Plant Cell* 35, 738–755. https://doi.org/10.1093/plcell/koac335

Chen, C., Xu, L., Xia, W., Qi, C., Liu, J., Hua, B., et al. (2023). Metabolomics and transcription profiling of pumpkin fruit reveals enhanced bioactive flavonoids and coumarins in giant pumpkin (*Cucurbita maxima*). *J. Agric. Food Chem.* 71, 27, 10459–10469. https://doi.org/10.1021/acs.jafc.3c01883

Chen, L., Duan, L., Sun, M., Yang, Z., Li, H., Hu, K., et al. (2022). Current trends and insights on EMS mutagenesis application to studies on plant abiotic stress tolerance and development. *Front. Plant Sci.* 13, 1052569. https://doi.org/10.3389/fpls.2022.1052569

Cheng, Z., Liu, Z., Xu, Y., Ma, L., Chen, J., and Gou, J. (2021). Fine mapping and identification of the candidate gene *BFS* for fruit shape in wax gourd (*Benincasa hispida*). *Theor. Appl. Genet.* 134, 3983–3995. https://doi.org/10.1007/s00122-021-03942-8

Chomicki, G., Schaefer, H., and Renner, S.S. (2020). Origin and domestication of Cucurbitaceae crops: Insights from phylogenies, genomics and archaeology. *New Phytol.* 226, 1240–1255. https://doi.org/10.1111/nph.16015

Colle, M., Weng, Y., Kang, Y., Ophir, R., Sherman, A., and Grumet, R. (2017). Variation in cucumber (*Cucumis sativus* L.) fruit size and shape results from multiple components acting pre-anthesis and post-pollination. *Planta* 246, 641–658. https://doi.org/10.1007/s00425-017-2721-9

Cui, J., Zhou, Y., Zhong, J., Feng, C.C., Hong, Y., Hu, K.L., et al. (2022). Genetic diversity among a collection of bittergourd (*Momordica charantia* L.) cultivars. *Genet. Resour. Crop Evol.* 69, 729–735. https://doi.org/10.1007/s10722-021-01258-6

Dhillon, N.P.S., Laenoi, S., Srimat, S., Pruangwitayakun, S., Mallappa, A., Kapur, A., et al. (2020). Sustainable cucurbit breeding and production in Asia using public–private partnerships by the World Vegetable Center. *Agronomy* 10, 1171. https://doi.org/10.3390/agronomy10081171

Dhillon, N.P.S., Sanguansil, S., Singh, S.P., Masud, M.A.T., Kumar, P., Bharathi, L.K., et al. (2017). Gourds: Bitter, bottle, wax, snake, sponge and ridge. In: Grumet, R., et al. (eds.) *Genetics and Genomics of Cucurbitaceae, Plant Genetics and Genomics: Crops and Models.* Springer International Publishing. https://doi.org/10.1007/7397_2016_24

Ding, W., Luo, Y., Li, W., Chen, F., Wang, C., Xu, W., et al. (2024). Fine mapping and transcriptome profiling reveal *CpAPRR2* to modulate immature fruit rind color formation in zucchini (*Cucurbita pepo*). *Theor. Appl. Genet.* 137, 167. https://doi.org/10.1007/s00122-024-04676-z

Fang, H., Huang, S., Li, R., Wang, P., Jiang, Q., Zhong, C., et al. (2024). Combined BSA-Seq and RNA-Seq to identify potential genes regulating fruit size in bottle gourd (*Lagenaria siceraria* L.). *Plants (Basel)* 13, 2154. https://doi.org/10.3390/plants13152154

Feng, Z., Wu, X., Wang, J., Wu, X., Wang, B., Lu, Z., et al. (2022). Identification of bottle gourd (*Lagenaria siceraria*) *OVATE* family genes and functional characterization of *LsOVATE1*. *Biomolecules* 13, 85. https://doi.org/10.3390/biom13010085

Ferriol, M., and Picó, B. (2008). Pumpkin and winter squash. In: Prohens, J., and Nuez, F. (eds.) *Vegetables I. Handbook of Plant Breeding*, Vol. 1. Springer, New York. https://doi.org/10.1007/978-0-387-30443-4_10

García, A., Aguado, E., Parra, G., Manzano, S., Martínez, C., Megías, Z., et al. (2018). Phenomic and genomic characterization of a mutant platform in *Cucurbita pepo*. *Front. Plant Sci.* 9, 1049. https://doi.org/10.3389/fpls.2018.01049

Gebretsadik, K., Chen, W., Duan, Y., Sun, Y., He, Y., Liao, Q., et al. (2024). Map-based cloning reveals *Cpgp* gene encoding an APRR2 protein to regulate the green fruit peel formation in *Cucurbita pepo*. *Mol. Breed.* 44, 53. https://doi.org/10.1007/s11032-024-01492-7

Gidamo, G.H. (2022). Genomic selection: A faster strategy for plant breeding. In: Wang, H. (ed.) *Case Studies of Breeding Strategies in Major Plant Species*, pp. 1–15. https://doi.org/10.5772/intechopen.105398

Goldman, I.L., Wang, Y., Alfaro, A.V., Brainard, S., Oravec, M.W., McGregor, C.E., et al. (2023). Form and contour: Breeding and genetics of organ shape from wild relatives to modern vegetable crops. *Front. Plant Sci.* 14, 1257707. https://doi.org/10.3389/fpls.2023.1257707

Hernandez, C.O., Labate, J., Reitsma, K., Fabrizio, J., Bao, K., Fei, Z., et al. (2023). Characterization of the USDA *Cucurbita pepo*, *C. moschata*, and *C. maxima* germplasm collections. *Front. Plant Sci.* 14, 1130814. https://doi.org/10.3389/fpls.2023.1130814

Hernandez, C.O., Wyatt, L.E., and Mazourek, M.R. (2020). Genomic prediction and selection for fruit traits in winter squash. *G3 (Bethesda)* 10, 3601–3610. https://doi.org/10.1534/g3.120.401215

Huang, H., Yang, Q., Zhang, L., Kong, W., Wang, H., Wei, A., et al. (2022b). Genome-wide association analysis reveals a novel QTL *CsPC1* for pericarp color in cucumber. *BMC Genom.* 23, 383. https://doi.org/10.1186/s12864-022-08606-5

Huang, X., Wu, W., Su, L., Lv, H., Cheng, Z., Yang, W., et al. (2022a). Development and application of InDel markers linked to fruit-shape and peel-colour genes in wax gourd. *Genes* 13, 1567. https://doi.org/10.3390/genes13091567

Huo, Y., Zhang, G., Yu, W., Liu, Z., Shen, M., Zhao, R., et al. (2023). Forward genetic studies reveal *LsAPRR2* as a key gene in regulating the green color of pericarp in bottle gourd (*Lagenaria siceraria*). *Front. Plant Sci.* 14, 1130669. https://doi.org/10.3389/fpls.2023.1130669

Jo, H.E., Song, K., Kim, J.-G., and Lee, C.H. (2022). Non-targeted metabolomic analysis for the comparative evaluation of volatile organic compounds in 20 globally representative cucumber lines. *Front. Plant Sci.* 13, 1028735. https://doi.org/10.3389/fpls.2022.1028735

Joshi, B.K., Shrestha, S., Adhikri, B., and Bhattarai, M. (2020). Traditional practices and genetic diversity on chayote landraces and their conservation. *Nat. Res. Sustain. Dev.* 10, 272–288. https://doi.org/10.31924/nrsd.v10i2.060

Junxing, L.I., Aiqing, M., Gangjun, Z., Xiaoxi, L., Haibin, W., Jianning, L., et al. (2022). Assessment of the 'taro-like' aroma of pumpkin fruit (*Cucurbita moschata* D.) via E-nose, GC-MS and GC-O analysis. *Food Chem. X.* 15, 100435. https://doi.org/10.1016/j.fochx.2022.100435

Kishor, D.S., Lee, H.-Y., Alavilli, H., You, C.-R., Kim, J.-G., Lee, S.-Y., et al. (2021b). Identification of an allelic variant of the *CsOr* gene controlling fruit endocarp color in cucumber (*Cucumis sativus* L.) using genotyping-by-sequencing (GBS) and whole-genome sequencing. *Front. Plant Sci.* 12, 802864. https://doi.org/10.3389/fpls.2021.802864

Kishore, D.S., Alavilli, H., Lee, S.-C., Kim, J.-G., and Song, K. (2021a). Development of SNP markers for white immature fruit skin color in cucumber (*Cucumis sativus* L.)

using QTL-seq and marker analyses. *Plants (Basel)* 10, 2341. https://doi.org/10.3390/plants10112341

Kumari, S.A.S.M., Nakandala, N.D.U.S., Nawanjana, P.W.I., Rathnayake, R.M.S.K., Senavirathna, T.M.T.N., Senevirathina, R.W.K.M., et al. (2019). The establishment of the species-delimits and varietal-identities of the cultivated germplasm of *Luffa acutangula* and *Luffa aegyptiaca* in Sri Lanka using morphometric, organoleptic and phylogenetic approaches. *PLoS ONE* 14, e0215176. https://doi.org/10.1371/journal.pone.0215176

Li, J., Gu, C., Yuan, Y., Gao, Z., Qin, Z., and Xin, M. (2024). Comparative transcriptome analysis revealed that auxin and cell wall biosynthesis play important roles in the formation of hollow hearts in cucumber. *BMC Genom.* 25, 36. https://doi.org/10.1186/s12864-024-09957-x

Li, S., Wang, C., Zhou, X., Liu, D., Liu, C., Luan, J., et al. (2020). The curvature of cucumber fruits is associated with spatial variation in auxin accumulation and expression of a *YUCCA* biosynthesis gene. *Hortic. Res.* 7, 135. https://doi.org/10.1038/s41438-020-00354-5

Li, Y., Li, D., Xiao, Q., Wang, H., Wen, J., Tu, J., et al. (2022). An in planta haploid induction system in Brassica napus. *J. Integr. Plant Biol.* 64, 1140–1144. https://doi.org/10.1111/jipb.13270

Liu, B., Guan, D., Zhai, X., Yang, S., Chen, S., Huang, J., et al. (2019). Selection footprints reflect genomic changes associated with breeding efforts in 56 cucumber inbred lines. *Hortic. Res.* 6, 127. https://doi.org/10.1038/s41438-019-0209-4

Liu, H., Meng, H., Pan, Y., Liang, X., Jiao, J., Li, Y., et al. (2015). Fine genetic mapping of the white immature fruit color gene w to a 33.0-kb region in cucumber (*Cucumis sativus* L.). 128, 2375–2385. https://doi.org/10.1007/s00122-015-2592-z

Liu, J., Cui, J., Dong, J., Zhong, J., Zhong, C., Yuan, F., et al. (2024). A 1-bp deletion in the *MC04g1399* is highly associated with failure to produce fruit wart in bitter gourd. *Hortic. Plant J.* 10, 171–180. https://doi.org/10.1016/j.hpj.2022.12.017

Liu, X., Ge, X., An, J., Liu, X., and Ren, H. (2023). *CsCER6* and *CsCER7* influence fruit glossiness by regulating fruit cuticular wax accumulation in cucumber. *Int. J. Mol. Sci.* 24, 1135. https://doi.org/10.3390/ijms24021135

Liu, X., Pan, Y., Liu, C., Ding, Y., Wang, X., Cheng, Z., et al. (2020). Cucumber fruit size and shape variations explored from the aspects of morphology, histology, and endogenous hormones. *Plants* 9, 772. https://doi.org/10.3390/plants9060772

Luo, C., Yan, J., He, C., Liu, W., Xie, D., and Jiang, B. (2022). Genome-wide identification of the *SAUR* gene family in wax gourd (*Benincasa hispida*) and functional characterization of *BhSAUR60* during fruit development. *Int. J. Mol. Sci.* 23, 14021. https://doi.org/10.3390/ijms232214021

Ma, L., Wang, Q., Zheng, Y., Guo, J., Yuan, S., Fu, A., et al. (2022). Cucurbitaceae genome evolution, gene function and molecular breeding. *Hortic. Res.* 9, uhab057. https://doi.org/10.1093/hr/uhab057

Mahapatra, S., Sureja, A.K., Behera, T.K., and Verma, M. (2022). Assessment of genetic diversity of ninety-one bottle gourd [Lagenaria siceraria (Mol.) Standl.] genotypes from fourteen different agro-climatic zones of India using agro-morphological traits and SSR markers. *Mol. Biol. Rep.* 49, 6367–6383. https://doi.org/10.1007/s11033-022-07446-6

Mallikarjuna, K.N., Tomar, B.S., Mangal, M., Singh, N., Singh, D., Kumar, S., et al. (2023). Genetic diversity and population structure analyses in bitter gourd (*Momordica charantia* L.) based on agro-morphological and microsatellite markers. *Plants (Basel)* 12, 3512. https://doi.org/10.3390/plants12193512

Mashilo, J., Shimelis, H., and Odindo, A.O. (2015). Genetic diversity of bottle gourd (*Lagenaria siceraria* (Molina) Standl.) landraces of South Africa assessed by morphological traits and simple sequence repeat markers. *S. Afr. J. Plant Soil.* 33. https://doi.org/10.1080/02571862.2015.1090024

Min, K., Song, K., Lim, S., Yi, G., and Lee, E.J. (2023). Cucurbitacin and volatile compound profiling reveals independent domestication of cucumber (*Cucumis sativus* L.) fruit. *Food Chem.* 405 Part B, 135006. https://doi.org/10.1016/j.foodchem.2022.135006

Mkhize, P., Mashilo, J., and Shimelis, H. (2021). Progress on genetic improvement and analysis of bottle gourd [*Lagenaria siceraria* (Molina) Standl.] for agronomic traits, nutrient compositions, and stress tolerance: A review. *Front. Sustain. Food Syst.* 5, 683635. https://doi.org/10.3389/fsufs.2021.683635

Montero, M.L., Colonna, A.E., Gallardo, R.K., and Ross, C.F. (2024). Sensory profiling of pears from the Pacific Northwest: Consumers' perspective and descriptive analysis. *J. Food Sci.* 89, 1225–1242. https://doi.org/10.1111/1750-3841.16913

Moreno, S.R., Sims, C.A., Odabasi, A., Simmone, A., Gao, Z., Chase, C.A., et al. (2023). Chemical and physical properties of winter squash and their correlation with liking of their sensory attributes. *J. Food Sci.* 88, 4440–4456. https://doi.org/10.1111/1750-3841.16771

Niu, J., Chen, Q., Lu, X., Wang, X., Tang, Z., Liu, Q., et al. (2023). Fine mapping and identifying candidate gene of *Y* underlying yellow peel in *Cucurbita pepo*. *Front. Plant Sci.* 14, 1159937. https://doi.org/10.3389/fpls.2023.1159937

Nong, L., Wang, P., Yang, W., Liu, T., Su, L., Cheng, Z., et al. (2023). Analysis of *BhAPRR2* allele variation, chlorophyll content, and chloroplast structure of different peel colour varieties of wax gourd (*Benincasa hispida*) and development of molecular markers. *Euphytica* 219, 107. https://doi.org/10.1007/s10681-023-03233-x

Oliver, P., Cicerale, S., Pang, E., and Keast, R. (2018). Identifying key flavors in strawberries driving liking via internal and external preference mapping. *J. Food Sci.* 83, 1073–1083. https://doi.org/10.1111/1750-3841.14109

Pan, L., Wang, M., Yang, Y., Chen, C., Dai, H., Zhang, Z., et al. (2022b). Whole-genome resequencing identified QTLs, candidate genes and Kompetitive Allele-Specific PCR markers associated with the large fruit of Atlantic Giant (*Cucurbita maxima*). *Front. Plant Sci.* 13, 942004. https://doi.org/10.3389/fpls.2022.942004

Pan, Y., Chen, B., Qiao, L., Chen, F., Zhao, J., Cheng, Z., et al., (2022a). Phenotypic characterization and fine mapping of a major-effect fruit shape QTL FS5.2 in cucumber, *Cucumis sativus* L., with near-isogenic line-derived segregating populations. *Int. J. Mol. Sci.* 23, 13384. https://doi.org/10.3390/ijms232113384

Paris, H.S. (2001). History of the cultivar-groups of *Cucurbita pepo*. *Hortic. Rev.* 25, 71–170. https://doi.org/10.1007/BF02935916

Paris, H.S., and Brown, R.N. (2005). The genes of pumpkin and squash. *HortScience* 40, 1620–1630.

Paris, H.S., Lebeda, A., Křistkova, E., Andres, T.C., and Nee, M.H. (2012). Parallel evolution under domestication and phenotypic differentiation of the cultivated subspecies of *Cucurbita pepo* (Cucurbitaceae). *Econ. Bot.* 66, 71–90. https://doi.org/10.1007/s12231-012-9186-3

Pradeepkumara, N., Prakash, C., Bhatia, R., Munshi, A.D., Rao, M., Subhasmita, S., et al. (2024). An efficient and easy-to-use protocol for induction of haploids in cucumber through parthenogenic embryo development. *Bot. Stud.* 65, 30. https://doi.org/10.1186/s40529-024-00436-w

Pu, Y.T., Luo, Q., Wen, L.-H., Li, Y.-R., Meng, P.-H., Wang, X.-J., et al. (2021). Origin, evolution, breeding, and omics of chayote, an important Cucurbitaceae vegetable crop. *Front. Plant Sci.* 12, 739091. https://doi.org/10.3389/fpls.2021.739091

Reddy, U.K., Lopez-Ortiz, C., Talavera-Caro, A.G., Natarajan, P., Tomason, Y., Alaparthi, S., et al. (2023). GWAS resolves molecular mechanisms underlying natural variation for carotenoids in *Cucurbita maxima* Duchesne. *Sci. Hortic.* 312, 111881. https://doi.org/10.1016/j.scienta.2023.111881

Ruangnam, S., Wanchana, S., Phoka, N., Saeansuk, C., Mahatheeranont, S., and de Hoop, S.J. (2017). A deletion of the gene encoding amino aldehyde dehydrogenase enhances the 'pandan-like' aroma of winter melon (*Benincasa hispida*) and is a functional marker for the development of the aroma. *Theor. Appl. Genet.* 130, 2557–2565. https://doi.org/10.1007/s00122-017-2976-3

Saensuk, C., Ruanggnam, S., Pitaloka, M.K., Dumhai, R., Mahatheeranont, S., de Hoop, S.J., et al. (2022). A SNP of betaine aldehyde dehydrogenase (BADH) enhances an aroma (2-acetyl-1-pyrroline) in sponge gourd (*Luffa cylindrica*) and ridge gourd (*Luffa acutangula*). *Sci. Rep.* 12, 3718. https://doi.org/10.1038/s41598-022-07478-9

Savage, J.A., Haines, D.F., and Holbrook, N.M. (2015). The making of giant pumpkins: How selective breeding changed the phloem of *Cucurbita maxima* from source to sink. *Plant Cell Environ.* 38, 1543–1554. https://doi.org/10.1111/pce.12502

Sidhu, M.K., and Kaur, J. (2021). Interspecific hybridization between sponge gourd (*Luffa cylindrica* L.) and 'Satputia' (*Luffa hermaphrodita* Singh & Bhandari) for preintrogression of cluster bearing, high yield and gynoecism. *J. Genet.* 100, 73. https://doi.org/10.1007/s12041-021-01323-0

Sun, H., Wu, S., Zhang, G., Jiao, C., Guo, S., Ren, Y., et al. (2017). Karyotype stability and unbiased fractionation in the paleo-allotetraploid cucurbita genomes. *Mol. Plant* 10, 1293–1306. https://doi.org/10.1016/j.molp.2017.09.003

Sun, W., Li, X., Huang, H., Wei, J., Zheng, F., Huang, Y., et al. (2023). Mutation of *CsARC6* affects fruit color and increases fruit nutrition in cucumber. *Theor. Appl. Genet.* 136, 111. https://doi.org/10.1007/s00122-023-04337-7

Sun, Y., Li, X., Ma, Z., and Chen, S. (2022). Quantitative trait locus mapping of fruit aroma compounds in cucumber (*Cucumber sativus* L.) based on a recombinant inbred line population. *Hortic. Res.* 9, uhac151. https://doi.org/10.1093/hr/uhac151

Tang, H.-Y., Dong, X., Wang, J.-K., Xia, J.-H., Xie, F., Zhang, Y., et al (2018). Fine mapping and candidate gene prediction for white immature fruit skin in cucumber (*Cucumis sativus* L.). *Int. J. Mol. Sci.* 19, 1493. https://doi.org/10.3390/ijms19051493

Tian, S., Yang, J., Fu, Y., Zhang, X., Zhang, J., Zhao, H., et al. (2023a). *McAPRR2*: The key regulator of domesticated pericarp color in bitter gourd. *Plants (Basel)* 12, 3585. https://doi.org/10.3390/plants12203585

Tian, S., Zhang, J., Zhao, H., Zong, M., Li, M., Gong, G., et al. (2023b). Production of double haploid watermelon via maternal haploid induction. *Plant Biotechnol. J.* 21,1308–1310. https://doi.org/10.1111/pbi.14045

Tyagi, R., Bhardwaj, R., Suneja, P., Sureja, A.K., Munshi, A.D., Arya, L., et al. (2023). Harnessing sponge gourd: An alternative source of oil and protein for nutritional security. *Front. Nutr.* 10, 1158424. https://doi.org/10.3389/fnut.2023.1158424

Tyagi, R., Sharma, V., Sureja, A.K., Munshi, A.D., Arya, L., Saha, D., et al. (2020). Genetic diversity and population structure detection in sponge gourd (*Luffa cylindrica*) using ISSR, SCoT and morphological markers. *Physiol. Mol. Biol. Plants* 26, 119–131. https://doi.org/10.1007/s12298-019-00723-y

Wakte, K., Zanan, R., Hinge, V., Khandagale, K., Nadaf, A., and Henry, R. (2017). Thirty-three years of 2-acetyl-1-pyrroline, a principal basmati aroma compound in scented rice (*Oryza sativa* L.): A status review. *J. Sci. Food Agric.* 97, 384–395. https://doi.org/10.1002/jsfa.7875

Wang, C., Ding, W., Chen, F., Zhang, K., Hou, Y., Wang, G., et al. (2024). Mapping and transcriptomic profiling reveal that the KNAT6 gene is involved in the dark green peel colour of mature pumpkin fruit (*Cucurbita maxima* L.). *Theor. Appl. Genet.* 137, 225. https://doi.org/10.1007/s00122-024-04741-7

Wang, H., Sun, J., Yang, F., Weng, Y., Chen, P., Du, S., et al. (2021). *CsKTN1* for a katanin p60 subunit is associated with the regulation of fruit elongation in cucumber (*Cucumis sativus* L.). *Theor. Appl. Genet.* 134, 2429–2441. https://doi.org/10.1007/s00122-021-03833-y

Wang, Y., Li, Y., Wu, X., Wu, X., Feng, Z., Wang, J., et al. (2022). Elucidation of the flavor aspects and flavor-associated genomic regions in bottle gourd (*Lagenaria siceraria*) by metabolomic analysis and QTL-seq. *Foods* 11, 2450. https://doi.org/10.3390/foods11162450

Williams, C., Weng, Y., and Du, X. (2022). Sensory profiles of 10 cucumber varieties using a panel trained with chemical references. *ACS Food Sci. Technol.* 2, 815–824. https://doi.org/10.1021/acsfoodscitech.1c00453

Wu, W., Wang, P., Huang, X., Su., L., Lv, H., Gou, J., et al. (2022). Fine mapping and functional analysis of major regulatory genes of soluble solids content in wax gourd (*Benincasa hispida*). *Int. J. Mol. Sci.* 23, 6999. https://doi.org/10.3390/ijms23136999

Wu, X., Wu, X., Wang, Y., Wang, B., Lu, Z., Xu, P., et al. (2019). Molecular genetic mapping of two complementary genes underpinning fruit bitterness in the bottle gourd (*Lagenaria siceraria* [Mol.] Standl.). *Front. Plant Sci.* 10, 1493. https://doi.org/10.3389/fpls.2019.01493

Xanthopoulou, A., Montero-Pau, J., Mellidou, I., Kissoudis, Blanca, J., Picó, B., et al. (2019). Whole-genome resequencing of *Cucurbita pepo* morphotypes to discover genomic variants associated with morphology and horticulturally valuable traits. *Hortic. Res.* 6, 94. https://doi.org/10.1038/s41438-019-0176-9

Xie, L., Wang, J., Liu, F., Zhou, H., Chen, Y., Pan, L., et al. (2022). Integrated analysis of multi-omics and fine-mapping reveals a candidate gene regulating pericarp color and flavonoids accumulation in wax gourd (*Benincasa hispida*). *Front. Plant Sci.* 13, 1019787. https://doi.org/10.3389/fpls.2022.1019787

Xie, Y., Liu, X., Sun, C., Song, X., Li, X., Cui, H., et al. (2023). *CsTRM5* regulates fruit shape via mediating cell division direction and cell expansion in cucumber. *Hortic. Res.* 10, uhad007. https://doi.org/10.1093/hr/uhad007

Xing, Y., Cao, Y., Ma, Y., Wang, F., Xin, S., and Zhu, W. (2023). QTL mapping and transcriptomic analysis of fruit length in cucumber. *Front. Plant Sci.* 14, 1208675. https://doi.org/10.3389/fpls.2023.1208675

Xu, P., Wang, Y., Sun, F., Wu, R., Du, H., Wang, Y., et al. (2021). Long-read genome assembly and genetic architecture of fruit shape in the bottle gourd. *Plant J.* 107, 956–968. https://doi.org/10.1111/tpj.15358

Xu, X., Zhu, Y., Yuan, Y., Sohail, H., He, S., Ye, Y., et al. (2024). R2R3-MYB transcription factor CsMYB60 controls mature fruit skin color by regulating flavonoid accumulation in cucumber. *Plant J.* 119, 796–813. https://doi.org/10.1111/tpj.16797

Xue, S., Wan, X., Lu, S., hong, Y., and Xie, D. (2022). A time-course transcriptome analysis of wax gourd fruit development reveals predominant genes regulating taste and nutrition. *Front. Plant Sci.* 13, 971274. https://doi.org/10.3389/fpls.2022.971274

Yan, J., Chen, F., Sun, P., Liu, W., Xie, D., Qian, Y., et al. (2022). Genome-wide association study and genetic mapping of BhWAX conferring mature fruit cuticular wax in wax gourd. *BMC Plant Biol.* 22, 539. https://doi.org/10.1186/s12870-022-03931-z

Yan, J., Cheng, J., Xie, D., Wang, Y., Wang, M., Yang, S., et al. (2024). A nonsynonymous mutation in BhLS, encoding an acyl-CoA N-acyltransferase leads to fruit and seed size variation in wax gourd (*Benincasa hispida*). *Theor. Appl. Genet.* 137, 100. https://doi.org/10.1007/s00122-024-04604-1

Yang, J., Weng, Y., Li, H., Kong, Q., Wang, W., Yan, C., et al. (2022b). *Epidermal Patterning Factor 2-like* (*McEPFL2*): A putative candidate for the continuous ridge (cr) fruit skin locus in bitter Gourd (*Momordica charantia* L.). *Gene (Basel)* 13, 1148. https://doi.org/10.3390/genes13071148

Yang, X., Zhang, W., He, H., Nie, J., Bie, B., Zhao, J., et al. (2014). Tuberculate fruit gene *Tu* encodes a C_2H_2 zinc finger protein that is required for the warty fruit phenotype in cucumber (*Cucumis sativus* L.). *Plant J.* 78, 1034–1046. https://doi.org/10.1111/tpj.12531

Yang, Y., Cai, C., Wang, Y., Wang, Y., Ju, H., Chen, X., et al. (2022a). Cucumber *glossy fruit 1* (*CsGLF1*) encodes the zinc finger protein 6 that regulates fruit glossiness by enhancing cuticular wax biosynthesis. *Hortic. Res.* 10, uhac237. https://doi.org/10.1093/hr/uhac237

Yin, S., Li, S., Sun, L., Shi, K., Fan, S., Liu, X., et al. (2024). Mutating the maternal haploid inducer gene *CsDMP* in cucumber produces haploids *in planta*. *Plant Physiol.* 194, 1282–1285. https://doi.org/10.1093/plphys/kiad600

Yu, R., Niu, Y., Wang, X., Yang, K., Han, X., Liu, Z., et al. (2022). Construction of a density mutant collection in bitter gourd *via* new germplasms innovation and gene functional study. *Front. Plant Sci.* 13, 1069750. https://doi.org/10.3389/fpls.2022.1069750

Zhai, X., Wu, H., Wang, Y., Zhang, Z., Shan, L., Zhao, X., et al. (2022). The fruit glossiness locus, *dull fruit* (*D*), encodes a C_2H_2-type zinc finger transcription factor, CsDULL, in cucumber (*Cucumis sativus* L.). *Hortic. Res.* 9, uhac146. https://doi.org/10.1093/hr/uhac146

Zhang, H., Wang, Y., Tan, J., and Weng, Y. (2022). Functional copy number variation of *CsSHINE1* is associated with fruit skin netting intensity in cucumber, *C. Sativus.* 135, 2101–2119. https://doi.org/10.1007/s00122-022-04100-4

Zhang, J., Feng, S., Yuan, J., Wang, C., Lu, T., Wang, H., et al. (2021). The formation of fruit quality in *Cucumis sativus* L. *Front. Plant Sci.* 12, 729448. https://doi.org/10.3389/fpls.2021.729448

Zhang, J., Gu, X., Yan, W., Lou, L., Xu, X., and Chen, X. (2022a). Characterization of differences in the composition and content of volatile compounds in cucumber fruit. *Foods* 1, 1101. https://doi.org/10.3390/foods11081101

Zhang, K., Yao, D., Chen, Y., Wen, H., Pan, J., Xiao, T., et al. (2023). Mapping and identification of *CsSF4*, a gene encoding a UDP-N-acetyl glucosamine-peptide N-acetylglucosaminyltransferase required for fruit elongation in cucumber (*Cucumis sativus* L.). *Theor. Appl. Genet.* 136, 54. https://doi.org/10.1007/s00122-023-04246-9

Zhang, T., Hong, Y., Zhang, X., Yuan, X., and Chen, S. (2022b). Relationship between key environmental factors and the architecture of fruit shape and size in near-isogenic lines of cucumber (*Cucumis sativus* L.). *Int. J. Mol. Sci.* 23, 14033. https://doi.org/10.3390/ijms232214033

Zhong, Y., Chen, B., Wang, D., Zhu, X., Li, M., Zhang, J., et al. (2022). In vivo maternal haploid induction in tomato. *Plant Biotechnol. J.* 20, 250–252. https://doi.org/10.1111/pbi.13755

Zhou, G., Chen, C., Liu, X., Yang, K., Wang, C., Lu, X., at al. (2022). The formation of hollow trait in cucumber (*Cucumis sativus* L.) fruit is controlled by *CsALMT2*. *Int. J. Mol. Sci.* 23, 6173. https://doi.org/10.3390/ijms23116173

Zhu, P., Meng, Y., Zhang, K., Wang, X., Liang, K., Wang, T., et al. (2022). Mapping of fruit apex shape related QTLs across multi-genetic backgrounds in cucumber (*Cucumis sativus* L.). *Hortic. Plant J.* 8, 328–340. https://doi.org/10.1016/j.hpj.2021.12.001

Zia, B., Chanda, B., Bai, J., Gilliard, A., and Ling, K.-S. (2023). Comparative evaluation of volatile organic compounds in two bottle gourd accessions with distinct fruit shapes. *Foods* 12, 3921. https://doi.org/10.3390/foods12213921

14 Genetic and Molecular Basis of Horticultural and Sensory Traits in Melon Species

Sangam L. Dwivedi and Autar K. Mattoo

14.1 INTRODUCTION

The Genesys system (https://www.genesys-pgr.org; accessed on March 07, 2024) contains 9986 and 6836 accessions of melon (*Cucumis melo*) and watermelon (*Citrullus lanatus*) genetic resources, respectively. Other countries holding significant germplasm repositories of these crops are China, India, Japan, Korea, Russia, Uzbekistan, and Taiwan (World Vegetable Centre) (Grumet et al., 2021). India is the primary center of diversity of the melon. It is genetically distinct from African germplasm, supporting independent domestication of the melon in Asia and Africa (Endl et al., 2018; Gonzalo et al., 2019). Melons are widely grown worldwide, particularly throughout warmer regions. Ripe melons are prized for their sweetness and are eaten raw as a cool dessert. Sweet melons are in demand in Europe, North and South America, Australia, and most parts of Asia. Three melon cultivar groups are the muskmelons (or Reticulatus group), the cantaloupes (or Cantalupensis group), and the casabas (or Inodorous group). Muskmelons have a reticulate (corky or netted) rind, while cantaloupes have a smooth or warty rind. Both groups of melons are aromatic and richly flavored, though with a short (about a week) shelf-life. Casabas have a smooth or wrinkled rind, inconspicuously lobed or non-lobed, neither aromatic nor richly flavored, but have the greatest sweetness of melons and a long (a month or more) shelf-life (Paris et al., 2012).

Contrarily, watermelons are native to Africa, where they have been in cultivation since the ancient times. Wild watermelon fruits are watery and typically hard-textured, pale-colored, and bland or bitter-tasting, while the dessert watermelons are sweet and tender and have colored flesh. The narrow genetic base of dessert watermelons suggests that it originated from a series of selection events in a single ancestral population. Multiple pieces of evidence suggest that Northeastern Africa is the center of origin of the dessert watermelon, domesticated for water and food over 4000 years ago. From there it spread to the Mediterranean approximately 2000 years ago (Paris, 2015). The progenitor of the watermelon can be found in Sudan. The red fruit flesh color became fixed through the domestication of watermelon,

DOI: 10.1201/9781003467335-14

while its fruit sweetness gradually increased over the course of domestication (Renner et al., 2021).

14.2 NATURAL VARIATION ASSOCIATED WITH HORTICULTURAL AND SENSORY TRAITS

14.2.1 Melons

Cultivated melons exhibit an abundant diversity in fruit morphotypes from oblate, spherical, oval, pyriform, and fusiform to extremely long and serpentine, with or without stripes, and with lobes and furrows ('sutures') or wrinkles or warts, with landraces highly variable in fruit morphotypes. The fruits are either highly aromatic or not at all, with relatively thin or thick fruit flesh. The flesh color at maturity can be orange, green, cream, or white and remains bland or becomes sweet or sour (Burger et al., 2010; Flores-León et al., 2022; Girelli et al., 2018; Moing et al., 2020). The melon genetic resources also have substantial genetic variation and high heritability for total soluble solid (TSS) content. Melons with greater than 9.0% of TSS are preferred for commercial production. Higher TSS also relate with greater sweetness and good flavor (Andrade et al., 2021). Intra-accession variability also exists within landraces. The landraces are more variable for sugars, acid, and volatile organic compounds (VOC). On-farm conservation and exploitation of landraces may provide additional opportunity to establish new breeding targets to harness such variability (Esteras et al., 2020; Flores-León et al., 2022; Shi et al., 2020). The melon fruit quality is highly influenced by variation in aroma and carotenoids. A huge amount of untapped aroma diversity has been discovered, both in germplasm and breeding lines, which may provide the basis for identifying specific genes/alleles contributing to melon flesh quality (Esteras et al., 2018, 2020; Manjithia et al., 2021).

14.2.2 Watermelons

Abundant phenotypic variation has been reported in watermelon germplasm for fruits of different sizes, shapes, weights, rind colors and thicknesses, rind stripes and grooving, aroma, flesh colors, TSS, sugar, lycopene, and carotenoids (Gong et al., 2022b; Gusmini and Wehner, 2006; Noh et al., 2020). The cultivated watermelons synthesize various kinds of carotenoids in the fruit, which are responsible for the vivid flesh colors, including white, yellow, orange, pink, red, and mixed color (Gusmini and Wehner, 2006; Levi et al., 2013; Nimmakayala et al., 2014; Noh et al., 2020; Wu et al., 2019). The US National Plant Germplasm System's watermelon collection consists of three *Citrullus* species: *C. lanatus*, *C. mucosospermus*, and *C. amarus*. The first two species are valuable source of resistance to many diseases. They are interfertile with *C. lanatus*, which may provide valuable alleles to broaden the genetic base of cultivated watermelons (Levi et al., 2017).

Moreover, both melon and watermelon seeds are rich sources of oil and protein, minerals, amino acids, phytochemicals, and antioxidants. The seeds have an abundant variation in size, shape, weight, and seed coat color in germplasm accessions (Guo et al., 2020; Saeed et al., 2023).

14.3 GENES AND MARKERS ASSOCIATED WITH HORTICULTURAL AND SENSORY TRAITS

14.3.1 Melon

Table 14.1. lists candidate genes and QTLs associated with climactic ripening of fruits, fruit aroma, fruit shape, fruit skin netting and surface groove, rind color, sucrose level, and starch content. A few functionally characterized genes and molecular markers have also been reported. For example, overexpression of *CmSUN23–24* and *CmSUN25-26-27c* increases fruit shape variation (Ma et al., 2022), whereas a KASP marker (*CmAPRR2*$^{SNP(G/T)}$) co-segregate with rind color (Shen et al., 2021). SNP-CAPS markers determine variation in TSS content (Zhao et al., 2023). A golden SNP in *CmOr* is linked with non-orange and orange melon fruit phenotypes (Tzuri et al., 2015). All these may be deployed in melon breeding.

TABLE 14.1
Candidate Genes/QTLs Associated with Horticultural and Sensory Traits in Melon (*Cucumis melo*)

Gene/QTL	Phenotype	Reference
sfw 2.2 (*MELO3C029669*)	Single fruit weight	Cai et al. (2024)
SSC6.1 and *SSC11.1*	Soluble solid content	Zhao et al. (2023)
EF12.1	Exocarp firmness	Zhao et al. (2023)
EPF3.1, *EPF3.2*, *EPF7.1*	Edible pericarp firmness	Zhao et al. (2023)
CmAPS2–2	Starch accumulation, impacting flavor	Wang et al. (2023)
CmMYB44	Increase sucrose levels	Gao et al. (2023)
CmSUN23–24, *CmSUN25-26-27c*	Fruit shape variation	Ma et al. (2022)
MELO3C019694.2 (*Cmfsg* locus)	Fruit surface groove	Du et al. (2022)
MELO3C010288 (*CmSN* locus)	Fruit skin netting	Liang et al. (2023)
CmCLV3	Carpel numbers in fruit	Wang et al. (2022)
CmMt1 and *CmMt2*	Regulate mottled rind with evidence of epistasis	Shen et al. (2021)
EYELLQW1.1, *EYELLQW8.1*	Fruit rind color—yellow	Domingo et al. (2022b)
ETHQV8.1 and *ETHQV6.3*	Fruit aroma	Mayobre et al. (2024)
EAROQW8.3	Fruit aroma	Mayobre et al. (2021)
ETHQB3.5, *ETHQV6.3*, *ETHQV8.1*	Climacteric fruit ripening and aroma	Domingo et al. (2022a)
ETHQB3.5	Climacteric fruit ripening and aroma	Dos-Santos et al. (2023)
CmERF024	A causal gene governing *ETHQV8.1* (climacteric fruit ripening)	Domingo et al. (2024)
ETHQV8.1	Climacteric fruit and aroma	Pereira et al. (2020)
CYPs (*Cyclophilin*)	Characteristic (strong) fragrance	Wibowo et al. (2021)
CmThAT1; CmPPR1	Fruit aroma and white flesh color, respectively	Galpaz et al. (2018)
CmOr	Flesh color (orange, white, or green)	Tzuri et al. (2015)

14.3.2 Watermelon

Table 14.2 summarizes candidate genes/QTLs associated with horticultural and sensory traits in watermelon, including fruit bitterness, fruit morphotypes (length, diameter, and index), fruit peel crack resistance, peel stripe margin, rind thickness and color, rind stripe and inter-stripe color, fruit texture and aroma, TSS content, sugar (including sucrose), β-carotene, and lycopene. Functionally characterized genes and markers, for example, a tightly linked SNP marker with yellow skin phenotype (Dou et al., 2018a), allelic variation at *C2* locus associated with varying flesh colors, canary yellow, incomplete canary yellow, and red flesh (Park et al., 2023), an InDel marker co-segregate with fruit shape (Dou et al., 2018b). nsSNP in *ClPSY1* linked with β-carotene (Jin et al., 2023), flesh color specific CAPS markers linked with red or yellow flesh color (Wang et al., 2019), or SNPs in *ClTST2* associated with sweetness (Ren et al., 2018) etc., are ideal resource to breed watermelon with specific fruit characteristics. A comparison of a wild reference genome (P84) with sequences of 1175 watermelon accessions detected 10,589 structural variants (SVs), and 18 GWAS signals associated with fruit acidity, fruit length, fruit weight, fruit color and sex determination. A total of 3317 highly diverged SVs differentiated the wild from cultivated melons, which could be associated with domestication related traits. Further analysis detected novel SVs and a key gene *CmPH* regulating fruit acidity through domestication and selection of different populations (Chen et al., 2024). A recessive gene *Clpf* regulates pink flesh (pf) trait in watermelon, which was further narrowed down to a 55.26-kb region on chromosome 6, harboring two candidate genes. Furthermore, *Cla97C06G122120*, annotated as a pentatricopeptide repeat (PPR), as the prime candidate gene associated with pink flesh trait (Zhang et al., 2024).

TABLE 14.2
Candidate Genes/QTLs Associated with Horticultural and Sensory Traits in Watermelon (*Citrullus lanatus*)

Gene/QTL/marker	Phenotype	Reference
Cla011508	Bitterness	Gong et al. (2022a)
MELO3C019694.2	Fruit surface grove	Du et al. (2022)
SNPs	Fruit shape, TSS, flesh color	Katuuramu et al. (2023)
ClFS1	Fruit shape	Dou et al. (2018b)
CmEXPB1	Fruit peel crack resistance	Hu et al. (2024)
QTL (41–61% PVE)	Fruit cracking	Zhan et al. (2023)
Cla016033	Fruit firmness	Sun et al. (2020)
qFPL4.1, qFPW4.1, qFPD2.1, qFPD2.2, qFPD8.1, qFPD10.1	Fruit pedicle	Amanullah et al. (2022)
Cla97C02G044160	Rind thickness	Gong et al. (2022b)
Cla97C09G175170, Cla97C09G175150	Rind stripe and inter-stripe color	Maragal et al. (2022)

(*Continued*)

Table 14.2 ***(Continued)***
Candidate Genes/QTLs Associated with Horticultural and Sensory Traits in Watermelon (Citrullus lanatus)

Gene/QTL/marker	Phenotype	Reference
Clyr	Rind color (yellow)	Liu et al. (2020)
SNPs (functional)	Rind color (yellow)	Dou et al. (2018a)
Cla97C06G126680	Peel stripe margin	Yang et al. (2024)
Cla97C01G008760	Flesh color	Nie et al. (2023)
C_2	Flesh with mixed color (red and canary yellow)	Park et al. (2023)
Clwf2	White flesh color	Yi et al. (2023)
ClCG01G008470 within *ClPSY1* locus	Orange-β flesh	Jin et al. (2023)
Cla97C06G121860, Cla97C06G121880, Cla97C06G121890, Cla97C06G121900	White flesh color	Yi et al. (2023)
Cla97C06G122050, Cla97C06G122120 (*Cyf* locus)	Canary yellow flesh color	Liu et al. (2023)
Cla97C10G200950 (*Clorf* locus)	Orange flesh color	Fang et al. (2023)
Cla97C01G008760 (*ClPsy1* locus)	Golden flesh color	Liu et al. (2022)
Cla003760, Cla021635	Yellow flesh color	Yuan et al. (2021)
Five genes (Y^{scr} locus)	Scarlet red and coral red flesh color	Li et al. (2020b)
Cla005011	Lycopene and red flesh color	Wang et al. (2019)
Cla018406	β-carotene accumulation, highly expressed in orange flesh	Yuan et al. (2021)
Cla007686	Lycopene accumulation, highly expressed in red flesh	Yuan et al. (2021)
Eight genes (cell wall biosynthesis/ ethylene pathway as hub genes)	Flesh firmness	Anees et al. (2021)
Cla97C01G013600, Cla97C05G089700, Cla97C01G001290, Cla97C05G095170, Cla97C06G118330	Fruit aroma	Gong et al. (2021)
ClTST2	Sugar accumulation in flesh cells	Ren et al. (2018)
ClVST1	Sucrose accumulation	Ren et al. (2020)
Numerous DEGs and DMs as hub genes associated with fruit quality	Flesh sweetness: 6 DEGs with 4 DMs; fruit bitterness: 6 DEGs with 13 DMs; flesh color: 9 DEGs with 10 DMs; flesh texture: 9 DEGs with 4 DMs	Chu et al. (2022)

14.4 GENES AND MARKERS ASSOCIATED WITH SEED TRAITS

14.4.1 Melon

Sufficient variability exists for seed size (seed length and width) and shape. Candidate genes and selective sweep signals (152 sweeps, spread 34.46 Mb across the 12 chromosomes) for seed size, including the QTL regions of qSL3 and qSWD3, were reported melon (Zhang et al., 2022). In addition, two SNPs mapped on chromosome 6 were significantly associated with seed traits, 1000-seed weight, and seed length and width (Hyun et al., 2021).

14.4.2 Watermelon

Two pleiotropic QTLs, *qSS4* and *qSS6*, control seed size. *qSS6* is a major QTL, explained ~93%, 94%, and 95% of the phenotypic variation associated with 1000-seed weight, seed length, and seed width, respectively. A newly developed PCR-based markers narrowed *qSS6* region to 55.5 kb and three candidates, including a known seed size regulator in rice, *SRS3* (Li et al., 2018). Both flesh and seeds of watermelon are edible. For seed consumption, large seeds are preferred, whereas for flesh consumption, small seeds are preferred. Seed traits (100-seed weight, seed hilum length, seed length, seed width, and seed thickness) are positively and strongly correlated with each other. *C. mucosospermu* and *C. lanatus* species are predominantly used for seed purposes. GWAS detected four SNPs (32250307, 32250454, 32256177, 32260870) on chromosome 5, significantly associated with seed size traits, which could be used to breed watermelon cultivars with target seed size (Gong et al., 2022c).

Seed coat color is an important quality trait in watermelon. A single dominant gene (*Cla019481*, named *ClCS1*), mapped on chromosome 3, controls the black seed coat (Li et al., 2020). Negative epistasis, mapped on chromosomes 6 and 5, respectively, regulates white and red seed coat color, and an SNP marker perfectly differentiated white and red seed coat color segregants (Jang et al., 2024). An earlier report confirmed four genes, *R*, T^l, *W*, and *D* loci model, mapped on chromosomes 3, 5, 6, and 8, respectively. KASP™ assays and SNPs (UGA3_5820134, UGA5_4591722, UGA6_7076766, and UGA8_22729513) linked to the four loci may facilitate genomic assisted breeding for seed coat color in watermelon (Paudel et al., 2019).

Watermelon seeds are a powerhouse for value-added traits, including proteins, free amino acids, vitamins, and essential minerals. Sufficient genetic variation for seed-bound free amino acids and protein content is reported across accessions and geographic regions. GWAS detected 188 SNPs coincided with 167 candidate genes associated with watermelon seed-bound amino acids and total protein. They are involved in metabolic pathways associated with amino acid metabolism; for example, the functionally characterized *argininosuccinate synthase* gene, explaining 7% of the variation in arginine content, may be deployed in breeding programs to develop value-added food/feed products (Joshi et al., 2021).

14.5 GERMPLASM DEVELOPMENT WITH SPECIFIC HORTICULTURAL AND SENSORY TRAITS

14.5.1 Melon

Advances in genetic and genomic resources have set the pace for the use of marker-assisted genetic enhancement for variation in fruit size and shape, fruit skin netting, fruit rind and flesh color, TSS content, carotenoids, and sugar content. It is achievable to develop melon cultivars with specific market demand and consumer preferences, combining best agronomic practices and sensory profiles (Shahwar et al., 2023).

Introgression lines (ILs) are useful genetic resources to identify, fine map and positional cloning, as well introduce favorable alleles into elite lines. A novel collection of ILs with 91–99% of genome coverage, involving two commercial European cultivars with different ripening process, segregate for multiple fruit quality traits. Such ILs are useful genomic resource to pyramid genes for important complex traits such as fruit weight and climacteric ripening (Domingo et al., 2022a; Pereira et al., 2021). ILs involving wild (Ames 24297) and cultivated (Piel de Sapo) genetic background, with 75% coverage of wild genome, has impressive variability for fruit size related traits, with fruit weight ranging from 800 to 4100 g, which reflects strong effect of the wild genome traits, likely due to new epistatic interactions in Piel de Sapo genetic background (Campos et al., 2023). NILs segregating for volatile organic compounds (acetate esters, non-acetate esters, and sulfur-derived families or those of alkanes, aldehydes, and ketones) that affect fruit aroma are also available for breeding melons with specific fruit aroma (Dos-Santos et al., 2023). ILs involving Japanese cultivar Ginsen Makuwa in recurrent genetic background of Vedrantais have characteristic VOC, not detected in either parent, may optimize use of these resources in melon breeding with novel aroma profiles (Perpiñá et al., 2021).

Ripening influences the quality and shelf-life of the fruits. Stacking of three climacteric QTLs (*ETHQB3.5*, *ETHQV6.3*, *ETHQV8.1*) in the non-climacteric cultivar Piel de Sapo interacted epistatically for ethylene production and aroma profile. The QTLs independently activate the production of esters changing the aroma profile of the fruits, with no significant effects on fruit size and firmness or TSS content, thus indicating a powerful tool to modify and fine-tune climacteric fruit ripening in melon-breeding programs (Domingo et al., 2022b). The exploitation of dudaim germplasm in breeding programs may enhance volatile profile of odorless Persian melons (Soltani et al., 2023). The identification of a golden SNP in *CmOr*, tightly linked with non-orange and orange melon fruit phenotypes, may provide additional opportunities for nutritional biofortification of melons (Tzuri et al., 2015).

The pangenome refers to the entire set of genes within a species, consisting of a core genome (containing sequences shared between all individuals of the species) and the dispensable genome. Assembling and assessing pangenomes reveals that relying on a single-reference genome may not represent full spectrum of genetic variability for a given species. Melon pangenome consists of de novo assembled genomes of a core subset of 25 diverse founder lines, consisting of 24 accessions from the two widely cultivated subspecies of *C. melo*, encompassing 12 horticultural groups and one feral accession, with substantial variation in genome size and structure, including ~300,000 structural variants and ~9 million SNPs. Furthermore, the availability of a half-diallel set of 300 F_2 populations, representing 25 founder lines, may enhance genetic dissection of horticulturally important traits to accelerate melon breeding (Oren et al., 2022).

The use of double haploid (DH) technology is one of the most efficient breeding strategies for developing new cultivars or inbred lines for hybrids. Identifying putative haploids is the key issue in haploid breeding. While in some crops, such as barley, maize and rice, DH technology is perfected and used routinely in cultivar development, in other crops, the technology has not reached the stage leading to its application in crop improvement (Dwivedi et al., 2015, and references therein). Parthenogenesis has been used to produce doubled haploid lines in melon. An improved procedure for recovery of haploid and double haploid melon plants through *in situ* parthenogenesis via pollination with irradiated pollen have been used to produce doubled haploid lines using seven melon donor genotypes (Hooghvorst et al., 2020, 2021). Parthenogenesis is a genotype-dependent process, and melon has a huge genotypic diversity. More research is required to increase the parthenogenetic efficiency and making it genotype-independent procedure to exploit double haploid technology in melon breeding.

14.5.2 Watermelon

Conventional crossing and selection have led to the development and release of several advanced lines and cultivars with excellent fruit characteristics. However, narrow genetic base and repeated use of a limited number of genotypes as breeding parents poses a serious genetic bottleneck in breeding. Advances in genetic mechanisms involved in the biosynthesis of phytochemicals, phytohormones, and fruit flesh chemistry elicited during watermelon fruit development and ripening together with the progress on gene(s) discovery and associated marker technology have recently paved the way for marker-assisted genetic improvement in watermelon.

A sequence comparison of 'Charleston Gray' (American dessert watermelon) and 97103 (East Asian cultivar), both of which are sequenced, revealed genetic variants underlying phenotypic differences between the two cultivars. Further, genotyping of 1,365 USDA watermelon collection of three *Citrullus* species (*C. lanatus*, *C. mucosospermus*, *C. amarus*), involving 25,000 high-quality SNPs obtained from the Charleston Gray genome, indicated a close relationship between *C. lanatus* and *C. mucosospermus*, and four distinct groups correlated to their geographic locations. *Citrullus amarus* is genetically distinct compared to *C. lanatus* and *C. mucosospermus*. Furthermore, SNP-based GWAS discovered genomic regions associated with fruit quality and disease resistance (Wu et al., 2019). Parallel to this, another group, re-sequenced 414 accessions representing all *Citrullus* species revealed independent evolution in *C. amarus* and another lineage containing *C. lanatus* and *C. mucosospermus*. Different loci affecting fruit size were under selection during speciation, domestication, and improvement. A non-bitter allele, originating in the progenitor of sweet watermelon, is largely fixed in *C. lanatus*. Selection for flesh sweetness started in the progenitor of *C. lanatus* and continues through modern breeding on loci controlling raffinose catabolism and sugar transport. Fruit flesh color and sugar accumulation co-evolved through shared genetic components including a sugar transporter gene (Guo et al., 2019). Resequencing of another set of 125 *C. amarus* accessions resulted in 2,126,759 genome-wide SNPs, of which 37 SNPs and many candidate genes (mapped on chromosomes Ca01, Ca03, Ca06, and Ca07) associated with fruit morphotypes, TSS content, and flesh color were uncovered (Katuuramu et al., 2023). *Citrullus* pangenome and super-pangenome serve as a comprehensive resource to mine and utilize genes in cultivated and wild species for improving the productivity and fruit quality of watermelons (Sun et al., 2023; Wu et al., 2023). Thus, these are powerful genomic resources, which, after functional validation, may facilitate marker-assisted genetic improvement in watermelon.

The disruption of the *DMP* (Zhong et al., 2020) gene induces single fertilization of the central cell, which can lead to the formation of haploids. A detailed method for creation of a haploid inducer line via *ClDMP3* mutation has been recently perfected in watermelon. The *cldmp3* mutant induces haploids in multiple watermelon genotypes at rates of up to 1.12%, which may greatly advance watermelon breeding (Chen et al., 2023).

14.6 CONCLUSIONS

Advances in the development of genetic and genomic resources, in both crops, are providing new opportunities to tailor improved germplasm combining fruit yield, nutritional and flavor related traits. The use of double haploid technology and exploitation of pangenome heat map (variability) may provide additional opportunity to accelerate development of inbred lines and hybrid cultivars with specific fruit characteristics.

REFERENCES

Amanullah, S., Osae, B.A., Yang, T., Li, S., Abbas, F., Liu, S., et al. (2022). Development of whole genome SNP-CAPS markers and preliminary QTL mapping of fruit pedicel traits in watermelon. *Front. Plant Sci.* 13, 879919. https://doi.org/10.3389/fpls.2022.879919

Andrade, I.S., de Melo, C.A.S., de Sousa Nunes, G.H., Holanda, I.S.A., Grangeiro, L.C., and Corrěa, R.X. (2021). Phenotypic variability, diversity and genetic-population structure in melon (*Cucumis melo* L.) associated with total soluble solids. *Sci. Hortic.* 278, 109844. https://doi.org/10.1016/j.scienta.2020.109844

Anees, M., Gao, L., Umer, M.J., Yuan, P., Zhu, H., Lu, X., et al. (2021). Identification of key gene networks associated with cell wall components leading to flesh firmness in watermelon. *Front. Plant Sci.* 12, 630243. https://doi.org/10.3389/fpls.2021.630243

Burger, Y., Paris, H.S., Cohen, R., Katzir, N., Tadmor, Y., Lewinsohn, E., et al. (2010). Genetic diversity of *Cucumis melo*. In: Janick, J. (ed.) *Horticultural Reviews*, Vol. 36. Wiley, New York, pp. 165–198. https://doi.org/10.1002/9780470527238.ch3

Cai, Y., Wang, D., Che, Y., Wang, L., Zhang, F., Liu, T., et al. (2024). Multiple localization analysis of the major QTL—*sfw 2.2* for controlling single fruit weight traits in melon based on SLAF sequencing. *Genes* 15, 1138. https://doi.org/10.3390/genes15091138

Campos, M., Gonzalo, M.J., Diaz, A., Picó, B., Gómez-Guillamón, M.L., Monforte, A.J., and Esteras, C. (2023). A novel introgression line library derived from a wild melon gives insights into the genetics of melon domestication, uncovering new genetic variability useful for breeding. *Int. J. Mol. Sci.* 24, 10099. https://doi.org/10.3390/ijms241210099

Chen, X., Li, H., Dong, Y., Xu, Y., Xu, K., Zhang, Q., et al. (2024). A wild melon reference genome provides novel insights into the domestication of a key gene responsible for melon fruit acidity. *Theor. Appl. Genet.* 137, 144. https://doi.org/10.1007/s00122-024-04647-4

Chen, X., Li, Y., Ai, G., Chen, J., Guo, D., Zhu, Z., et al. (2023). Creation of a watermelon haploid inducer line via *ClDMP3*-mediated single fertilization of the central cell. *Hortic. Res.* 10, uhad081. https://doi.org/10.1093/hr/uhad081

Chu, S., Wang, S., Zhang, R., Yin, M., Yang, X., Shi, Q., et al. (2022). Integrative analysis of transcriptomic and metabolomic profiles reveals new insights into the molecular foundation of fruit quality formation in *Citrullus lanatus* (Thunb.) Matsum. & Nakai. *Food Qual. Saf.* 6, fyac015. https://doi.org/10.1093/fqsafe/fyac015

Domingo, M.S., Areco, L., Mayobre, C., Valverde, L., Martín-Hernández, M., Pujol, M., et al. (2022a). Modulating climacteric intensity in melon through QTL stacking. *Hortic. Res.* 9, uhac131. https://doi.org/10.1093/hr/uhac131

Domingo, M.S., Mayobre, C., Pereira, L., Argyris, J., Valverde, L., Martín-Hernández, A.M., et al. (2022b). Fruit morphology and ripening-related QTLs in a newly developed introgression line collection of the elite varieties 'Védrantais' and 'Piel de Sapo'. *Plants (Basel)* 11, 3120. https://doi.org/10.3390/plants11223120

Domingo, M.S., Orduña, L., Navarro, D., Mayobre, C., Santiago, A., Valverde, L., et al. (2024). The *ethylene-responsive transcription factor ERF024* is a novel regulator of climacteric fruit ripening in melon. *Plant J.* 119, 1844–1858. https://doi.org/10.1111/tpj.16889

Dos-Santos, N., Bueso, M.C., Díaz, A., Moreno, E., Garcia-Mas, J., Monforte, A.J., et al. (2023). Thorough characterization of *ETHQB3.5*, a QTL involved in melon fruit climacteric behavior and aroma volatile composition. *Foods* 12, 376. https://doi.org/10.3390/foods12020376

Dou, J., Lu, X., Ali, A., Zhao, S., Zhang, L., He, N., et al. (2018a). Genetic mapping reveals a marker for yellow skin in watermelon (*Citrullus lanatus* L.). *PLoS ONE* 13, e0200617. https://doi.org/10.1371/journal.pone.0200617

Dou, J., Zhao, S., Lu, X., He, N., Zhang, L., Ali, A., et al. (2018b). Genetic mapping reveals a candidate gene (*ClFS1*) for fruit shape in watermelon (*Citrullus lanatus* L.). *Theor. Appl. Genet.* 131, 947–958. https://doi.org/10.1007/s00122-018-3050-5

Du, X., Liu, H., Zhu, Z., Liu, S., Song, Z., Xia, L., et al. (2022). Identification of candidate chromosome region related to melon (*Cucumis melo* L.) fruit surface groove trait through biparental genetic mapping and genome-wide association study. *Front. Plant Sci.* 13, 828287. https://doi.org/10.3389/fpls.2022.828287

Dwivedi, S.L., Britt, A.B., Tripathi, L., Sharma. S., Upadhyaya, H.D., and Ortiz, R. (2015). Haploids: Constraints and opportunities in plant breeding. *Biotechnol. Adv.* 33, 812–829. https://doi.org/10.1016/j.biotechadv.2015.07.001

Endl, J., Achigan-Dako, E.G., Pandey, A.K., Monforte, A.K., Picó, B., Schaefer, H. (2018). Repeated domestication of melon (*Cucumis melo*) in Africa and Asia and a new close relative from India. *Am. J. Bot.* 105, 1662–1671. https://doi.org/10.1002/ajb2.1172

Esteras, C., Rambla, J.L., Sánchez, G., Granell, A., and Picó, M.B. (2020). Melon genetic resources characterization for rind volatile profile. 10, 1512. https://doi.org/10.3390/agronomy10101512

Esteras, C., Rambla, J.L., Sánchez, G., López-Gresa, M.P., González-Mas, M.C., Fernández-Trujillo, J.P., et al. (2018). Fruit flesh volatile and carotenoid profile analysis within the *Cucumis melo* L. species reveals unexploited variability for future genetic breeding. *J. Sci. Food Agric.* 98, 3915–3925. https://doi.org/10.1002/jsfa.8909

Fang, X., Li, S., Zhu, Z., Zhang, X., Xiong, C., Wang, X., et al. (2023). *Clorf* encodes carotenoid isomerase and regulates orange flesh color in watermelon (*Citrullus lanatus* L.). *J. Agric. Food Chem.* 71, 15445–15455. https://doi.org/10.1021/acs.jafc.3c02122

Flores-León, A., Moro, C.P., Marti, R., Beltrn, J., Roselló, S., Cebolla-Cornejo, J., et al. (2022). Spanish melon landraces: Revealing useful diversity by genomic, morphological, and metabolomic analysis. *Int. J. Mol. Sci.* 23, 7162. https://doi.org/10.3390/ijms23137162

Galpaz, N., Gonda, I., Shem-Tov, D., Barad, O., Tzuri, G., Lev, S., et al. (2018). Deciphering genetic factors that determine melon fruit-quality traits using RNA-Seq-based high-resolution QTL and eQTL mapping. *Plant J.* 94, 169–191. https://doi.org/10.1111/tpj.13838

Gao, G., Yang, F., Wang, C., Duan, X., Li, M., Ma, Y., et al. (2023). The transcription factor CmERFI-2 represses CmMYB44 expression to increase sucrose levels in oriental melon fruit. *Plant Physiol.* 192, 1378–1395. https://doi.org/10.1093/plphys/kiad155

Girelli, C.R., Accogli, R., Coco, L.D., Angilé, F., de Bellis, L., and Fanizzi, F.P. (2018). ^{1}H-NMR-based metabolomic profiles of different sweet melon (*Cucumis melo* L.) Salento varieties: Analysis and comparison. *Food Res. Intl.* 114, 81–89. https://doi.org/10.1016/j.foodres.2018.07.045

Gong, C., Diao, W., Zhu, H., Umer, M.J., Zhao, S., He, N., et al. (2021). Metabolome and transcriptome integration reveals insights into flavor formation of 'Crimson' watermelon flesh during fruit development. *Front. Plant Sci.* 12, 629361. https://doi.org/10.3389/fpls.2021.629361

Gong, C., Li, B., Anees, M., Zhu, H., Zhao, S., He, N., et al. (2022a). Fine mapping reveals that the bHLH gene *Cla011508* regulates the bitterness of watermelon fruit. *Sci. Hortic.* 292, 110626. https://doi.org/10.1016/j.scienta.2021.110626

Gong, C., Lu, X., Zhu, H., Anees, M., He, N., and Liu, W. (2022b). Genome-wide association study provides genetic insights into natural variation in watermelon rind thickness and single fruit weight. *Front. Plant Sci.* 13, 1074145. https://doi.org/10.3389/fpls.2022.1074145

Gong, C., Zhao, S., Yang, D., Lu, X., Anees, M., He, N., et al. (2022c). Genome-wide association analysis provides molecular insights into the natural variation of watermelon seed size. *Hortic. Res.* 9, uhab074. https://doi.org/10.1093/hr/uhab074

Gonzalo, M.J., Díaz, A., Dhillon, N.P.S., Reddy, U.K., Picó, B., and Monforte, A.J. (2019). Re-evaluation of the role of Indian germplasm as center of melon diversification based on genotyping-by-sequencing analysis. *BMC Genom.* 20, 448. https://doi.org/10.1186/s12864-019-5784-0

Grumet, R., McCreight, J.D., McGregor, C., Weng, Y., Mazourek, M., Reitsma, K., et al. (2021). Genetic resources and vulnerabilities of major cucurbit crops. *Genes (Basel)* 12, 1222. https://doi.org/10.3390/genes12081222

Guo, S., Zhao, S., Sun, H., Wang, X., Wu, S., Lin, T., et al. (2019). Resequencing of 414 cultivated and wild watermelon accessions identifies selection for fruit quality traits. *Nat. Genet.* 51, 1616–1623. https://doi.org/10.1038/s41588-019-0518-4

Guo, Y., Gao, M., Liang, X., Xu, M., Liu, X., Zhang, Y., et al. (2020). Quantitative trait loci for seed size variation in Cucurbits—A review. *Front. Plant Sci.* 11, 304. https://doi.org/10.3389/fpls.2020.00304

Gusmini, G., and Wehner, T.C. (2006). Qualitative inheritance of rind pattern and flesh color in watermelon. *J. Hered.* 97, 177–185. https://doi.org/10.1093/jhered/esj023

Hooghvorst, I., Torrico, O., Hooghvorst, S., and Nogués, S. (2020). *In situ* parthenogenetic doubled haploid production in melon "Piel de Sapo" for breeding purposes. *Front. Plant Sci.* 11, 378. https://doi.org/10.3389/fpls.2020.00378

Hooghvorst, I., Torrico, O., and Nogués, S. (2021). Doubled haploid parthenogenetic production of melon 'Piel de Sapo'. *Methods Mol. Biol.* 2289, 87–95. https://doi.org/10.1007/978-1-0716-1331-3_5

Hu, Y., Li, Y., Zhu, B., Huang, W., Chen, J., Wang, F., et al. (2024). Genome-wide identification of the expansin gene family in netted melon and their transcriptional responses to fruit peel cracking. *Front. Plant Sci.* 15, 1332240. https://doi.org/10.3389/fpls.2024.1332240

Hyun, D.Y., Sebastin, R., Lee, G.-A., Lee, K.J., Kim, S.-H., Yoo, E., et al. (2021). Genome-wide SNP markers for genotypic and phenotypic differentiation of melon (*Cucumis melo* L.) varieties using genotyping-by-sequencing. *Int. J. Mol. Sci.* 22, 6722. https://doi.org/10.3390/ijms22136722

Jang, Y.J., Kim, K.R., Ryu, J., Kim, Y., Chung, S.-M., and Lee, G.P. (2024). Markers for seed coat color in watermelon. *Hort. Sci. Technol.* 42, 480–491. http://www.hst-j.org

Jin, B., Jang, G., Park, G., Shahwar, D., Shin, J., Kwon, G., et al. (2023). Development of a bene-Based marker set for orange-colored watermelon flesh with a high β-carotene content. *Int. J. Mol. Sci.* 25, 210. https://doi.org/10.3390/ijms25010210

Joshi, V., Nimmakayala, P., Song, Q., Abburi, V., Natarajan, P., Levi, A., et al. (2021). Genome-wide association study and population structure analysis of seed-bound amino acids and total protein in watermelon. *PeerJ.* 9, e12343. https://doi.org/10.7717/peerj.12343

Katuuramu, D.N., Levi, A., and Wechter, W.P. (2023). Genome-wide association study of soluble solids content, flesh color, and fruit shape in citron watermelon. *Plant Genome* 16, e20391. https://doi.org/10.1002/tpg2.20391

Levi, A., Jarret, R., Kousik, S., Patrick Wechter, W., Nimmakayala, P., and Reddy, U.K. (2017). Genetic resources of watermelon. In: Grumet, R., Katzir, N., and Garcia-Mas, J. (eds.) *Genetics and Genomics of Cucurbitaceae*. Springer, Cham, pp. 87–110. https://doi.org/10.1007/7397_2016_34

Levi, A., Thies, J.A., Wechter, W.P., Harrison, H.F., Simmons, A.M., Reddy, U.K., et al. (2013). High frequency oligonucleotides: Targeting active gene (HFO-TAG) markers revealed wide genetic diversity among *Citrullus* spp. accessions useful for enhancing disease or pest resistance in watermelon cultivars. *Genet. Resour. Crop Evol.* 60, 427–440. https://doi.org/10.1007/s10722-012-9845-3

Li, B., Lu, X., Gebremeskel, H., Zhao, S., He, N., Yuan, P., et al. (2020). Genetic mapping and discovery of the candidate gene for black seed coat color in watermelon (*Citrullus lanatus*). *Front. Plant Sci.* 10, 1689. https://doi.org/10.3389/fpls.2019.01689

Li, N., Shang, J., Wang, J., Zhou, D., Li, N., and Ma, S. (2018). Fine mapping and discovery of candidate genes for seed size in watermelon by genome survey sequencing. *Sci. Rep.* 8, 17843. https://doi.org/10.1038/s41598-018-36104-w

Liang, X., Li, Q., Cao, L., Du, X., Qiang, J., Hou, J., et al. (2023). Natural allelic variation in the EamA-like transporter, *CmSN*, is associated with fruit skin netting in melon. *Theor. Appl. Genet.* 136, 192. https://doi.org/10.1007/s00122-023-04443-6

Liu, D., Yang, H., Yuan, Y., Zhu, H., Zhang, M., Wei, X., et al. (2020). Comparative transcriptome analysis provides insights into yellow rind formation and preliminary mapping of the *Clyr* (*Yellow Rind*) gene in watermelon. *Front. Plant Sci.* 11, 192. https://doi.org/10.3389/fpls.2020.00192

Liu, S., Gao, Z., Wang, X., Luan, F., Dai, Z., Yang, Z., et al. (2022). Nucleotide variation in the phytoene synthase (*ClPsy1*) gene contributes to golden flesh in watermelon (*Citrullus lanatus* L.). *Theor. Appl. Genet.* 135, 185–200. https://doi.org/10.1007/s00122-021-03958-0

Liu, S., Liu, M., Cao, Y., Xu, Y., Liu, H., Zhu, Q., et al. (2023). Identification of chromosome region and candidate genes for *canary-yellow flesh* (*Cyf*) locus in watermelon (*Citrullus lanatus*). *Plant Sci.* 329, 111594. https://doi.org/10.1016/j.plantsci.2023.111594

Ma, M., Liu, S., Wang, Z., Shao, R., Ye, J., Yan, W., et al. (2022). Genome-wide identification of the *SUN* gene family in melon (*Cucumis melo*) and functional characterization of two *CmSUN* genes in regulating fruit shape variation. *Int. J. Mol. Sci.* 23, 16047. https://doi.org/10.3390/ijms232416047

Manjithia, D., Metrani, R., Dhowlaghar, N., Crosby, K.M., and Patil, B.S. (2021). Assessment and classification of volatile profiles in melon breeding lines using headspace solid-phase microextraction coupled with gas chromatography-mass spectrometry. *Plants (Basel)* 10, 2166. https://doi.org/10.3390/plants10102166

Maragal, S., Nagesh, G.C., Reddy, D.C.L., and Rao, E.S. (2022). QTL mapping identifies novel loci and putative candidate genes for rind traits in watermelon. *3 Biotech.* 12, 46. https://doi.org/10.1007/s13205-022-03112-7

Mayobre, C., Domingo, M.S., Özkan, E.N., Fernández-Borbolla, A., Ruiz-Lasierra, J., Garcia-Mas, J., et al. (2024). Genetic regulation of volatile production in two melon introgression line collections with contrasting ripening behavior. *Hortic. Res.* 11, uhae020. https://doi.org/10.1093/hr/uhae020

Mayobre, C., Pereira, L., Eltahiri, A., Bar, E., Lewinsohn, E., Garcia-Mas, J., et al. (2021). Genetic dissection of aroma biosynthesis in melon and its relationship with climacteric ripening. *Food Chem.* 353, 129484. https://doi.org/10.1016/j.foodchem.2021.129484

Moing, A., Allwood, J.W., Aharoni, A., Baker, J., Beale, M.H., Ben-Dor, S., et al. (2020). Comparative metabolomics and molecular phylogenetics of melon (*Cucumis melo*, Cucurbitaceae) biodiversity. *Metabolites* 10, 121. https://doi.org/10.3390/metabo10030121

Nie, H., Kim, M., Lee, S., Lim, S., Lee, M.S., Kim, J.H., et al. (2023). High-quality genome assembly and genetic mapping reveal a gene regulating flesh color in watermelon (*Citrullus lanatus*). *Front. Plant Sci.* 14, 1142856. https://doi.org/10.3389/fpls.2023.1142856

Nimmakayala, P., Levi, A., Abburi, L., Abburi, V., Tomason, Y.R., Saminathan, T., et al. (2014). Single nucleotide polymorphisms generated by genotyping by sequencing to characterize genome-wide diversity, linkage disequilibrium, and selective sweeps in cultivated watermelon. *BMC Genom.* 15, 767. https://doi.org/10.1186/1471-2164-15-767

Noh, J.-J., Hur, O.-S., Ro, N.-Y., Lee, J.-E., Hwang, A.-J., Kim, B.-S., et al. (2020). Lycopene content and fruit morphology of red, pink, orange, and yellow fleshed watermelon (*Citrullus lanatus*) germplasm collections. *Korean J. Plant Res.* 33, 624–637. https://doi.org/10.7732/kjpr.2020.33.6.624

Oren, E., Dafna, A., Tzuri, G., Halperin, I., Isaacson, T., Elkabetz, M., et al. (2022). Pan-genome and multi-parental framework for high-resolution trait dissection in melon (*Cucumis melo*). *Plant J.* 112, 1525–1542. https://doi.org/10.1111/tpj.16021

Paris, H.S. (2015). Origin and emergence of the sweet dessert watermelon, *Citrullus lanatus*. *Ann. Bot.* 116, 133–148. https://doi.org/10.1093/aob/mcv077

Paris, H.S., Amar, Z., and Lev, E. (2012). Medieval emergence of sweet melons, *Cucumis melo* (Cucurbitaceae). *Ann. Bot.* 110, 23–33. https://doi.org/10.1093/aob/mcs098

Park, G., Shahwar, D., Jang, G., Shin, J., Kwon, G., Kim, Y., et al. (2023). Identification of a novel locus C2 controlling canary yellow flesh color in watermelons. *Front. Genet.* 14, 1256627. https://doi.org/10.3389/fgene.2023.1256627

Paudel, L., Clevenger, J., and McGregor, C. (2019). Chromosomal locations and interactions of four loci associated with seed coat color in watermelon. *Front. Plant Sci.* 10, 788. https://doi.org/10.3389/fpls.2019.00788

Pereira, L., Domingo, M.S., Argyris, J., Mayobre, C., Valverde, L., Martín-Hernández, A.M., et al. (2021). A novel introgression line collection to unravel the genetics of climacteric ripening and fruit quality in melon. *Sci. Rep.* 11, 11364. https://doi.org/10.1038/s41598-021-90783-6

Pereira, L., Domingo, M.S., Ruggieri, V., Argyris, J., Phillips, M.A., Zhao, G., et al. (2020). Genetic dissection of climacteric fruit ripening in a melon population segregating for ripening behavior. *Hortic. Res.* 7, 187. https://doi.org/10.1038/s41438-020-00411-z

Perpiñá, G., Roselló, S., Esteras, C., Beltrán, J., Monforte, A.J., Cebolla-Cornejo, J., et al. (2021). Analysis of aroma-related volatile compounds affected by 'Ginsen Makuwa' genomic regions introgressed in 'Vedrantais' melon background. *Sci. Hortic.* 276, 109664. https://doi.org/10.1016/j.scienta.2020.109664

Ren, Y., Guo, S., Zhang, J., He, H., Sun, H., Tian, S., et al. (2018). A tonoplast sugar transporter underlies a sugar accumulation QTL in watermelon. *Plant Physiol.* 176, 836–850. https://doi.org/10.1104/pp.17.01290

Ren, Y., Sun, H., Zong, M., Guo, S., Ren, Z., Zhao, J., et al. (2020). Localization shift of a sugar transporter contributes to phloem unloading in sweet watermelons. *New Phytol.* 227, 1858–1871. https://doi.org/10.1111/nph.16659

Renner, S.S., Wu, S., Pérez-Escobar, O.A., Silber, M.V., Fei, Z., and Chomicki, G. (2021). A chromosome-level genome of a Kordofan melon illuminates the origin of domesticated watermelons. *Proc. Natl. Acad. Sci. U. S. A.* 118, e2101486118. https://doi.org/10.1073/pnas.2101486118

Saeed, F., Afzaal, M., Niaz, B., Hussain, M., Rasheed, A., Raza, M.A., et al. (2023). Comparative study of nutritional composition, antioxidant activity and functional properties of *Cucumis melo* and *Citrullus lanatus* seeds powder. *Cogent Food Agric.* 10, Article: 2293517. https://doi.org/10.1080/23311932.2023.2293517

Shahwar, D., Khan, Z., and Park, Y. (2023). Molecular marker-assisted mapping, candidate gene identification, and breeding in melon (*Cucumis melo* L.): A review. *Int. J. Mol. Sci.* 20, 15490. https://doi.org/10.3390/ijms242015490

Shen, J., Xu, X., Zhang, Y., Niu, X., and Shou, W. (2021). Genetic mapping and identification of the candidate genes for mottled rind in *Cucumis melo* L. *Front. Plant Sci.* 12, 769989. https://doi.org/10.3389/fpls.2021.769989

Shi, J., Wu, H., Xiong, M., Chen, Y., Chen, J., Zhou, B., et al. (2020). Comparative analysis of volatile compounds in thirty-nine melon cultivars by headspace solid-phase microextraction and gas chromatography-mass spectrometry. *Food Chem.* 316, 126342. https://doi.org/10.1016/j.foodchem.2020.126342

Soltani, F., Nouri, M., Sokrpour, M., Alabboud, M., Erban, A., Fehrle, I., et al. (2023). Breeding of Persian melon: Purpose to increase volatile compounds in melon (*Cucumis melo*) intergroup F1 cross population. *Plant Breed.* 142, 372–387. https://doi.org/10.1111/pbr.13089

Sun, L., Zhang, Y., Cui, H., Zhang, L., Sha, T., Wang, C., et al. (2020). Linkage mapping and comparative transcriptome analysis of firmness in watermelon (*Citrullus lanatus*). *Front. Plant Sci.* 11, 192. https://doi.org/10.3389/fpls.2020.00192

Sun, Y., Kou, D.-R., Li, Y., Ni, J.-P., Wang, J., Zhang, Y.-M., et al. (2023). Pan-genome of *Citrullus* genus highlights the extent of presence/absence variation during domestication and selection. *BMC Genom.* 24, 332. https://doi.org/10.1186/s12864-023-09443-w

Tzuri, G., Zhou, X., Chayut, N., Yuan, H., Portnoy, V., Meir, A., et al. (2015). A 'golden' SNP in CmOr governs the fruit flesh color of melon (*Cucumis melo*). *Plant J.* 82, 267–279. https://doi.org/10.1111/tpj.12814

Wang, C., Jiang, H., Gao, G., Yang, F., Guan, J., and Qi, H. (2023). *CmMYB44* might interact with *CmAPS2–2* to regulate starch metabolism in oriental melon fruit. *Plant Physiol. Biochem.* 196, 361–369. https://doi.org/10.1016/j.plaphy.2023.01.047

Wang, C., Qiao, A., Fang, X., Sun, L., Gao, P., Davis, A.R., et al. (2019). Fine mapping of lycopene content and flesh color related gene and development of molecular marker–assisted selection for flesh color in watermelon (*Citrullus lanatus*). *Front. Plant Sci.* 10, 1240. https://doi.org/10.3389/fpls.2019.01240

Wang, L., Wang, Y., Luan, F., Zhang, X., Zhao, J., Yang, Z., et al. (2022). Biparental genetic mapping reveals that *CmCLAVATA3* (*CmCLV3*) is responsible for the variation in carpel number in melon (*Cucumis melo* L.). *Theor. Appl. Genet.* 135, 1909–1921. https://doi.org/10.1007/s00122-022-04083-2

Wibowo, W.A., Maryanto, S.D., and Daryono, S.D. (2021). Phenotypic characters and identification *CYPs* (*Cyclophilin*) gene in *Cucumis melo* L. cv. Gama Melon Parfum. *Biodiversitas* 22, 3007–3014. https://doi.org/10.13057/biodiv/d220601

Wu, S., Sun, H., Gao, L., Branham, S., McGregor, C., Renner, S.S., et al. (2023). A *Citrullus* genus super-pangenome reveals extensive variations in wild and cultivated watermelons and sheds light on watermelon evolution and domestication. *Plant Biotechnol. J.* 21, 1926–1928. https://doi.org/10.1111/pbi.14120

Wu, S., Wang, X., Reddy, U., Sun, H., Bao, K., Gao, L., et al. (2019). Genome of 'Charleston Gray', the principal American watermelon cultivar, and genetic characterization of 1,365 accessions in the U.S. National Plant Germplasm System watermelon collection. *Plant Biotechnol. J.* 17, 2246–2258. https://doi.org/10.1111/pbi.13136

Yang, S., Amanullah, S., Duan, Y., Guo, Y., Xu, M., Bao, X., et al. (2024). Fine genetic mapping and transcriptomic analysis revealed major gene modulating the clear stripe margin pattern of watermelon peel. *Front. Plant Sci.* 15, 1462141. https://doi.org/10.3389/fpls.2024.1462141

Yi, L., Zhou, W., Zhang, Y., Chen, Z., Wu, N., Wang, Y., et al. (2023). Genetic mapping of a single nuclear locus determines the white flesh color in watermelon (*Citrullus lanatus* L.). *Front. Plant Sci.* 14, 1090009. https://doi.org/10.3389/fpls.2023.1090009

Yuan, P., Umer, M.J., He, N., Zhao, S., Lu, X., Zhu, H., et al. (2021). Transcriptome regulation of carotenoids in five flesh-colored watermelons (*Citrullus lanatus*). *BMC Plant Biol.* 21, 203. https://doi.org/10.1186/s12870-021-02965-z

Zhan, Y., Hu, W., He, H., Dang, X., Chen, S., and Bie, Z. (2023). A major QTL identification and candidate gene analysis of watermelon fruit cracking using QTL-seq and RNA-seq. *Front. Plant Sci.* 14, 1166008. https://doi.org/10.3389/fpls.2023.1166008

Zhang, H., Li, M., Yang, Y., Li, Z., Xu, Y., Wang, H., et al. (2022). Molecular mapping for fruit-related traits, and joint identification of candidate genes and selective sweeps for seed size in melon. *Genomics* 114, 110306. https://doi.org/10.1016/j.ygeno.2022.110306

Zhang, W., Ji, Z., Hu, G., Yuan, L., Liu, M., Zhang, X., et al. (2024). *Clpf* encodes pentatricopeptide repeat protein (PPR5) and regulates pink flesh color in watermelon (*Citrullus lanatus* L.). *Theor. Appl. Genet.* 137, 126. https://doi.org/10.1007/s00122-024-04619-8

Zhao, H., Zhang, T., Meng, X., Song, J., Zhang, C., and Gao, P. (2023). Genetic mapping and QTL analysis of fruit traits in melon (*Cucumis melo* L.). *Curr. Issues Mol. Biol.* 45, 3419–3433. https://doi.org/10.3390/cimb45040224

Zhong, Y., Chen, B., Li, M., Wang, D., Jiao, Y., Qi, X., et al. (2020). A DMP-triggered in vivo maternal haploid induction system in the dicotyledonous Arabidopsis. *Nat. Plants* 6, 466–472. https://doi.org/10.1038/s41477-020-0658-7

15 Genetics, Genomics, and Breeding of Health-Relative Bioactive Compounds in Bitter Melons

Sangam L. Dwivedi and Chittaranjan Kole

15.1 BIOACTIVE COMPOUNDS IN BITTER MELON AND HUMAN HEALTH

The pre-clinical efficacy studies revealed that regular consumption of bitter melon (*Momordica charantia*), also known as bitter gourd, is beneficial for managing obesity and type 2 diabetes associated metabolic syndromes and cancers. Plant extract possesses antioxidant, anti-inflammatory, anticancer, antidiabetic, antibacterial, anti-obesity, and immunomodulatory activities. The bioactive compounds in bitter melon are cucurbitane-type triterpenoids, triterpene glycosides, phenolic acids, flavonoids, essential oils, saponins, fatty acids, and proteins. The isolated compounds from these bioactive molecules include kuguacin J, karaviloside XI, kuguaglycoside C, momordicoside Q–U, charantin, α-eleostearic acid, and proteins (α-momorcharin, RNase MC2, MAP30), all possessing potential biological activity (Dandawate et al., 2016; Fang et al., 2019; Gayathry and John, 2022; Raina et al., 2016). Bitter melons are known for their blood-glucose-lowering potential. *In vitro* studies involving wild bitter melon (WBM) extract effectively prevent diabetes and related metabolic abnormalities in HFD/STZ-induced type 2 diabetic rats (Sun et al., 2023). Fruits improve glycemic control while stems/leaves of bitter melons increase plasma insulin concentrations in obese pig models with mild diabetes. The effect of bitter melon fruit on glycemic control is small but consistent and cultivar-specific (Koopmans et al., 2024). The main bioactive compound with anti-hyperglycemic activities of bitter melon is saponin that increases insulin secretion via the PI3K/Akt/FoxO1 signaling pathway (Liu et al., 2021). Type 2 diabetes is also linked to the onset of Alzheimer's disease, a gradual neurodegenerative brain disorder characterized by the chronic loss of cognitive and behavioral functions. The evidence to date suggests that bioactive compounds, including those from bitter melons, may have a direct effect on the treatment of metabolic disorders and neurodegenerative diseases (Richter et al., 2023). Bitter

DOI: 10.1201/9781003467335-15

melon extract has also been found effective in ameliorating high fat diet induced obesity and hyperlipidemia in male Sprague Dawley rats (Mahwish et al., 2017).

The anticancer effects of bitter melon have also been noted to minimize the risk of breast, ovarian, and cervical cancer cells (Psilopatis et al., 2023). The bioactive compounds isolated from bitter melon effectively inhibited cancer development and progression via regulation of protein network, modulating AKT/mTOR/p70S6K signaling, p38MAPK-MAPKAPK-2/HSP-27 pathway, cell cycle regulatory proteins, and apoptosis-associated proteins in cancer cells (Farooqi et al., 2018).

Antiaging effects of bitter melon fruit extract (BMFE) were also noted in mice. Fourteen triterpenoids were simultaneously identified in BMFE, activated by PI3K/AKT signaling pathway protein genes. It improved the learning and memory ability, and significantly improved the activity of antioxidant enzymes and decreased the accumulation of lipid peroxides in aging mice of subacute mice (Wang et al., 2022). Antiaging effects of bitter melon fruit extract were found to extend the lifespan of K6001 yeast. A novel cucurbitane-type triterpenoid significantly increases the survival rate of yeast under oxidative stress and decreases reactive oxygen species (ROS) level (Cao et al., 2018).

Age-related neurological disorders are a common problem as the person advances to ageing. Bitter melon is a rich source of luteolin, which increases acetylcholine in neurons by binding to enzymes in acetylcholine metabolism pathways. This binding inhibits the hyperphosphorylation of tau protein by restraining its kinase enzyme. A cost-effective supplement of bitter melon, with minimal side effects, reduces the risk of age-related neurological disorders (Adarmanabadi et al., 2024).

The various bioactive compounds of bitter melons have also been effective in managing overweight and obesity. Proteins, triterpenoids, saponins, phenolics, and conjugated linolenic acids are the major bioactive components with anti-obesity activities. The mechanisms included inhibition of fat synthesis, promotion of glucose utilization, and stimulation of auxiliary lipid-lowering activity (Fan et al., 2019).

Oxidative stress induced infertility, a serious medical and social human problem, can be overcome using medicinal plants. Bitter melons have many medicinal uses, including overcoming the oxidative-stress induced infertility. Pentacyclic triterpene 13, 14-epoxyoleanan-3-ol-acetate is the most active n-hexane fraction, isolated for the first time from aerial parts of bitter melons. *In vitro* application of 10 mg mL^{-1} of the purified compound equipotent with 25 mg mL^{-1} of vitamin E has been found effective in reducing peroxidation of polyunsaturated fatty acids and improving sexual activity of normal male rats, indicating that it can be used as fertility enhancer in men (Adedokun et al., 2022).

Cutibacterium acne, a key bacterial species, is involved in the development and progression of acne (*Acne vulgaris*) inflammation. Ethanol-soluble fraction of WBM leaf extract modulates *in vitro* and *in vivo C. acnes*–induced inflammatory responses in humans and mice (Chuang et al., 2020). Further investigation has revealed that 5β,19-epoxycucurbita-6,23-dien-3β,19,25-triol (Kuguacin R; KR) and 3β,7β,25-trihydroxycucurbita-5,23-dien-19-al (TCD) are major cucurbitane-type triterpenoids associated with acne-induced inflammation (Chuang et al., 2021).

A 12-week randomized placebo-controlled clinical trial involving bitter melon extract exhibited glucose-lowering effects through suppression of glucagon levels in Korean people with prediabetes (Kim et al., 2022). Meta-analysis involving 414 patients in total and 4–16 weeks of follow-up revealed the effect of bitter melon in treatment (randomized control trials) over placebo had no significant effect on fasting blood glucose level, HbA1c level, lipid profiles, management of body weight or on blood (systolic and diastolic) pressure, though small changes in values were noted, between the treatments that were too short to reliably detect long-term effects. The effect of bitter melon treatment as compared to placebo was not associated with a notable change in alanine aminotransferase (ALT), aspartate aminotransferase (AST), and creatinine levels (Laczkó-Zöld et al., 2023). Additional research is therefore recommended involving placebo and controlled human trials for a longer duration.

15.2 CUCURBITANE-TYPE TRITERPENOIDS FROM BITTER MELON FOR DEVELOPING FUNCTIONAL AND THERAPEUTIC FOOD

Bitter melons, though not liked by many because of its fruit bitterness, is a storehouse of several key nutrients, with more than 60 bioactive compounds (alkaloids, carotenoids, cucurbitane triterpenoids, phenolics, and phytosterols) active against more than 30 diseases, including cancer and diabetes. The incorporation of these bioactive compounds is a potential area for developing functional and therapeutic foods or alternative medicines by including plant-based molecules. Bitter melons have antidiabetic, anticancer, antioxidant, antidementia, antimicrobial, anthelmintic, hypolipidemic, and hypotensive activities (Gayathry and John, 2022).

The two chemotypes identified in bitter melons, cucurbitane-type triterpene glycosides (CTGs) and oleanane-type triterpene glycosides (OTGs), differ in their richness in cucurbitane versus oleanane triterpenoid glycosides. CTGs chemotypes are rich in cucurbitane-type triterpene glycosides, while OTGs in oleanane-type triterpene glycosides. The discovery of two chemotypes may play a crucial role in the selection of bitter melon varieties rich in bioactive components for the development of diabetes therapeutics or dietary supplements (Zhou et al., 2019).

Twenty-one types of cucurbitane triterpenoids, including two new cucurbitane glycosides (charantoside IX and X), were reported from bitter melon fruits (Yue et al., 2017). Seven cucurbitane-type triterpenoids (which include triterpenes and saponins), including six undescribed compounds, kuguaovins H–M, and a known compound, momordicoside K, were isolated from wild bitter melon vines. Kuguaovins H, J, and L variants exhibited cytotoxicity against various types of cancers (Liaw et al., 2022a). The same group isolated four novel triterpene glycosides, taimordisins A-D (1–4), from fresh fruits, which possess anti-inflammatory biological activities and are safe to use (Liaw et al., 2022b). Another natural cucurbitane-type triterpenoid, charantin (a mixture of β-sitosterol and stigmastadienol glucosides), has been reported in maturing bitter melon fruits (Cuong et al., 2017). Cucurbitane triterpenoids regulate type 2 diabetes by exhibiting α-glucosidase inhibitory activity (Yue et al., 2017). 7β,23(S)-dimethoxycucurbita-5,24-dien-3β-ol and 3β-hydroxycucurbita-6,23(E),25-trien-5β,19-olide are two new cucurbitane-type triterpenoids, in addition to two known cucurbitane-type triterpenoids, which were reported from the methanol based dried fruit pulp extract (Chen et al., 2017). Differences in cucurbitane-type triterpenoids were also noted among root, vine, and fruit tissues of bitter melons (Chen et al., 2008, Chang et al., 2008, Akihisa et al., 2007).

Artificial neural network (ANN) and stepwise regression analysis are used to optimize and predict the accumulation of total carotenoids, phenols, flavonoids, and diosgenin contents in the leaves and the unripe and ripe fruits of the bitter melon plant in response to elicitors (phytohormones). The study revealed ANN models ($R^2 = 0.60–0.98$) are more accurate and sensitive modeling tools for optimization and prediction than regression models ($R^2 = 0.24–0.69$). The plant organs have a pivotal role in the amount of total carotenoid, phenols, and flavonoids contents. The elicitor type is the main determinant factor in diosgenin content. The application of 40.5 mM β-aminobutyric acid (BABA), 3.99 mM salicylic acid (SA), 1.84 mM methyl jasmonate acid (MeJA), and 4.4 mM SA, respectively, produced highest amounts of total carotenoids, phenols, and flavonoids in leaves and diosgenin in unripe fruits. Thus, ANN is an accurate, low-cost, time-saving, and powerful tool to predict and optimize the production of secondary metabolite (Jandoust et al., 2023).

15.3 NATURAL VARIATION FOR COMPOUNDS ASSOCIATED WITH BITTERNESS IN BITTER MELONS

Significant differences in total phenols, flavonoids, organic acid, and triterpenoids as well on antioxidative activity were reported in bitter melon cultivars. Metabolic analysis detected 370 secondary metabolites, with flavonoids and phenolics being predominant. The differentially accumulated metabolites were mainly associated with secondary metabolic pathways, including pathways of flavonoid, flavonol, isoflavonoid, flavone, folate, and phenylpropanoid biosynthesis. Twenty-seven of these metabolites were significantly or negatively correlated with antioxidative capacity ($r \geq 0.7$), indicating the bitter melon varieties contain plethora of bioactive compounds that can be used to develop novel molecules with significant medicinal value (Zhang et al., 2023). Another study differing in fruit pericarp color, dark green (mo), pale green (lv), and white (bai), detected 93 metabolites, including 90 flavonoids and three tannins. The flavonoids mainly included isoflavones, dihydroflavones, flavanols, flavonols, flavonoids, chalcones, flavonoid carbonosides, and dihydroflavonols. Both 'lv' and 'bai' fruit types relative to 'mo' fruit type had 21 and 25 different metabolites, respectively, while only nine metabolites differentiated 'lv' and 'bai' fruits. The 'mo' type fruits showed greater vitexin and isovitexin contents. This suggests that different metabolites in colored bitter melons are mainly involved in the biosynthesis of flavonols, flavonoid carbonosides, and flavonoids (Zhang et al., 2022). Bitter melon accessions with diverse pericarp fruit colors (grayish orange, grayish yellow, very soft green, dark grayish cyan) exhibited 112 significant metabolite compounds. The variation in six anthocyanin contents drives the differences in pericarp color (Yang et al., 2024).

Analysis of seed metabolites of 55 bitter melon samples from different parts of the China revealed two major chemotypes, CTGs and OTGs. These two chemotypes differ in their richness in cucurbitane versus oleanane triterpenoid glycosides (Zhou et al., 2019).

Bitter melon accessions had significant differences in morphological (leaf color, tendril character, sex form, fruit color, fruit shape, tubercle character, fruit length, fruit girth, fruit weight), nutritional (β-carotene, ascorbic acid) and nutraceutical (chlorophyll, total phenols, phenolic acids, flavonoids, radical scavenging activities, charantin) contents. Indian-origin accessions with dark green, spindle-shaped fruits with sharp tubercles were associated with high nutrient and most nutraceutical contents but were comparatively low in charantin content. Those with whitish yellow, yellowish white, obtuse, or roundish-oblong fruit with light ridges or roundish blunt tubercles from East Asia region were associated with comparatively low nutrient content and high nutraceutical content but high in charantin content as well. Morphological characters influenced the phytochemical contents. Hence, combining morphological characteristics with nutrient and nutraceutical contents may be used to develop functionally rich bitter melon cultivars (Hazra et al., 2022). Indian-origin bitter melon accessions, involving wild species, landraces, primitives, and other germplasms, had significant genetic variation for ascorbic acid, chlorophyll content, Ca, Mg, P, Fe, and Zn contents. Unique accessions high in ascorbic acid and Zn or in Mg, Ca, P, and Fe contents were also identified (Meghashree et al., 2024). The differences in ascorbic acid, amino acids, and phenolic compounds were also reported among bitter melon cultivars (Perez et al., 2018). Large variability in charantin content (15.1 to 161.4 $\mu g \cdot g^{-1}$ dry-weight basis), an antidiabetic compound, and accessions with high charantin content were reported in Korean bitter melon germplasms (Kim et al., 2013).

15.4 GENETIC AND MOLECULAR BASIS OF BITTERNESS IN BITTER MELON

A large number of SSR, InDels, SNPs, CNVs, and SVs were reported from bitter melon genomes (Cui et al., 2017, 2021; Fu et al., 2023) that paved the way to construct varying density of genetic maps (Cui et al., 2021; Kaur et al., 2021).

The bitter melon fruit pericarp color (dark green vs. white) has a significant impact on consumer preferences. *McPRR2,* a homolog of *Arabidopsis pseudo response regulator* 2 (*APRR2*), is the key gene determining pericarp color in bitter melons. A 15-bp nucleotide insertion in the coding region of *McPRR2* confers with bitter gourd pericarp color. Marker, Mc958, linked to pericarp color may be deployed to breed bitter gourd populations with specific pericarp fruit color (Guo et al., 2024). Furthermore, *McPRR2* (*McAPRR2*) revealed three primary haplotypes: *Hap1*, a wild type with a green pericarp; *Hap2*, a South Asia (SA) and Southeast Asia (SEA) type with a green pericarp; and *Hap3*, a light green pericarp primarily from SEA. *McAPRR2* haplotypes are significantly correlated with both pericarp color and ecological adaptation (Tian et al., 2023).

Ten putative phenylpropanoid and flavonoid biosynthetic genes are reported, with rutin and epicatechin being the most abundant in bitter melon. While the rutin is abundant in leaves, epicatechin is highly accumulated in flowers and fruits. The trans-cinnamic acid, p-coumaric acid, ferulic acid, kaempferol, and rutin accumulation coincide, respectively, with the expression patterns of *McPAL*, *McC4H*, *McCOMT*, *McFLS*, and *Mc3GT* (Cuong et al., 2018).

Fifteen candidate genes encoding enzymes related to triterpenoid biosynthesis were unlocked, with most genes highly expressed in flower and/or fruit from the ripening stages. The accumulation of charantin was highest in fruits from the ripening stage and coincided with the expression pattern of *McSE* and *McCAS1*, suggesting that both genes play important roles in charantin biosynthesis in bitter melon (Cuong et al., 2017).

The bitter melon seed oil is a rich source of conjugated fatty acids (polyunsaturated fatty acids containing at least one pair of conjugated double bonds, mainly α-eleostearic acid), with a significant therapeutic value to treat anti-ageing and obesity associated with metabolic syndromes. Fruit transcriptome analysis of green rind, white rind, pulp, immature seeds, and mature seeds unlocked 26, 3, and 10 full-length genes coding for enzymes associated with synthesis of conjugated fatty acids and triacylglycerol (Revichandiran and Parani, 2022). This suggests that plants deficient in conjugated fatty acids can be metabolically engineered to produce a higher amount of these fatty acids by transferring the relevant genes.

15.5 NAVIGATING GENOMES TO ACCELERATE DEVELOPMENT OF TAILOR-MADE BITTER MELON CULTIVARS WITH ENHANCED NUTRITIONAL AND NUTRACEUTICAL VALUES

Momordica balsamina is the closest wild relative species of cultivated bitter melon (*M. charantia*). It has multiple resistance to major biotic and abiotic stresses. The species is easily hybridized with cultivated bitter melon. The high quality of chromosome assembly genome of *M. balsamina* is of 384.90 Mb and N50 30.96 Mb in size. It contains 6,32,098 transposons elements (TEs), 215,379 simple sequence repeats (SSRs), 567,483 TF binding sites, 3,376 noncoding RNA genes and 41,652 protein-coding genes, 4,347 disease resistance-related, 67 heat-stress-related, 15 salt-stress-related, 5 carotenoid-related, 229 cucurbitacin-related, 19 terpene-related, 37 antioxidant activity, and 6 sex-determination-related genes. This information is web-catalogued at http://webtom.cabgrid.res.in/mbger/, accessed on May 12, 2024, and may facilitate introgression of desirable traits from *M. balsamina* to cultivated bitter melon (Vinay et al., 2023). The de novo assembly of draft genome sequence of a monoecious bitter gourd inbred line OHB3–1 contains scaffolds of 285.5 Mb in length, equivalent to estimated 84% of the of bitter melon genome, 339 Mb. The genome contains 5,859 protein-coding genes and 15.3% of TEs. The genome is close to watermelon (*Citrullus lanatus*) than to cucumber (*Cucumis sativus*) or melon (*C. melo*). The presence of putative trypsin-inhibitor and ribosome-inactivating genes are indicative of the fact that bitter melon is a medicinal plant (Urasaki et al., 2017).

The whole-genome sequencing of a bitter melon cultivar Dali-11, a wild small-fruited line TR, and re-sequencing of diverse bitter melon germplasms unlocked the major gene (*Bi*) clusters involved in biosynthesis of cucurbitane triterpenoids causing bitter taste. This gene cluster is highly conserved in cucumber, melon, and watermelon (Cui et al., 2020). An assembled 295.6-Mb telomere-to-telomere high-quality bitter melon genome is anchored to 11 chromosomes comprising of 98 contigs and 95-scaffolds. It harbors 19,895 protein-coding genes, 45.59% of which constitute predicted repeat sequences, 211,329 SNPs, 100,187 InDels, 210 copy number variations (CNVs), and 374 structural variations (SVs). Synteny analysis revealed variations involved in fruit quality during the divergence of bitter melons. The mechanisms of pigment accumulation and cucurbitacin biosynthesis in bitter melons have been unearthed. A large number of differentially expressed genes (DEGs) and differentially expressed metabolites (DEMs) were unfolded in unripe and matured fruit peel and pulp tissues (Fu et al., 2023). Such genomic information may provide new avenues to breed bitter melon cultivars with targeted fruit metabolites for developing nutritionally enhanced and therapeutic functional foods.

15.6 BREEDING TO DEVELOP NUTRITIONALLY AND THERAPEUTICALLY ENHANCED BITTER MELON CULTIVARS

Sufficient variation exists for agro-morphological and fruit-yield traits in bitter melon germplasms (Alhariri et al., 2021; Bhati et al., 2023; Mallikarjuna et al., 2023; Tomar et al., 2023). In contrast, only limited germplasms that were screened showed sufficient variability for macro- and micro-minerals (Singla et al., 2023) and for bioactive compounds to effect targeted improvement (Hazra et al., 2022; Kim et al., 2013; Mahwish et al., 2018; Perez et al., 2018; Zhou et al., 2019). Using conventional crossing and selection, largely for yield and adaptation traits, several breeding lines and cultivars with improved fruit yield and/or bioactive compounds have been developed (Trierweiler et al., 2019). There is a growing interest in plant secondary metabolites conferring attractive fruits and vegetables colors, rich in nutritional and nutraceutical properties. Some groups of vegetables such as those from the family Cucurbitaceae, Solanaceae, or Brassicaceae are consumed worldwide. Naturally these groups of vegetables are attractive targets of genetic improvement by exploiting allelic variation associated with secondary metabolites (bioactive compounds such as carotenoids, flavonoids, and phenolics). Advances in development of genetic and genomic resources, including development of a reference genome or resequencing of large collection of germplasm, are paving the way to accelerate development of new germplasm rich in bioactive compounds on these crops, including bitter melons (Bassolino et al., 2022; Mattoo et al., 2022).

A rapid, easily accessible, and cost-effective high-throughput phenotyping is needed to screen segregants with specific metabolites in large breeding populations. Green extraction methodology using ultrasonically assisted extraction and choline chloride–acetic acid (CHAC)–based natural deep eutectic solvent has been optimized to enhance the extraction of metabolites from bitter melon leaves. CHAC-based extraction significantly improved the extraction of total phenolics, flavonoids, and individual phenolic compounds compared to conventional solvents (water, ethanol, or methanol). A molar ratio of 1:4.35 CHAC with 20.68% water content at 75°C for 21.23 min has been optimized for extraction of total phenols, flavonoids, antioxidant activity, gallic acid, chlorogenic acid, vanillic acid, epicatechin, syringic acid, quercetin-3-glucoside, and salicylic acid, respectively, of which, vanillic acid, salicylic acid and chlorogenic acid being highly bio-accessible (Zannou et al., 2022). This technology has also been optimized in diverse crops for extraction of phenolics and flavonoids in crops as diverse as black mulberry (*Morus nigra*) or jackfruit (*Artocarpus heterophyllus*) (Ly and Sothornvit, 2024; Vo et al., 2023) suggesting that it may be adopted in other crops as well with some modifications. Despite advances in extraction technology, further work is needed to develop a reliable, simple, and cost-effective assay for screening of early generation breeding populations to select segregants with specific bioactive compounds.

15.7 CONCLUSION

All parts of the bitter melon are rich sources of nutrients and bioactive (secondary metabolite) compounds. The evidence to date suggests variability of nutritional and nutraceutical compounds in germplasms. Targeted evaluation of the gene pool may provide additional and novel sources of variation. The sequencing of a reference genome, including a close relative of the bitter melon, resulted in identification of several putative genes associated with bioactive compounds, which should be functionally characterized to develop genetic tags (marker) linked with nutrients or bioactive compounds to tailor germplasm/cultivars with specific plant-based molecules for drug therapy or for developing nutritious and therapeutic foods.

REFERENCES

Adarmanabadi, S.M.H.H., Abadi, O.J.K., Amiri, A., Tamannaeifar, R., Balanian, S., Rasekhejam, M., et al. (2024). Pharmacotherapeutic potential of bitter gourd (*Momordica charantia*) in age-related neurological diseases. *J. Integr. Neurosci.* 23, 86. https://doi.org/10.31083/j.jin2304086

Adedokun, O., Gbolade, A., and Ayinde, B. (2022). 13, 14-Epoxyoleanan-3-ol-acetate: A male fertility-enhancing constituent from hexane fraction of *Momordica charantia* L. (*Cucurbitaceae*). *Turk. J. Pharm Sci.* 19, 180–186. https://doi.org/10.4274/tjps.galenos.2021.48264

Akihisa, T., Higo, N., Tokuda, H., Ukiya, M., Akazawa, H., Tochigi, Y., et al. (2007). Cucurbitane-type triterpenoids from the fruits of *Momordica charantia* and their cancer chemopreventive effects. *J. Nat. Prod.* 70, 1233–1239. https://doi.org/10.1021/np068075p

Alhariri, A., Beera, T.K., Jat, G.S., Devi, MB., Boopalakrishnan, G., Teleb, A.A., et al. (2021). Analysis of genetic diversity and population structure in bitter gourd (*Momordica charantia* L.) using morphological and SSR markers. *Plants* 10, 1860. https://doi.org/10.3390/plants10091860

Bassolino, L., Petroni, K., Plito, A., Marinelli, A., Ferrari, E.A.M., Ficco, D.B.M., et al. (2022). Does plant breeding for antioxidant-rich foods have an impact on human health? *Antioxidants* 11, 794. https://doi.org/10.3390/antiox11040794

Bhati, S., Prakash, S., Kumar, V., Singh, B., Singh, M.K., Gangwar, L.K., et al. (2023). Genetic variability, heritability and genetic advance for yield and yield contributing characters in bitter gourd (*Momordica charantia* L.) genotypes. *Int. J. Environ. Clim.* 13, 1867–1875. https://doi.org/10.9734/ijecc/2023/v13i92418

Cao, X., Sun, Y., Lin, Y., Pan, Y., Farooq, U., Xiang, L., et al. (2018). Antiaging of cucurbitane glycosides from fruits of *Momordica charantia* L. *Oxid. Med. Cell. Longe.* 2018, Article ID 1538632. https://doi.org/10.1155/2018/1538632

Chang, C.-I., Chen, C.R., Liao, Y.-W., Cheng, H.-L., Chen, Y.-C., and Chou, C.-H. (2008). Cucurbitane-type triterpenoids from the stems of *Momordica charantia. J. Nat. Prod.* 71, 1327–1330. https://doi.org/10.1021/np070532u

Chen, C.R., Liao, Y.-W., Kuo, Y.-H., Hsub, J.-L., and Chang, C.-I. (2017). Cucurbitane-type triterpenoids from *Momordica charantia. Nat. Prod. Commun.* 12, 889–892.

Chen, J., Tian, R., Qiu, M., Lu, L., Zheng, Y., and Zhang, Z. (2008). Trinorcucurbitane and cucurbitane triterpenoids from the roots of *Momordica charantia*. *Phytochemistry* 69, 1043–1048. https://doi.org/10.1016/j.phytochem.2007.10.020

Chuang, L.-T., Huang, W.-C., Hou, Y.-C., Chyuan, J.-H., Chang, H., Chang, C.-I., et al. (2021). Suppressive effect of two cucurbitane-type triterpenoids from *Momordica charantia* on *Cutibacterium acnes*-induced inflammatory responses in human THP-1 monocytic cell and mouse models. *Molecules* 26, 579. https://doi.org/10.3390/molecules26030579

Chuang, L.-T., Shih, Y.-H., Huang, W.-C., Lin, L.-C., Hsu, C., Chyuan, J.-H., et al. (2020). In vitro and in vivo screening of wild bitter melon leaf for anti-inflammatory activity against *Cutibacterium acnes*. *Molecules* 25, 4277. https://doi.org/10.3390/molecules25184277

Cui, J., Cheng, J., Nong, D., Peng, J., Hu, Y., He, Y., et al. (2017). Genome-wide analysis of simple sequence repeats in bitter gourd (*Momordica charantia*). *Front. Plant Sci.* 8, 1103. https://doi.org/10.3389/fpls.2017.01103

Cui, J., Peng, J., Cheng, J., and Hu, K. (2021). Development and validation of genome wide InDel markers with high levels of polymorphism in bitter gourd (*Momordica charantia*). *BMC Genom.* 22, 190. https://doi.org/10.1186/s12864-021-07499-0

Cui, J., Yang, Y., Luo, S., Wang, L., Huang, R., Wen, Q., et al. (2020). Whole-genome sequencing provides insights into the genetic diversity and domestication of bitter gourd (*Momordica* spp.). *Hortic. Res.* 7, 85. https://doi.org/10.1038/s41438-020-0305-5

Cuong, D.M., Jeon, J., Morgan, A.M.A., Kim, C., Kim, J.K., Lee, S.Y., et al. (2017). Accumulation of charantin and expression of triterpenoid biosynthesis genes in bitter melon (*Momordica charantia*). *J. Agric. Food Chem.* 65, 7240–7249. https://doi.org/10.1021/acs.jafc.7b01948

Cuong, D.M., Kwon, S.-J., Jeon, J., Park, Y.J., Park, J.S., and Park, S.U. (2018). Identification and characterization of phenylpropanoid biosynthetic genes and their accumulation in bitter melon (*Momordica charantia*). *Molecules* 23, 469. https://doi.org/10.3390/molecules23020469

Dandawate, P.R., Subramaniam, D., Padhye, S.B., and Anant, S. (2016). Bitter melon: A panacea for inflammation and cancer. *Chin. J. Nat. Med.* 14, 81–100. https://doi.org/10.1016/S1875-5364(16)60002-X

Fan, M., Kim, E.-K., Choi, Y.-J., Tang, Y., and Moon, S.-H. (2019). The Role of *Momordica charantia* in resisting obesity. *Int. J. Environ. Res. Public Health* 16, 3251. https://doi.org/10.3390/ijerph16183251

Fang, E.F., Froetscher, L., Scheibye-Knudsen, M., Bohr, V.A., Wong, J.H., Wong, J.H., et al. (2019). Emerging antitumor activities of the bitter melon (*Momordica charantia*). *Curr. Protein Pept. Sci.* 20, 296–301. https://doi.org/10.2174/1389203719666180622095800

Farooqi, A.A., Khalid, S., Tahir, F., Sabitaliyevich, U.Y., Yalim, I., Attar, R., et al. (2018). Bitter gourd (*Momordica charantia*) as a rich source of bioactive components to combat cancer naturally: Are we on the right track to fully unlock its potential as inhibitor of deregulated signaling pathways. *Food Chem. Toxicol.* 119, 98–105. https://doi.org/10.1016/j.fct.2018.05.024

Fu, A., Zheng, Y., Guo, J., Grierson, D., Zhao, X., Wen, C., et al. (2023). Telomere-to-telomere genome assembly of bitter melon (*Momordica charantia* L. var. *abbreviata* Ser.) reveals fruit development, composition and ripening genetic characteristics. *Hortic. Res.* 10, uhac228. https://doi.org/10.1093/hr/uhac228

Gayathry, K.S., and John, J.A. (2022). A comprehensive review on bitter gourd (*Momordica charantia* L.) as a gold mine of functional bioactive components for therapeutic foods. *Food Prod. Process Nutr.* 4, 10. https://doi.org/10.1186/s43014-022-00089-x

Guo, J., Han, X., Wu, T., Wang, R., Zhao, J., Wang, R., et al. (2024). Potential locus *W* and candidate gene *McPRR2* associated with pericarp pigment accumulation in bitter gourd (*Momordica charantia* L.) revealed via BSA-seq analysis. *Plant Physiol. Biol.* 208, 108515. https://doi.org/10.1016/j.plaphy.2024.108515

Hazra, P., Hazra, S., Acharya, B., Dutta, S., Saha, S., Mahapatra, P., et al. (2022). Diversity of nutrient and nutraceutical contents in the fruits and its relationship to morphological traits in bitter gourd (*Momordica charantia* L.). *Sci. Hortic.* 305, 111414. https://doi.org/10.1016/j.scienta.2022.111414

Jandoust, S., Shojaeiyan, A.A., Ayyari, M., Tohidfar, M., Ahmadi, H., and Ebrahimi, S.N. (2023). Using artificial neural network for prediction of accumulation of bioactive compounds in *Momordica charantia* through application of different elicitors. *Ind. Crop. Prod.* 192, 115984. https://doi.org/10.1016/j.indcrop.2022.115984

Kaur, G., Pathak, M., Singla, D., Sharma, A., Chhuneja, P., and Sarao, N.K. (2021). High-density GBS-based genetic linkage map construction and QTL identification associated with yellow mosaic disease resistance in bitter gourd (*Momordica charantia* L.). *Front. Plant Sci.* 12, 671620. https://doi.org/10.3389/fpls.2021.671620

Kim, A.K., Lee, H.J., Oh, M.M., Lee, W.M., Lee, S.G., Chae, W.B., et al. (2013). Selection of Bitter Gourd (*Momordica charantia* L.) germplasm for improvement anti-diabetic compound contents. *Korean J. Breed. Sci.* 45, 332–338. http://dx.doi.org/10.9787/KJBS.2013.45.4.332

Kim, B., Lee, H.S., Kim, H.-J., Lee, H., Lee, I.-Y., Ock, S., et al. (2022). *Momordica charantia* (bitter melon) efficacy and safety on glucose metabolism in Korean prediabetes participants: A 12-week, randomized clinical study. *Food Sci. Biotechnol.* 32, 697–704. https://doi.org/10.1007/s10068-022-01214-9

Koopmans, S.J., Binnendijk, G., Ledoux, A., Choi, Y.H., Mes, J.J., Guan, X., et al. (2024). *Momordica charantia* fruit reduces plasma fructosamine whereas stems and leaves increase plasma insulin in adult mildly diabetic obese Göttingen minipigs. *PLoS ONE* 19, e0298163. https://doi.org/10.1371/journal.pone.0298163

Laczkó-Zöld, E., Csuper-Löffler, B., Kolcsár, E.-B., Ferenci, T., Nan, M., Tóth, B., et al. (2023). The metabolic effect of *Momordica charantia* cannot be determined based on the available clinical evidence: A systematic review and meta-analysis of randomized clinical trials. *Front. Nutr.* 10, 1200801. https://doi.org/10.3389/fnut.2023.1200801

Liaw, C.-C., Huang, H.-T., Liu, H.-K., Lin, Y.-C., Zhang, L-J., and Wei, W.-C. (2022a). Cucurbitane-type triterpenoids from the vines of *Momordica charantia* and their anti-inflammatory, cytotoxic, and antidiabetic activity. *Phytochem.* 195, 113026. https://doi.org/10.1016/j.phytochem.2021.113026

Liaw, C.-C., Lo, I.-Q., Lin, Y.-C., Huang, H.-T., Zhang, L.-J., Hsiao, P.-C., et al. (2022b). Four cucurbitane glycosides taimordisins A-D with novel furopyranone skeletons isolated from the fruits of *Momordica charantia*. *Food Chem. X* 14, 100286. https://doi.org/10.1016/j.fochx.2022.100286

Liu, Y., Mu, S., Chen, W., Liu, S., Cong, Y., Liu, J., et al. (2021). Saponins of *Momordica charantia* increase insulin secretion in INS-1 pancreatic β-cells via the PI3K/Akt/FoxO1 signaling pathway. *Endocrinol Diabetes Nutr. (Engl Ed)* 68, 329–337. https://doi.org/10.1016/j.endien.2021.08.004

Ly, L., and Sothornvit, R. (2024). Novel sequential microwave-ultrasonic-assisted extraction of phenolic compounds from Jackfruit (*Artocarpus heterophyllus* Lam.) peel waste using choline chloride-lactic acid deep eutectic solvent. *Food Bioprocess. Technol.* https://doi.org/10.1007/s11947-024-03393-1

Mahwish, M., Saeed, F., Nisa, M.U., and Nadeem, M.T. (2018). Minerals and phytochemical analysis of bitter melon fruits and its components in some indigenous and exotic cultivars. *Bioscience J.* 34, 1622–1631. https://doi.org/10.14393/BJ-v34n6a2018-39827

Mahwish, Saeed, F., Arshad, M.S., Nisa, M.U., Nadeem, M.T., and Arshad, M.U. (2017). Hypoglycemic and hypolipidemic effects of different parts and formulations of bitter gourd (*Momordica Charantia*). *Lipids Health Dis.* 16, 211. https://doi.org/10.1186/s12944-017-0602-7

Mallikarjuna, K.N., Tomar, B.S., Mangal, M., Singh, N., Singh, D., Kumar, S., et al. (2023). Genetic diversity and population structure analyses in bitter gourd (*Momordica charantia* L.) based on agro-morphological and microsatellite markers. *Plants* 12, 3512. https://doi.org/10.3390/plants12193512

Mattoo, A.K., Dwivedi, S.L., Dutt, S., Singh, B., Garg, M., and Ortiz, R. (2022). Anthocyanin-rich vegetables for human consumption—Focus on potato, sweetpotato and tomato. *Int. J. Mol. Sci.* 23, 2634. https://doi.org/10.3390/ijms23052634

Meghashree, J.R., Ganiger, V.M., Kumar, J.S.A., Bhuvaneshwari, G., Gopali, J.B., Evoor, S., et al. (2024). Genetic diversity and population structure assessment of Indian bitter gourd accessions using nutritional content and molecular markers. *Genet. Resour. Crop Evol.* 71, 1557–1573. https://doi.org/10.1007/s10722-023-01709-2

Perez, J.L., Jayaprakashs, G.K., Crosby, K., and Patil, B.S. (2018). Evaluation of bitter melon (*Momordica charantia*) cultivars grown in Texas and levels of various phytonutrients. *J. Sci. Food Agric.* 99, 379–390. https://doi.org/10.1002/jsfa.9199

Psilopatis, I., Vrettou, K., Giaginis, C., and Theocharis, S. (2023). The role of bitter melon in breast and gynecological cancer prevention and therapy. *Int. J. Mol. Sci.* 24, 8918. https://doi.org/10.3390/ijms24108918

Raina, K., Kumar, D., and Agarwal, R. (2016). Promise of bitter melon (*Momordica charantia*) bioactives in cancer prevention and therapy. *Semin Cancer Biol.* 40–41, 116–129. https://doi.org/10.1016/j.semcancer.2016.07.002

Ravichandiran, K., and Parani, M. (2022). Transcriptome analysis of five different tissues of bitter gourd (*Momordica charantia* L.) fruit identifies full-length genes involved in seed oil biosynthesis. *Sci. Rep.* 12, 15374. https://doi.org/10.1038/s41598-022-19686-4

Richter, E., Geetha, T., Burnett, D., Broderick, T.L., and Babu, J.R. (2023). The effects of *Momordica charantia* on type 2 diabetes Mellitus and Alzheimer's disease. *Int. J. Mol. Sci.* 24, 4643. https://doi.org/10.3390/ijms24054643

Singla, D., Sangha, M.K., Singh, M., Pathak, M., and Bala, M. (2023). Variation of mineral composition in different fruit parts of bitter gourd (*Momordica charantia* L.). *Biol. Trace. Elem. Res.* 201, 4961–4971. https://doi.org/10.1007/s12011-022-03546-3

Sun, K., Ding, M., Fu, C., Li, P., Li, T., Fang, L., et al. (2023). Effects of dietary wild bitter melon (*Momordica charantia* var. *abbreviate* Ser.) extract on glucose and lipid metabolism in HFD/STZ-induced type 2 diabetic rats. *J. Ethnopharmacol.* 306, 116154. https://doi.org/10.1016/j.jep.2023.116154

Tian, S., Yang, J., Fu, Y., Zhang, X., Zhang, J., Zhao, H., et al. (2023). *McAPRR2*: The key regulator of domesticated pericarp color in bitter gourd. *Plants (Basel)* 12, 3585. https://doi.org/10.3390/plants12203585

Tomar, B.S., Mangal, M., Singh, N., Mallikarjuna, K.N., Singh, D., Kumar, S., et al. (2023). Genetic diversity and population structure analyses in bitter gourd (*Momordica charantia* L.) based on agro-morphological and microsatellite markers. *Plants* 12, 3512. https://doi.org/10.3390/plants12193512

Trierweiler, B., Frechen, M.A., Soukup, S.T., Egert, B., and Baldermann, S. (2019). Bitter gourd, *Momordica charantia* L., breeding lines differ in secondary metabolite content according to market type. *J. Appl. Bot. Food Qual.* 92, 106–115. https://doi.org/10.5073/JABFQ.2019.092.015

Urasaki, N., Takagi, H., Natsume, S., Uemura, A., Taniai, N., Miyagi, N., et al. (2017). Draft genome sequence of bitter gourd (*Momordica charantia*), a vegetable and medicinal plant in tropical and subtropical regions. *DNA Res.* 24, 51–58. https://doi.org/10.1093/dnares/dsw047

Vinay, N.D., Singh, K., Ellur, R.K., Chinnusamy, V., Jaiswal, S., Iquebal, M.A., et al. (2023). High-quality *Momordica balsamina* genome elucidates its potential use in improving stress resilience and therapeutic properties of bitter gourd. *Front. Plant Sci.* 14, 1258042. https://doi.org/10.3389/fpls.2023.1258042

Vo, T.P., Pham, T.V., Weina, K., Tran, T.N.H., Vo, L.T.V., Nguyen, P.T., et al. (2023). Green extraction of phenolics and flavonoids from black mulberry fruit using natural deep eutectic solvents: Optimization and surface morphology. *BMC Chem.* 17, 119. https://doi.org/10.1186/s13065-023-01041-x

Wang, D., Wang, E., Li, Y., Teng, Y., Li, H., Jiao, L., et al. (2022). Anti-aging effect of *Momordica charantia* L. on d-galactose-induced subacute aging in mice by Activating PI3K/AKT signaling pathway. *Molecules* 27, 4502. https://doi.org/10.3390/molecules27144502

Yang, L., Li, Z., Li, J., Ma, Y., Miao, M., Long, H., et al. (2024). Integrative metabolome and transcriptome analyses reveal the pericarp coloration mechanisms in bitter melon (*Momordica charantia* L.). *Horticulturae* 10, 291. https://doi.org/10.3390/horticulturae10030291

Yue, J., Xu, J., Cao, J., Zhang, X., and Zhao, Y. (2017). Cucurbitane triterpenoids from *Momordica charantia* L. and their inhibitory activity against α-glucosidase, α-amylase and protein tyrosine phosphatase 1B (PTP1B). *J. Funct. Foods* 37, 624–631. https://doi.org/10.1016/j.jff.2017.07.041

Zannou, O., Pashazadeh, H., Ghellam, M., Redha, A.A., and Koca, I. (2022). Enhanced ultrasonically assisted extraction of bitter melon (Momordica charantia) leaf phenolic compounds using choline chloride-acetic acid–based natural deep eutectic solvent: An optimization approach and in vitro digestion. *Biomass Convers. Bioref.* 14, 11491–11503. https://doi.org/10.1007/s13399-022-03146-0

Zhang, H., Du, X., Yu, J., Jin, H., and Liu, N. (2022). Comparative metabolomics study of flavonoids in the pericarp of different coloured bitter gourds (*Momordica charantia* L.). *Physiol. Mol. Biol. Plants* 28, 1347–1357. https://doi.org/10.1007/s12298-022-01210-7

Zhang, Y., Lu, P., Jin, H., Cui, J., Miao, C., He, L., et al. (2023). Integrated secondary metabolomic and antioxidant ability analysis reveals the accumulation patterns of metabolites in *Momordica charantia* L. of different cultivars. *Int. J. Mol. Sci.* 24, 14495. https://doi.org/10.3390/ijms241914495

Zhou, S., Allard, PM., Wolfrum, C., Ke, C., Tang, C., Ye, Y., et al. (2019). Identification of chemotypes in bitter melon by metabolomics: A plant with potential benefit for management of diabetes in traditional Chinese medicine. *Metabolomics* 15, 104. https://doi.org/10.1007/s11306-019-1565-7

16 Profiling Heirloom Tomato Germplasm
Metabolites, Textural Traits, and Consumer Preferences

Rodomiro Ortiz and Sangam L. Dwivedi

16.1 INTRODUCTION

While there exist numerous informal definitions for heirloom tomatoes (*Solanum lycopersicum* L.), it is important to note that none of these definitions constitute an official botanical classification or certification scheme. Unlike official classifications or certifications regulated by recognized authorities, such as botanical organizations or agricultural agencies, the designation of being an 'heirloom' variety is typically conferred to traditional, open-pollinated cultivars that have been passed down through generations, often prized for their unique flavors, shapes, and colors. However, the lack of a standardized classification or certification means that the term 'heirloom' can vary widely in its interpretation and application within the tomato-growing community.

Heirloom tomatoes, renowned for their diverse flavors and unique appearances, have garnered considerable attention not only for their taste but also for their distinct textural attributes. The profiling of heirloom tomato germplasm for metabolites and textural traits holds immense significance as it unveils the intricate relationship between biochemical compounds, physical characteristics, and consumer preferences. This chapter provides an overview of the heirloom tomato germplasm, exploring the metabolites and textural traits that contribute to their distinct profiles and how these factors intersect with consumers' textural preferences.

16.2 HEIRLOOM TOMATO DIVERSITY

Heirloom tomatoes represent a diverse array of traditional cultivars that have been passed down through generations, each with its unique combination of flavors, colors, and shapes. These tomatoes, which are cultivated for specific regional or familial purposes, showcase a wealth of genetic diversity that contributes to their distinctive quality fruit (Bauchet and Causse, 2012). In this regard, Tripodi et al. (2023) conducted an extensive evaluation of 60 heirloom tomato accessions sourced from a gene bank in Italy across two seasons, scrutinizing the primary chemical and biochemical traits of the fruits. Their diverse samples encompassed various varietal

DOI: 10.1201/9781003467335-16

types, including beefsteak, globe, oxheart, plum, and cherry tomatoes. This research also revealed a medium to high level of heritability for all traits, ranging from 0.52 for soluble solids to an impressive 0.99 for fruit weight. Across both seasons, the average content of ascorbic acid remained consistent at approximately 31 mg per 100 g of fresh weight, while the greatest variability was observed in carotenoid levels, with peaks reaching 245.65 mg per g of fresh weight for total lycopene and 32.29 mg per g of fresh weight for β-carotene. Furthermore, by dissecting the genotypic (G) and seasonal (Y) factors, it was found that the genotype was the primary source of variation for all traits. Interestingly, no significant effect of seasonal variation or G × Y interaction was observed for ascorbic acid and fruit weight, respectively, while lycopene variation was significantly influenced by seasonal changes. Moreover, after incorporating molecular fingerprinting via the 10K SolCAP array, the study was able to identify a total of 7,591 single-nucleotide polymorphisms (SNPs). Analysis of population structure, phylogenetic relationships, and principal components underscored a clear differentiation between plum and cherry genotypes compared to beefsteak and globe types. Likewise, marker-trait association analysis revealed 11 significantly associated loci for β-carotene and fruit weight. Notably, a single variant on chromosome 8, located 12 kb from CCD8, a cleavage dioxygenase pivotal in apocarotenoid biosynthesis, was associated with β-carotene levels. For fruit weight, a single association was found less than 3 Mbp from SLSUN31 and fw11.3, two candidate genes implicated in augmenting fruit mass. These findings highlight the potential of heirloom tomato germplasm for genetic enhancement and the identification of candidate genes crucial for target fruit quality traits.

16.3 METABOLITE PROFILING

Metabolites, the small molecules involved in various biochemical pathways, play a pivotal role in shaping the flavor and nutritional profile of tomatoes. Metabolite profiling involves comprehensive analysis of these compounds, providing insights into the chemical composition of heirloom tomatoes. Phenolic compounds, flavonoids, carotenoids, and volatile compounds are among the key metabolites that contribute to the unique taste and aroma of heirlooms.

The complexity of tomato fruit aroma is underscored by the presence of an extensive array of over 400 volatile organic compounds (VOCs) released by these fruits (Martina et al., 2021). Among these compounds, 21 emerge as the primary contributors to the perceived tomato aroma, delineating the intricate interplay of various chemical components in shaping the characteristic scent of tomatoes. The VOCs belong to five distinct clusters based on their biosynthetic origins. This classification offers valuable insights into the underlying biochemical pathways responsible for the generation of tomato aroma, providing a framework for understanding the molecular basis of scent production in tomatoes. By unraveling the complexities of VOC biosynthesis and categorization, this study contributes to a deeper understanding of the factors influencing tomato aroma, facilitating efforts to enhance and manipulate aroma profiles in tomato breeding programs and food processing industries. Furthermore, Martina et al. indicated that there are 359 quantitative trait loci (QTL) associated with tomato fruit VOCs, which were then physically mapped onto the genome. This mapping effort enabled the investigation of potential candidate genes underlying the variation in VOC production.

Bauchet et al. (2017) conducted an extensive metabolic profiling of 300 tomato accessions and related wild species, spanning a diverse array of genetic backgrounds. Their study involved the quantification of 60 primary and secondary metabolites, encompassing volatile organic compounds, over a two-year period. Notably, they observed a wide variation in metabolite content and genetic inheritance, both within individual accessions and across different genetic groups. Through a comprehensive genome-wide association study utilizing 10,000 SNPs, the researchers identified 79 associations that influenced 13 primary and 19 secondary metabolites, with considerable effects at high resolution. Remarkably, their analysis pinpointed four genome regions housing clusters of associations governing the variation in several metabolites. Furthermore, employing local linkage disequilibrium analysis and allele mining techniques, Bauchet et al. (2017) successfully identified potential candidate genes implicated in modulating the content of metabolites crucial for human diet and fruit consumption. These findings represent a significant advancement in understanding the genetic mechanisms underlying metabolic diversity in tomatoes, with implications for breeding programs aimed at enhancing nutritional quality and flavor profiles in the fruit of this vegetable crop.

16.3.1 Phenolic Compounds and Flavonoids

Improving tomato flavor poses significant challenges due to several factors, as highlighted by Klee and Tieman (2013). One major obstacle is the complexity of assessing flavor phenotype accurately. Additionally, there is a dearth of fundamental knowledge concerning the specific chemicals driving consumer preferences, the intricate pathways responsible for their synthesis, and the genes that regulate the output of these pathways. Heirloom tomatoes are often rich in phenolic compounds and flavonoids, contributing not only to their flavor but also to their antioxidant properties. The specific types and concentrations of these compounds vary among heirloom cultivars, influencing the perceived taste and potential health benefits. They are described in the following sections.

While recent advances, including systematic analyses and the availability of the tomato genome sequence, have propelled understanding in this area, delivering enhanced flavor while maintaining high yield and long postharvest shelf-life remains a daunting task. Balancing these conflicting objectives continues to present formidable challenges in tomato quality improvement efforts.

16.3.1.1 Carotenoids

They play a pivotal role in defining the vibrant colors characteristic of heirloom tomatoes. These pigments, which include lycopene, beta-carotene, and lutein, among others, are responsible for the rich hues ranging from deep reds to vibrant oranges and golden yellows observed in heirloom tomato varieties. Beyond their visual appeal, carotenoids contribute significantly to both the flavor profile and nutritional composition of heirloom tomatoes. Lycopene, a predominant carotenoid found in tomatoes, not only imparts the characteristic red color but also contributes to the distinctively rich and savory taste of ripe tomatoes. Similarly, beta-carotene, another important carotenoid, adds vibrant orange tones to certain heirloom varieties while enhancing their sweetness and providing essential precursor compounds for vitamin A synthesis in the body. Additionally, lutein, though less conspicuous in color, offers valuable antioxidant properties and supports eye health. The presence of these diverse carotenoids not only enhances the sensory experience of consuming heirloom tomatoes but also underscores their nutritional value. As potent antioxidants, carotenoids help combat oxidative stress in the body, thereby promoting overall health and well-being. Moreover, their abundance in heirloom tomatoes reflects the heritage and genetic diversity of these traditional cultivars, highlighting their significance beyond mere aesthetic appeal.

16.3.1.2 Anthocyanins

Tomato lines carrying the genetic variants *Aft*, *atv*, *Abg*, *hp-1*, and an unspecified gene derived from line LA2099 exhibited a remarkable trait, i.e., elevated anthocyanin levels in their fruits (Mes and Myers, 2004). This combination of genetic factors resulted in the production of tomatoes characterized by heightened concentrations of anthocyanins, which are pigments responsible for bestowing red, purple, or blue hues upon various plant tissues, including fruits. Furthermore, the increased abundance of flavonoids observed in these tomatoes with elevated anthocyanin levels correlated with a rise in water-soluble antioxidant activity within the fruit, as confirmed through *in vitro* evaluations. This discovery underscores the potential health-enhancing properties associated with these anthocyanin-rich tomatoes, suggesting their utility as functional foods endowed with augmented antioxidant capacity. Overall, the identification of these genetic variants and their impact on anthocyanin accumulation not only sheds light on the genetic regulation of fruit pigmentation in tomatoes but also highlights the potential of these tomatoes as valuable resources for developing antioxidant-rich foods with potential health benefits.

16.3.1.3 Volatile Compounds

The aroma of heirloom tomatoes is shaped by a complex mixture of volatile compounds. These compounds, including terpenes and esters, are responsible for the characteristic scents that distinguish one variety from another. Metabolite profiling helps identify the specific compounds contributing to the aromatic nuances of each heirloom tomato. Over the past few decades, there has been a growing focus on enhancing the aroma and flavor of tomatoes among breeders, agronomists, and food technologists, recognizing it as a pivotal factor in solidifying tomato consumption in the global market. However, this pursuit poses a formidable challenge, given the complex nature of tomato aroma and flavor, which is influenced by over 400 VOCs in varying degrees. Recent advancements in molecular tools have enabled the identification of key alleles that impact the volatile composition of tomatoes. Yet the composition of these volatiles is governed by intricate polygenic systems, strongly shaped by a myriad of external, preharvest, and postharvest factors. Despite some exceptions, comprehensive knowledge regarding the effects of these determinants remains limited, hindering the ability to formulate robust conclusions to harness the full genetic potential. Distefano et al. (2022) did a thorough review of the existing knowledge on major VOCs that influence tomato organoleptic quality, delving into their classification, metabolic pathways, genetic regulation, and technical aspects affecting their concentrations in fruits. They highlighted that the expanding genetic insights into volatile compound expression are paving the way for the regulation of their biosynthetic pathways, thus facilitating the creation of novel products with unprecedented sensory profiles. Nevertheless, given the multifactorial nature of tomato volatiles, efforts to improve them must grapple with the influence of external factors, such as environmental conditions, crop management practices, postharvest technologies, and processing methods, all of which introduce considerable variability in key quality traits of tomatoes. To truly enhance tomato flavor, it is, therefore, imperative to comprehend these intricate interactions and devise a target breeding strategy, as well as pre- and postharvest management techniques or processing methods that align with consumers' preferences. In this context, integrated sensory analysis and metabolomic approaches, as applied to fresh fruits, have unveiled intricate molecular interactions among volatiles, challenging the conventional concentration/odor threshold theory that typically elucidates the aromatic sensation's molecular basis. Consequently, future research endeavors must navigate the intricate relationships between technical factors, volatilome remodeling, and consumers' responses, considering the inherent complexity in characterizing both fresh (e.g., cherry, plum, beef, and round tomatoes, among others) and processed products (e.g., inter alia, purees, juices, or sauces).

16.3.2 Textural Traits and Shelf-Life

In addition to flavor and aroma, the texture of heirloom tomatoes is a defining characteristic that influences consumer preferences. Textural traits encompass attributes, such as firmness, juiciness, and skin thickness. The interplay of genetic factors, environmental conditions, and cultivation practices contributes to the diverse factors affecting textures observed among heirloom cultivars, as indicated in the following sections.

16.3.2.1 Firmness

This characteristic is a critical attribute of tomato fruit quality that is influenced by various factors, such as cell wall composition and water content, ultimately shaping the overall mouthfeel experienced by consumers. In heirloom tomatoes, the spectrum of firmness ranges widely, offering consumers diverse textural experiences. The firmness of tomatoes is intricately linked to the composition and integrity of their cell walls, which provide structural support and determine the fruit's resistance to deformation upon handling. Additionally, water content plays a crucial role, with higher water content generally associated with softer textures. Some heirloom tomato cultivars are prized for their delicate and tender textures, characterized by a soft and melt-in-your-mouth consistency. These cultivars are often favored for fresh consumption and culinary applications where a smooth and luxurious mouthfeel is desired. Conversely, other heirloom tomatoes maintain a robust firmness, with denser and firmer flesh that offers a satisfying crunch or bite. These tomatoes are well-suited for slicing, grilling, or incorporating into dishes where a more substantial texture is preferred. The diversity in firmness observed among heirloom tomatoes not only reflects the genetic variability and heritage of these traditional cultivars but also provides consumers with a range of textural options to suit their preferences and culinary needs.

16.3.2.2 Juiciness

It is a defining characteristic of tomatoes, which is influenced by factors such as water content and cellular structure, playing a significant role in the perception of freshness and succulence. Heirloom tomatoes exhibit variability in juiciness, which directly influences the overall eating experience. The juiciness of tomatoes is primarily determined by their water content, with higher levels of moisture contributing to a juicier texture. Additionally, the cellular structure of the fruit, including the size and arrangement of cells, influences how readily the water is released upon consumption. Some heirloom tomato cultivars are known for their exceptionally juicy flesh, bursting with flavorful juices that impart a refreshing and invigorating sensation with each bite. These cultivars are often prized for their thirst-quenching qualities and are favored for fresh consumption, salads, and sandwiches. Other heirloom tomatoes may exhibit a firmer texture with less pronounced juiciness, thus offering a more subdued but still satisfying eating experience. These heirlooms are well-suited for cooking applications where excessive juiciness may lead to unwanted dilution of sauces or dishes. The variability in juiciness observed among heirloom tomatoes underscores the richness and diversity of these traditional cultivars, providing consumers with a range of sensory experiences to explore and enjoy.

16.3.2.3 Skin Thickness

This trait is a crucial attribute influencing the ease of biting into a tomato and can significantly impact the release of flavors during consumption. Among heirloom tomatoes, there is notable variation in skin thickness, which contributes to the overall eating experience. The thickness of the skin on tomatoes dictates the degree of resistance encountered when biting into the fruit. Heirloom tomatoes may exhibit thin, delicate skins that offer minimal resistance, allowing effortless bites and enhancing the perception of tenderness. In these cultivars, the thin skin also facilitates the release of flavors, allowing the taste of the tomato to be more readily experienced. Conversely, some heirloom tomatoes may feature thicker skins that provide a satisfying resistance when bitten into. While thicker skins may require a bit more effort to penetrate, they can contribute to a more substantial mouthfeel and offer a satisfying tactile sensation during consumption. Additionally, thicker skin may help protect the fruit during handling and transportation, reducing the risk of damage. The variability in skin thickness among heirloom tomatoes adds another dimension to their culinary appeal, thereby providing consumers with options to suit their preferences for texture and mouthfeel. Whether thin and delicate or thick and robust, the skin of heirloom tomatoes plays a vital role in shaping the overall eating experience and enjoyment of these traditional cultivars.

16.3.2.4 Shelf-Life

Mellidou et al. (2023) observed notable variations among Greek tomato accessions concerning key ripening parameters, such as color, firmness, and weight loss. Their investigation revealed a spectrum of postharvest performances, striking a balance between short and long shelf-life accessions (SSL and LSL, respectively, hereafter). Through metabolome analysis conducted on 14 carefully selected accessions exhibiting contrasting shelf-life potentials, a rich profile of 206 phytonutrients and volatile compounds was delineated. Moreover, transcriptome analysis conducted on fruits sourced from the most exemplary SSL and LSL accessions unveiled significant disparities in the expression patterns of transcripts pivotal to fundamental metabolic pathways governing fruit quality and postharvest longevity. Notably, pathways governing cell wall synthesis, polyamine synthesis, ABA catabolism, and steroidal alkaloid synthesis were predominantly induced in the LSL accession, whereas those associated with ethylene biosynthesis, cell wall degradation, isoprenoids, phenylpropanoids, ascorbic acid, and aroma (TomloxC) were notably stimulated in the SSL accession. This comprehensive understanding offers invaluable insights into the molecular mechanisms underpinning enhanced shelf-life and the augmentation of flavor and aroma in contemporary tomato cultivars.

16.3.3 Consumer Textural Preferences

Understanding the relationship between metabolites, textural traits, and consumer preferences is key for both growers and breeders. Consumer acceptance and enjoyment of heirloom tomatoes are influenced by individual preferences, regional culinary traditions, and cultural factors (Lázaro, 2018). Furthermore, consumers have shown a strong preference for heirloom tomatoes, particularly when they are cultivated in open fields near their selection areas. In such environments, heirloom tomatoes are thought to express their full potential in terms of nutrient synthesis and sensory properties. While the total soluble solid content remains, a key factor influencing consumer appreciation, other morphological traits also play a significant role in consumers' choices. Moreover, it is observed that certain consumers exhibit distinct preferences for various tomato typologies and nutritional characteristics, such as acidity levels. This diversity in consumer preferences underscores the importance of offering a wide range of tomato cultivars to cater to different tastes and preferences. Overall, the positive reception of heirloom tomatoes reflects their unique flavors, textures, and nutritional profiles, making them a sought-after choice among discerning consumers.

16.3.3.1 Culinary Applications

Heirloom tomatoes, with their diverse metabolite profiles and textural traits, offer a versatile ingredient for a wide range of culinary applications. Some consumers may prefer certain tomato cultivars for fresh salads, where juiciness and tenderness are prized, while others may favor firmer tomato cultivars for slicing and grilling.

16.3.3.2 Market Trends

Consumer trends and preferences in the culinary world often drive market demand for specific heirloom cultivars. Metabolite profiling and understanding textural traits allow growers to cater to evolving market preferences, ensuring a steady supply of sought-after tomatoes.

16.3.3.3 Culinary Education

Enlightening consumers about the nuanced qualities of heirloom tomatoes, including metabolite composition and textural attributes, can enhance their appreciation for these traditional cultivars. Initiatives that highlight the diversity within heirloom germplasm can foster a deeper understanding of flavor complexity.

16.4 CONCLUSION

Heirloom tomatoes are a category of open-pollinated cultivars renowned for their rich diversity in colors, shapes, and flavors (Constantino et al., 2022). This diversity makes heirlooms valuable resources for plant breeders seeking to enhance various physical and biochemical traits in tomato fruits (Wang et al., 2023). By tapping into the wide array of characteristics present in heirloom cultivars, plant breeders could introduce novel traits and qualities into modern tomato cultivars, thus contributing to the ongoing improvement of the tomato cultigen pool. The broadening of this cultigen pool may be further enhanced by using crop wild relatives (CWR) as shown by Li et al. (2023), who observed that the content of ascorbic acid was notably elevated in three F_2 hybrids resulting from the cross between tomato and its wild relative *Solanum pimpinellifolium*, surpassing the levels found in LA1585, the wild, red-fruited accession which exhibited higher parental levels. Similarly, two of the hybrid lines exhibited heightened lycopene content compared to LA1585. Additionally, the fructose and glucose contents were found to be elevated in five F_2 hybrids in contrast to e9292, the cultivated tomato *S. lycopersicum*. Moreover, the sugar/acid value was notably increased in three hybrids.

In recent years, clustered regularly interspaced short palindromic repeats (CRISPR)/Cas9 gene editing has emerged as a powerful tool in germplasm enhancement of tomatoes. As highlighted by Rothan et al. (2019), this technology has demonstrated remarkable efficiency in engineering favorable alleles and introducing new genetic diversity in tomatoes. Through techniques such as gene disruption, gene replacement, and precise base editing, CRISPR/Cas9 enables precise and targeted modifications to the tomato genome. The versatility and precision of CRISPR/Cas9 gene editing could revolutionize tomato breeding efforts, by allowing to introduce desired traits more efficiently than crossbreeding methods. By precisely altering specific genes, CRISPR/Cas9 offers the potential to enhance traits related to fruit quality in the tomato crop. Overall, the successful application of CRISPR/Cas9 gene editing in tomatoes holds great promise for accelerating its breeding and developing improved tomato cultivars that meet the evolving needs of end-users.

While the profiling of heirloom tomato germplasm for metabolites and textural traits offers valuable insights, challenges exist in reconciling the inherent variability of these tomatoes with consumer expectations and market demands. Additionally, the integration of advanced technologies, such as high-throughput metabolomics and precision breeding techniques, holds promise for accelerating the identification of key flavors and texture-associated genes. The profiling of heirloom tomato germplasm for metabolites and textural traits provides a holistic understanding of these traditional cultivars. From the intricate interplay of metabolites shaping flavor to the diverse textural experiences offered by different cultivars, heirloom tomatoes embody a rich tapestry of genetic diversity. As consumers increasingly seek unique and flavorful culinary experiences, the continued exploration of heirloom tomato germplasm will contribute to the preservation of agricultural heritage and the development of novel-bred cultivars that meet evolving taste preferences.

REFERENCES

Bauchet, G., and Causse, M.M. (2012). Genetic diversity in tomato (*Solanum lycopersicum*) and its wild relatives. In: *Genetic Diversity in Plants*. IN-TECH Education and Publishing, London.

Bauchet, G., Grenier, S., Samson, N., Segura, V., Kende, A., Beekwilder, J., et al. (2017). Identification of major loci and genomic regions controlling acid and volatile content in tomato fruit: Implications for flavor improvement. *New Phytol.* 215, 624–641.

Constantino, L.V., Shimizu, G.D., Macera, R., Fukuji, A.S.S., Zeffa, D.M., Koltun, A., et al. (2022). Genetic diversity and selection of heirloom tomato accessions based on the physical and biochemical fruit-related traits. *Bragantia* 81, e1422. https://doi.org/10.1590/1678-4499.20210193

Distefano, M., Mauro, D.P., Page, D., Giuffrida, F., Bertin, N., and Leonardi, C. (2022). Aroma volatiles in tomato fruits: The role of genetic, preharvest and postharvest factors. *Agronomy* 12, 376. https://doi.org/10.3390/agronomy12020376

Klee, H.J., and Tieman, D.M. (2013). Genetic challenges of flavor improvement in tomato. *Trends Genet.* 29, 257–262. http://doi.org/10.1016/j.tig.2012.12.003

Lázaro, A. (2018). Tomato landraces: An analysis of diversity and preferences. *Plant Genet. Resour.* 16, 315–324. https://doi.org/10.1017/S1479262117000351

Li, W., Li, Y., Liang, Y., Ni, L., Huang, H., Wei, Y., et al. (2023). Generating novel tomato germplasm using the ancestral wild relative of *Solanum pimpinellifolium*. *Horticulturae* 9, 34. https://doi.org/10.3390/horticulturae9010034

Martina, M., Tikunov, Y., Portis, E., and Bovy, A.G. (2021). The genetic basis of tomato aroma. *Genes* 12, 226. https://doi.org/10.3390/genes12020226

Mellidou, I., Koukounaras, A., Frusciante, S., Rambla, J.L., Patelou, E., and Ntoanidou, S. (2023). A metabolome and transcriptome survey to tap the dynamics of fruit prolonged shelf-life and improved quality within Greek tomato germplasm. *Front. Plant Sci.* 14, 1267340. https://doi.org/10.3389/fpls.2023.1267340

Mes, P.J., and Myers, J.R. (2004). Increased anthocyanin content in tomato fruit from improved antioxidant activity. *HortScience* 39, 869. https://doi.org/10.21273/HORTSCI.39.4.869D

Rothan, C., Diouf, I., and Causse, M. (2019). Trait discovery and editing in tomato. *Plant J.* 97, 73–90. https://doi.org/10.1111/tpj.14152

Tripodi, P., D'Alessandro, A., and Francese, G. (2023). An integrated genomic and biochemical approach to investigate the potentiality of heirloom tomatoes: Breeding resources for food quality and sustainable agriculture. *Front. Plant Sci.* 13, 1031776. https://doi.org/10.3389/fpls.2022.1031776

Wang, S., Qiang, Q., Xiang, L., Fernie, A.R., and Yang, J. (2023). Targeted approaches to improve tomato fruit taste. *Hortic. Res.* 10, uhac229. https://doi.org/10.1093/hr/uhac229

17 Genetic and Chemical Toolbox for Developing Tastier Tomato Germplasm and Reinstating Flavor to Commercially Grown Tomatoes

Autar K. Mattoo and Sangam L. Dwivedi

17.1 INTRODUCTION

Tomato (*Lycopersicon esculentum*) is an economically important crop, widely produced and consumed worldwide both as fresh and processed byproduct. It is widely recognized as an ideal model system for classical genetics, cytogenetics, molecular genetics, and fruit developmental biology (Stikić et al., 2015). Variation in environmental conditions, such as light intensity, cultivation practices, and temperature, may adversely impact fruit quality including its nutrients and flavor (Delgado-Vargas et al., 2023; Fan et al., 2023; Vinha et al., 2014).

Tomato consumption is associated with many health benefits to humans including lowered risk for certain types of cancers, protection against cardiovascular and neurodegenerative diseases and diabetes, reducing hepatitis-induced inflammation in livers, maintenance of a healthy gut microbiome including reducing the risk of inflammatory bowel diseases, and improved skin health (Collins et al., 2022a; Dzakovich et al., 2024; Tanambell et al., 2019). However, excessive consumption of tomato may cause adverse impact on human health including heartburn, irritable bowel syndrome, urinary problems, migraines, body aches, anaphylactic reactions, lycopenodermia, renal calculi, and hepatitis A (Collins et al., 2022b).

Quality attributes are complex and highly affected by genotype, environment, and management interactions. The stakeholders in the agri-food chain have different quality requirements. The producers, for example, want their crops to synchronize flowering/maturity, resist biotic and abiotic stresses, produce higher yields, and be easy to harvest and handle. The marketers (or the traders) want the produce to be

DOI: 10.1201/9781003467335-17

nutritious, have longer shelf-life, and less prone to handling and shipping damage. Consumers prefer wholesomeness in terms of appearance, color, nutrition, flavor, succulence, juiciness, crispness, chewiness, and oily (in case of fruits and vegetables), sustainable and affordable (Gascuel et al., 2017).

Most plant genetic interventions over the last hundred years have been driven by industry demand of larger fruit/seeds, improved yields, uniformity, stress tolerance, and good shipping quality to ensure profits on the farm and affordable food for consumers. These milestones were, however, achieved at the cost of sensory quality. With self-sufficiency in staple foods, the crop genetic enhancers are paying attention to improving taste and flavor, especially in fruits and vegetables. Consumer's willingness to pay more for safe and nutritious food with pleasing taste and aroma has put extra pressure on crop improvement community to combine enhanced yield with nutrition, color, aroma, and texture.

Advances in the development of a genetic (including germplasm and genomic resources) and chemical (metabolomic resources) toolbox to breed tomatoes with improved flavor and taste are highlighted here.

17.2 GERMPLASM RESOURCES

Over 82,400 tomato germplasms, including cultivated and wild tomato species, are maintained in ex-situ gene banks globally. The World Vegetable Center in Taiwan maintains the largest collection followed by gene banks in the United States, the Philippines, Germany, the Russian Federation, and Japan (Ebert and Chou, 2015). The Solanaceae database (http://www.bguard.science.ru.nl/) and the isogenic tomato mutation library (http://zamir.sgn.cornell.edu/mutant/) are other useful gene banks holding unique tomato germplasms (Gascuel et al., 2017). The Vavilov Research Institute of Plant Industry (VIR) in Russia maintains 7,678 accessions of cultivated and wild tomato species (Kurina et al., 2021). C.M. Rick Tomato Genetics Resource Center (TGRC) is a gene bank of tomato wild relatives, monogenic mutants, and miscellaneous genetic stocks. The wild species group includes representatives of each of the 13 species in the tomato clade (*Solanum* section *Lycopersicon*), as well as the related nightshades *S. lycopersicoides*, *S. sitiens*, *S. juglandifolium*, and *S. ochranthum* (https://tgrc.ucdavis.edu/about-us).

The whole-genome sequencing (WGS) of 132 doubly mutagenized lines of Arka Vikas genome by ethyl methanesulfonate (EMS) yielded ~41 million novel mutations and 5.5 million short InDels, not detected in the Arka Vikas genome. About 97% of the genome had mutations, including genes, promoters, UTRs, and introns. More than one-third of genes in the mutagenized population had one or more deleterious mutations. Nearly one-fourth of deleterious genes mapped on tomato metabolic pathways modulate multiple pathway steps (Gupta et al., 2024b). This could be a valuable resource to unlock functional characterization of genes associated with flavor related traits in tomatoes. *S. pimpinellifolium*, the closest relative of domesticated tomatoes and excellent source of flavor related traits, harbors alleles that could be used to increase glucose and fructose content and adjust acidity by altering malic and citric acid levels (Çolak et al., 2020). Both sugars (fructose and glucose) and organic acids (citric and malic acids) contribute to tomato flavor (Anđelini et al., 2023).

Neuronal network analysis detected unique volatile organic compounds (VOCs) among Andean tomato landraces that are significantly associated with the typical and atypical tomato odor and taste (Cortina et al., 2018). Heirloom cultivars, relative to modern tomatoes, possess unique flavor, and are grouped as those with better flavor, better taste, and better odor. Variation in total soluble solids, fructose, glucose, and citric acid positively contributes to tomato taste. Fifteen VOCs determine tomato odor, with many significantly and positively correlated ($P < 0.05$) with flavor scores (Cheng et al., 2020). USDA tomato core collection unfolded sufficient variability for total soluble solids (TSS) and total titratable acids (TTA). TSS ranges from 3.4 to 9.0%, while TTA ranges from 0.20 to 0.64%, with overall flavor score positively correlated with TTS ($r = 0.37$; $P < 0.05$) and TTA ($r = 0.33$; $P < 0.05$), indicating that both TSS and TTA play important roles in determining the overall flavor in tomato (Panthee et al., 2012).

Assessment of Italian tomato landraces with distinct fruit shape (flattened/ribbed, pear/oxheart, round/elongate) has shown that those with round/elongate types are rich in glycoalkaloids and flattened/ribbed types in phenolic compounds. The latter type is also rich in total amino acids, particularly in glutamic acid. Multivariate analysis of amino acid content clearly separated the three classes of fruit types. The changes in quality-related metabolites depend on the genetic background in traditional tomato germplasm, which matches with specific fruit shape categories (Baldina et al., 2016).

Traditional criollo tomato cultivars from the Andean region are appreciated for their excellent organoleptic fruit quality. Tomato aroma and sourness, involving criollo and modern tomatoes, correlate with citrate and several VOCs in criollo tomatoes. Two sensory attributes, sweetness and a not-acidic taste, correlate with the characteristic tomato taste, fructose, glucose, and two VOCs, suggesting that criollo tomatoes could be used in breeding programs for improving flavor and taste of modern tomatoes (D'Angelo et al., 2018). Likewise, VOCs significantly influence unique flavor characteristics of cherry tomatoes, *Solanum lycopersicum* var. *cerasiforme* (Guan et al., 2024).

Tomatoes contain many wild relatives. The variation in VOCs across closely related wild species is poorly understood. Assessment of phylogenetic patterns of these VOCs and their biosynthetic pathways may offer greater insights into the evolutionary drivers of fruit diversity. The VOCs content across the tomato clade involving 32 accessions indicates that each species in tomato clade (cultivated tomato and 12 related wild species differing in fruit color) can be differentiated by their unique VOCs, with *Solanum pennellii* as containing altogether distinct VOCs. The variation in VOCs among major clades corresponded to divergence of the brightly colored-fruited species from the green-fruited species. The nitrogen-containing compounds were higher in colored-fruited species, while esters were higher in green-fruited species, with the latter appearing to signal more sugars (Barnett et al., 2023).

Accumulation of VOCs during fruit ripening largely affects the tomato fruit flavor. Sensory analysis of three tomato subgroups (BIG, *S. lycopersicum*; CER, *S. lycopersicum* var. *cerasiforme*; and PIM, *S. pimpinellifolium*) have revealed that CER subgroup had the highest fruit-flavor score. Ninety-four volatile compounds were detected. Green fruits contained more pentanal, while red fruits more 1-pentanol.

1-Octen-3-ol caused bitterness in green tomatoes, abundant in PIM green fruits. Phenylethyl alcohol affected the acidity and sweetness of red tomatoes, which was most abundant in CER red fruits. The branched-chain volatiles were abundant in PIM and BIG red fruits, while apocarotenoids were abundant in CER red fruits. The apocarotenoids and branched-chain volatiles synergistically mediated aromatic flavors in red fruits (Zhang et al., 2024b).

Solanum pimpinellifolium (SP) and the weedy *S. lycopersicum* var. *cerasiforme* (SLC) are unexploited genetic reservoirs easily accessible to breeders and cross-compatible with cultivated tomato (*S. lycopersicum* var. *lycopersicum*). Four SP and four weedy SLC accessions were used as founder populations to develop the first interspecific multiparent advanced generation intercross (MAGIC) population, which showed immense diversity for plant, inflorescence, fruit, and agronomic traits. The whole-genome resequencing of these founder populations unlocked over 12 million variants, ranging from 1.2 to 1.9 million variants in SLC and from 3.1 to 4.8 million in SP, 4,897,803 being the private variants. Morpho-agronomic, chemical, and genetic characterization of these populations may provide valuable knowledge and tools to the tomato community for genetic and genomic studies and for breeding purposes, including flavor-related traits (Gramazio et al., 2020). MAGIC population consisted of 354 lines and 6,488 high-quality SNPs. They show a high degree of homozygosity, an absence of genetic structure, and a balanced representation of the founder genomes (Arrones et al., 2024). *Solanum pimpinellifolium* is a potential genetic resource to improve flavor of domesticated tomatoes (Çolak et al., 2020).

17.3 ADVANCES IN HIGH-THROUGHPUT ASSAYS FOR ASSESSING FLAVOR-RELATED TRAITS

Both sugar and organic acids are major factors that influence tomato fruit flavor. A simple method for quantification of glucose, fructose, citric acid, and malic acid is known in tomatoes. It efficiently removes interfering compounds by solid phase extraction, with high intra- and inter-day reproducibility (Agius et al., 2018).

The sugars and organic acids largely influence tomato fruit taste. Analysis of these compounds, individually, using high-performance liquid chromatography (HPLC) or capillary zone electrophoresis (CZE) are expensive and time-consuming, thereby not suited for screening of large number of breeding samples. The Fourier-transform mid-infrared (FT-MIR) method, aided with use of partial least squares (PLS) models, is efficient to identify samples with high contents of taste-related compounds, even in the absence of within assay calibration (Ibáñez et al., 2019). The development of portable spectrometers facilitate access to quality attributes within a few seconds during fruit ripening in the field. A portable/handheld (palm-sized) near-infrared (NIR) spectrometer (Neospectra, Si-ware), with spectral range of 1295–2611 nm and equipped with a micro-electro-mechanical system (MEMs), accurately predicted tomato fruit quality attributes non-destructively, with high reproducibility. For example, the reproducibility was 0.87 for TSS, 0.87 for fructose, 0.83 for glucose, 0.81 for ascorbic acid, and 0.86 for citric acid. Thus, a handheld NIR spectrometer is an ideal tool to simultaneously determine several quality attributes of different types of tomatoes in the field itself (Borba et al., 2021).

Visible and near-infrared (Vis-NIR) spectroscopy is another high-throughput method to assess fruit quality attributes of tomatoes. Use of a portable spectroradiometer RS-3500 (Spectral Evolution Inc.) and two calibration models (partial least squares regression, PLS, and multiple scatter correction, MSC) showed good prediction accuracies for lycopene and dry matter of intact tomatoes with a determination coefficient of 0.90 for both parameters. The reproducibility (R^2) was 0.86 for taste index, 0.84 for flavonoids, 0.82 for β-carotene, and 0.73 for total phenols, respectively, with many of these compounds affecting flavor-related traits in tomatoes (Duckena et al., 2023).

Headspace–solid-phase microextraction–gas chromatography–mass spectrometry (HS-SPME-GC-MS) is the most efficient method for extracting and identifying volatile compounds because of its high sensitivity and selectivity (Gupta et al., 2024a). A comparison of two detection technology platforms, i.e., headspace–solid-phase microextraction–gas chromatography–mass spectrometry (HS-SPME-GC-MS) and headspace–solid-phase microextraction—full two-dimensional gas chromatography–time-of-flight mass spectrometry (HS-SPME-GC×GC-TOFMS), and fruit samples from a common tomato cherry variety detected a total of 1,544 VOCs. HS-SPME-GC-MS detected 663 VOCs, HS-SPME-GC×GC-TOFMS 1026 VOCs, while both assays detected 145 VOCs, thereby the latter assay is more efficient in determining VOCs (Guan et al., 2024). A simple and rapid method is now available for quantifying non-polar pigments (carotenoids and chlorophyll) of tomato fruits (Aono et al., 2021).

17.4 PROTEOMIC AND METABOLOMIC DIVERSITY IMPACTING FRUIT QUALITY TRAITS

Modern tomato cultivars are less flavorful than traditional or heirloom cultivars. Quantifying flavor associated chemicals among large collection of modern, heirloom, and wild accessions revealed that modern cultivars contain significantly lower amounts of many of the flavor-associated chemicals (sugars, acids, and volatiles) than heirloom cultivars (Tieman et al., 2017). Carotenoids indirectly affect flavor as precursors of aroma compounds, while chlorophylls contribute to sugar production through photosynthesis. Assessment of large collection of tomato germplasm including *S. lycopersicum* and its wild relatives (*S. l.* var. *cerasiforme* and *S. pimpinellifolium*) reveals that chlorophyll-abundant accessions have relatively higher sugar accumulation, while prolycopene is associated with an abundance of linear carotenoid-derived aroma compounds in an orange-fruited tomato, the Dixie Golden Giant (Aono et al., 2021).

Transcriptome and metabolome analysis involving wild and cultivated tomatoes detected a total of 35 volatile flavor compounds. The volatile compounds affecting fruit flavor significantly increased in the transition from breaker to red ripe fruit stage, with total content of the volatiles in wild tomatoes much greater than that in the cultivated tomatoes. Gene expression levels are highly correlated with the content of volatile flavor compounds, involving two regulatory pathways for volatile flavors. Tetrahydrothiazolidine *N*-hydroxylase (*SlTNH1*)-dependent and lipoxygenase (*SlLOX*)-dependent pathways. *SlTNH1*-dependent pathway is regulated by two TFs from the bHLH and AP2/ERF families, controlling the synthesis of

2-isobutylthiazole in amino acid metabolism. *SlLOX*-dependent pathway is controlled by one TF from the HD-Zip family, regulating the synthesis of hexanal and (Z)-2-heptenal in fatty acid metabolism. bHLH and AP2/ERF TFs are bound to their structural genes. Clearly, these provide new insights into volatile flavor formation in tomato fruit, which could be exploited to improve tomato flavor (Liu et al., 2024).

Unlocking the molecular mechanism of flavor formation during fruit ripening can guide us to formulate breeding strategies balancing the consumer and producer demands. Cherry tomatoes possess a distinctive taste and flavor. Cherry tomatoes (*Solanum lycopersicum* var. *cerasiforme*) are a small-fruited variety of tomato named after cherry. The fruits can be red (the predominant type), yellow, orange, green, or almost black. The metabolomic and transcriptomic analyses of cherry tomato fruits at different fruit ripening stages revealed differentially accumulated metabolites and differential gene expression during fruit ripening, providing candidate metabolites and genes related to flavor. The key flavor-related metabolites reached a steady level at the mature green fruit stage, while the expression levels of the most flavor related genes stabilized at the pink stage. Gene enrichment analysis indicated that changes in metabolic and biosynthetic pathways are evident throughout the process of fruit ripening. The fruit flavor and firmness genes rather than disease resistance and fruit color genes contribute positively to the accumulation of metabolites related to unique cherry tomato flavor (Pan et al., 2023).

Two independent gene sets affecting fruit metabolites and 57 candidate genes, including *glucarate caffeoyltransferase* (*SlCGT*) related to polyphenol and polyamine biosynthesis were reported in consumer-preferred tomatoes. *SlCGT* affected fruit quality and diseases resistance in tomato. Selective sweeps and key metabolites controlled by multiple genes modified fruit quality through domestication and genetic improvement (Yang et al., 2022).

Tomato introgression lines (ILs) are the major source of variation for stress tolerance, fruit yield, and quality traits. IL4–4, in which the bottom of cultivated tomato chromosome 4 is replaced by the corresponding *S. pennellii* segment, has many primary and secondary metabolites changed, including those related to fruit flavor and nutritional quality. A comprehensive profiling of IL4–4 ripe fruit metabolites unfolded 327 metabolites, of which 185 were significantly changed compared to the control. Many of the altered metabolites belong to volatile organic compounds, and primary and secondary metabolites. A genomic region encompassing 20 putative open reading frames is responsible for most of the altered metabolites in IL4–4 fruit (Liu et al., 2016).

17.5 GENOMIC RESOURCES

17.5.1 High-Density Genetic Maps

Genetic maps characterize the genome and contribute to trait improvement in plant breeding. Genetic improvement is based on our knowledge of inheritance patterns of genes (or quantitative trait loci, QTLs) and DNA fragments (markers) associated with agronomically beneficial traits. Such genes and associated markers are then tagged along the physical map (i.e., the draft genome assembly or complete genome). A consensus genetic map depicts the position (both marker order and distance between flanking markers) of genetic markers involving multiple populations (biparental F_2's or recombinant inbred lines [RLSs] originating for biparental populations, multiparent advanced generation intercross populations [MAGIC], or nested association mapping [NAM] populations or RILs) of a single species could be a common framework population for developing consensus genetic map. Thus, genetic maps are a solid foundation for genetic enhancement in crop improvement programs.

Advances in high-throughput genotyping platforms and next-generation sequencing (NGS) has increased multifold chances of discovering sequence-based markers, improved efficiency of constructing genetic maps, and our ability to overlay recombination and physical maps of the genome. NGS-derived SNPs were used to develop high-density SNP genetic maps involving three interspecific tomato F_2 populations—EXPEN 2000 (*Solanum lycopersicum* LA0925 x *S. pennellii* LA0716), EXPEN 2012 (*S. lycopersicum* Moneymaker x *S. pennellii* LA0716), and EXPIM 2012 (*S. lycopersicum* Moneymaker x *S. pimpinellifolium* LA0121). EXPEN 2000-SNP and EXPEN 2012 maps consisted of 3,503 and 3,687 markers representing 1,076 and 1,229 unique map positions (genetic bins), respectively. EXPEN 2000-SNP and EXPEN 2012 maps, respectively, had an average marker bin interval of 1.6 cM and 0.9 cM. The EXPIM 2012 map with 4,491 markers (1,358 bins) had an average bin interval of 0.8 cM. All three maps revealed an uneven distribution of markers across the genome. EXPEN 2012 and EXPIM 2012 maps showed high levels of co-linearity across all 12 chromosomes and revealed evidence of small inversions between LA0716 and LA0121. The physical positions of 7,666 SNPs relative to the tomato genome sequence were identified, with genetic and physical positions being mostly consistent with a few exceptions. The genomic regions with high recombination rates were consistent with the known distribution of euchromatin across the 12 chromosomes, while those with very low recombination rates were observed in the heterochromatic regions (Sim et al., 2012). *S. pennellii* and *S. pimpinellifolium* have remarkable stress tolerance and/or intense fruit flavor, so this map is a useful resource to identify, clone, and functionally characterize genes/markers associated with stress tolerance, and fruit yield and/or flavor related traits (Çolak et al., 2020; Gur and Zamir, 2004; Bolger et al., 2014; Liu et al., 2016; Wang et al., 2020).

17.5.2 Genes and Markers Regulating Fruit Chemical Variations, Including Metabolites

The olfactory and gustatory systems in humans perceive flavor in food (Chen et al., 2023; Vincis and Fontanini, 2019). Flavor is one of the main determinants of fruit and vegetable quality. Sugars, acids, and volatile compounds determine tomato flavor. Several genes including *LIN5, ALMT9, AAT1, CXE1,* and *LoxC* control the levels of metabolite compounds in tomatoes. *LIN5* and *ALMT9* function in sugar and acid metabolism affecting taste, while *AAT1, CXE1,* and *LoxC* regulate volatile production affecting smell (Pereira et al., 2021).

Citric acid (CA) acts as a fruit flavor enhancer that mediates multiple metabolic pathways in the development of tomato fruit. Eleven candidate genes were found associated with CA accumulation in tomatoes. The gene expression data of certain candidate genes was significantly associated with changes in CA metabolism. Three genes exerted a positive regulatory effect on CA accumulation, while the remaining exhibited negative impacts associated with citric acids. The CA content of tomatoes is primarily influenced by improvement sweeps with minimal influence from domestication sweeps. Allelic variation within candidate genes caused differences in CA content among tomato accessions (Gai et al., 2024).

Malic acid (MA) mediates a series of metabolic pathways to influence tomato fruit flavor. The accumulation of MA is controlled by complex heredity and environmental factors. GWAS identified 15 high-confidence annotated genes, predicted based on linkage disequilibrium and lead SNPs. Transcriptome data of candidate genes, involving different tomato organs and fruit tissues, suggested specific expression patterns in fruit pericarp. MA content is highly influenced by domestication sweeps and less by genetic improvement sweeps. Additionally, genotype × environment interaction may have also contributed to the difference in MA contents in different environments (Gai et al., 2023).

Fatty-acid-derived volatile organic compounds (FA-VOCs) contribute to tomato fruit flavor and human preferences. Short-chain FA-VOCs (C5 and C6) are among the most abundant volatile compounds. *Sl-LIP8* is significantly associated with accumulation of short-chain FA-VOCs. *Sl-LIP8* cleaves linoleic acid (18:2) and linolenic acid (18:3) acyl groups from glycerolipids. CRISPR/Cas9-generated *Sl-LIP8* mutant had much lower content of multiple fruit short-chain FA-VOCs. *Sl-LIP8* RNA abundance correlated with FA-VOC content, indicating that glycerolipid turnover by *Sl-LIP8* is an important early step in the synthesis of multiple short-chain FA-VOCs (Li et al., 2020).

FLORAL4 regulates floral aroma volatile in tomato fruit. A QTL on chromosome 4 within *FLORAL4* affects the levels of the phenylalanine-derived volatiles (PHEVs) 2-phenylethanol, phenylacetaldehyde, and 1-nitro-2-phenylethane. Near-isogenic lines (NILs) fruits contrasting for *FLORAL4* and in the composition of PHEVs significantly differed in the perception of fruity and rose-hip-like aroma. Fine mapping of PHEV locus detected *FLORAL4* as a candidate gene for PHEV regulation, as also evidenced by reverse genetics (Tikunov et al., 2020).

A number of SNPs were significantly associated with tomatoes primary and secondary metabolites with large effects at high resolution. Four genome regions and putative candidate genes were significantly associated with variation in several metabolites, with two associations involved in fruit acidity and phenylpropanoid volatile production (Bauchet et al., 2017). A study involving breeders' sensory test in an F_2 population, originated from two superior flavor parents and evaluated under two contrasting production systems (organic low-input and hydroponic cultivation) identified ~27% of the loci co-localized between both cultivation systems. Nine distinct QTL clusters were identified for flavor-related traits, with a large cluster on chromosome 6 comprising five sensory and nine volatile QTLs (Hagenguth et al., 2024). A breeders' sensory test refers to a small team of human subjects and large number of small samples from breeding populations, as well as evaluation of physicochemical attributes (total soluble solids, titratable acidity, and dry matter) and volatile compounds. The physicochemical and sensory traits were significantly correlated, which facilitated selection of desirable segregants for most traits in early generation, indicating that a breeders' sensory test is an effective tool in breeding for flavor in tomato (Hagenguth et al., 2022).

17.5.3 Pangenome, Graph Pangenome, and Super-Pangenome

The tomato pangenome, involving genome sequences of 725 phylogenetically and geographically distinct accessions, unfolded 4,873 genes that are absent in the reference genome, including important genes associated with disease resistance and flavor. Domestication and genetic improvement resulted in substantial gene loss and intense negative selection of genes and promoters. A rare allele in the TomLoxC promoter is selected against during domestication. TomLoxC contributes to desirable tomato flavor, which is associated with apocarotenoid production. Orange-fruited accessions harboring both rare and common TomLoxC alleles (heterozygotes) have higher TomLoxC expression than those homozygous for either that are resurgent in modern tomatoes (Gao et al., 2019).

Missing heritability is one of the major biological constraints while dealing with improvement of complex traits. Identifying all casual genetic variants and their individual contributions may offer an opportunity to tap missing heritability. The recently constructed graph pangenome of tomato involves over 19 million variants from 838 genomes, including 32 new reference-level genome assemblies. GWAS-based heritability estimation using 20,323 gene-expression and metabolite traits recorded greater heritability, 0.41, compared to 0.33 when using single linear reference genome. Resolving incomplete linkage disequilibrium through the inclusion of additional causal structural variants in the graph pangenome improved heritability estimate. Moreover, the structural variants improved the power to identify genetic factors underlying agronomically important traits leading to identification of two new genes contributing to TSS content (Zhou et al., 2022).

Most of the accessions included in tomato pangenome (Gao et al., 2019) or graph-pangenome (Zhou et al., 2022) belong to domesticated tomatoes and their closely related progenitor species *Solanum pimpinellifolium*. A super-pangenome that includes genomic information of many diverse species, especially wild relatives within a genus, could further expedite crop improvement (Khan et al., 2020). The tomato super-pangenome, involving chromosome-scale genomes of nine wild species and two cultivated accessions of tomato clade, is the landscape of structural variants facilitating access to the genomic diversity among tomato wild relatives. It may facilitate access to wild tomato genes with the potential to increase yields and improve fruit quality including numerous signals impacting with tomato flavor-related traits and fruit metabolites (Li et al., 2023a). Clearly, the availability of pangenome or super-pangenome adds depth and completeness to the reference genomes and is a useful resource for future biological discovery and in crop improvement programs.

17.6 Advances in Genetic Transformation and Genome Editing in the Improvement of Tomato Fruit Flavor

Modern tomato cultivars have inferior flavor and quality compared to traditional (or heirloom) tomatoes. The sustained effort to improve tomato flavor via conventional crossing and selection has achieved only limited success to improve flavor of modern cultivars. Fatty acids, phenylpropanes, carotenoids, and branched-chain amino acids are the major sources of variation in volatile compounds, with fatty acids contributing almost 50% of volatile compounds. Transgenic tomato overexpressing *SlCCD1A* improved tomato fruit flavor. *SlCCD1A* cleaved lycopene, α-carotene, and β-carotene to produce 6-methyl-5-hepten-2-one, geranylacetone, α-ionone, and β-ionone to enhance, respectively, floral, fruity, fatty, and sweet-like aromas of tomato fruits (Cheng et al., 2021).

Overexpression of *SlMYB75* improves multiple fruit quality traits, including anthocyanin, phenolics, flavonoids, TSS, and aromatic volatile compounds in tomato. The anthocyanin content reached 1.86 mg g^{-1} fresh fruit weight, while phenolics, flavonoids, and TSS increased 2.6-, 4-, and 1.2-fold than those of wild-type (WT) fruits. Aldehyde, phenylpropanoid-derived, and terpene volatiles were significantly increased in transgenic fruit overexpressing *SlMYB75*, with some terpene volatiles showing more than 10 times higher levels than those in WT fruits (Jian et al., 2019).

Extending fruit shelf-life with no loss of quality is a significant challenge to crop genetic enhancers. Simultaneously knocking out *SlPG* and *SlPL* by CRISPR/Cas9 system improves tomato fruit firmness and extends shelf-life without negatively impacting consumer-based quality attributes, including flavor (Ortega-Salazar et al., 2023). CRISPR/Cas9-mediated multiplex gene-editing system using three fruit-color-related genes (*PSY1*, *MYB12*, and *SGR1*), experimented in red-fruited the Ailsa Craig tomato cultivar, generated lines with different fruit colors, including yellow, brown, pink, light yellow, pink-brown, yellow-green, and light green at the mature fruit stage. The transgene-free homozygous plants were obtained in less than one year, without any adverse impact on yield and fruit quality (Yang et al., 2024).

Sugar in fruits correlates with consumers' preference. It influences fruit flavor and taste. Modern tomatoes lost flavor and are less sugary. Sugar content is inversely correlated with fruit size and yield. Two genes, calcium-dependent protein kinase 27 (*SlCDPK27*; also known as *SlCPK27*) and its paralogue *SlCDPK26*, control fruit sugar content in tomato. Knockout mutant involving *SlCDPK27* and *SlCDPK26*, via CRISPR/Cas9-mediated gene editing, resulted in sweeter fruits with no penalty on fruit weight or fruit yield, but with fewer and lighter seeds, which exhibited normal germination (Zhang et al., 2024a). Japan has become the first country to release gene-edited tomatoes enriched with high amounts of γ-aminobutyric acid (GABA) for commercial production and marketed by Tokyo-based Sanatech Seed (waltz et al., 2021).

Clearly, adoption of transgenic route or CRISPR/Cas9-mediated gene-editing systems provide additional avenues for innovative breeding aimed at developing tomato cultivars with better aroma and nutrition.

17.7 ALLELIC VARIATIONS AND INTROGRESSION LINES (ILS) WITH IMPROVED FLAVOR FOR USE IN APPLIED BREEDING

Meta genome-wide association studies (mGWAS), involving 775 tomato accessions and 2,316,117 SNPs, discovered 305 significant associations related to sugars, acids, amino acids, and flavor-related volatiles in tomatoes. Domestication and genetic improvement adversely impacted citric and malic acid contents, while sugar content was less affected, clearly showing the influence of human selection on loss of flavor in modern tomatoes. However, it should be possible to significantly increase volatiles that positively contribute to consumer preferences while reducing unpleasant volatiles, by selecting the relevant allele combinations (Zhao et al., 2019).

Introgression lines with desirable horticultural quality traits and abiotic stress tolerance were established by crossing *S. pimpinellifolium* (a red-fruited LA1585 accession) with yellow-fruited tomato (*S. lycopersicum*) mutant, e9292. Many of these lines had higher concentration of ascorbic acid, lycopene, fructose, glucose, and sugar/acid ratios, indicating that novel genetic resources can be created to enhance tomato fruit flavor and stress tolerance, using wild relatives in breeding programs (Li et al., 2023b). ToMAGIC populations developed recently by intercrossing four accessions each of *Solanum lycopersicum* var. *cerasiforme* and *Solanum pimpinellifolium* generated a set of novel germplasm (354 lines) with immense variability for agronomic traits including fruit size, plant pigmentation, leaf morphology, and earliness, demonstrating that the newly developed ToMAGIC population is a valuable resource for genetic research in tomato, including for research on tomato flavor and taste (Arrones et al., 2024).

Variation in volatile organic compounds (VOCs) are major determinants of fruit flavor, including tomatoes. Introgression lines (ILs) carrying a single introgression from *Solanum pimpinellifolium* (*SP*), a wild species of tomato, maintained positive effect on tomato flavor, indicating SP alleles are valuable resources for improving the flavor of modern tomatoes (Rambla et al., 2017). The undesirable flavor in ILs involving *S. pennellii* alleles relates to a segment on the short arm of chromosome 8. Analysis of ripe fruits volatiles from ILs carrying this chromosome segment revealed up to 60-fold increase in the levels of 2-phenylethanol and phenylacetaldehyde, as compared to the modern tomato. Higher levels of 2-phenylacetaldehyde (>0.005 ppm) causes nauseating (unfavorable) aroma, while low levels of phenylacetaldehyde contribute to the desirable aroma in cultivated tomato (Tadmor et al., 2002). Phenylalanine-derived volatile compound produced by plants contributes to desirable flavor and fragrance in many plants, including tomato fruits (Tieman et al., 2006). The ripe fruits of IL4–4 containing a segment of *S. pennellii* in the genetic background of cv. M82 contain many metabolite compounds related to VOCs and those with primary and secondary metabolites significantly altered (Liu et al., 2016).

The latest advances in GMO and CRISPR-edited tomatoes with enhanced flavor have also been achieved; for example, introgression of a dominant phototropin 1 mutant enhances carotenoids and boosts flavor-related volatiles in genome-edited tomato RIN mutants (Nizampatnam et al., 2024).

17.8 CONCLUSIONS

Domestication and genetic improvement brought loss of allelic variation associated with improved flavor in modern tomatoes. The traditional cultivars (heirlooms), including landraces, and a few wild relatives (cross-compatible with cultivated tomatoes), have fruits with good taste and flavor. Current advances in the development of genetic and chemical toolboxes are providing new avenues to improve tomatoes' fruit flavor and taste. Genetic stocks with unique flavor and taste can be channeled through breeding programs to develop cultivars with improved fruit flavor and taste. The proof of concept has also been demonstrated to improve tomato fruit flavor and taste via genetic transformation or genome-editing approaches.

REFERENCES

Agius, C., von Tucher, S., Poppenberger, B., and Rozhon, W. (2018). Quantification of sugars and organic acids in tomato fruits. *MethodsX* 5, 537–550. https://doi.org/10.1016/j.mex.2018.05.014

Anđelini, M., Major, N., Išić, N., Kovačević, T.K., Ban, D., Palčić, I., et al. (2023). Sugar and organic acid content is dependent on tomato (*Solanum Lycoperiscum* L.) peel color. *Horticulturae* 9, 313. https://doi.org/10.3390/horticulturae9030313

Aono, Y., Asikin, Y., Wang, N., Tieman, D., Klee, H., and Kusano, M. (2021). High-throughput chlorophyll and carotenoid profiling reveals positive associations with sugar and apocarotenoid volatile content in fruits of tomato varieties in modern and wild accessions. *Metabolites* 11, 398. https://doi.org/10.3390/metabo11060398

Arrones, A., Antar, O., Peirera-Dias, L., Solana, A., Ferrante, P., Aprea, G., et al. (2024). A novel tomato inter-specific (*Solanum lycopersicum* var. *cerasiforme* and *S. pimpinellifolium*) MAGIC population facilitates trait association and candidate gene discovery in untapped exotic germplasm. *Hortic. Res.* uhae154. https://doi.org/10.1093/hr/uhae154

Baldina, S., Picarella, M.E., Troise, A.D., Pucci, A., Ruggieri, V., Ferracane, R., et al. (2016). Metabolite profiling of Italian tomato landraces with different fruit types. *Front. Plant Sci.* 7, 664. https://doi.org/10.3389/fpls.2016.00664

Barnett, J.R., Tieman, D.M., and Caicedo, A.L. (2023). Variation in ripe fruit volatiles across the tomato clade: An evolutionary framework for studying fruit scent diversity in a crop wild relative. *Ann. Bot.* 110, e16223. https://doi.org/10.1002/ajb2.16223

Bauchet, G., Grenier, S., Samson, N., Segura, V., and Kende, A. (2017). Identification of major loci and genomic regions controlling acid and volatile content in tomato fruit: Implications for flavor improvement. *New Phytol.* 215, 624–641. https://doi.org/10.1111/nph.14615

Bolger, A., Scossa, F., Bolger, M.E., Lanz, C., Maumus, F., Tohge, T., et al. (2014). The genome of the stress-tolerant wild tomato species *Solanum pennellii. Nat. Genet.* 46, 1034–1038. https://doi.org/10.1038/ng.3046

Borba, K.R., Aykas, D.P., Milani, M.I., Colnago, L.A., Ferreira, M.D., and Rodriguez-Saona, L.E. (2021). Portable near infrared spectroscopy as a tool for fresh tomato quality control analysis in the field. *Appl. Sci.* 11, 3209. https://doi.org/10.3390/app11073209

Chen, Y., Liu, Y., and Feng, X. (2023). Food perception: Taste, smell and flavour. *Foods* 12, 3628. https://doi.org/10.3390/foods12193628

Cheng, G., Chang, P., Shen, Y., Wu, L., El-Sappah, A.H., Zhang, F., et al. (2020). Comparing the flavor characteristics of 71 tomato (*Solanum lycopersicum*) accessions in central Shaanxi. *Front. Plant Sci.* 11, 586834. https://doi.org/10.3389/fpls.2020.586834

Cheng, G.-T., Li, Y.-S., Qi, S.-M., Wang, J., Zhao, P., Lou, Q.-Q., et al. (2021). *SlCCD1A* enhances the aroma quality of tomato fruits by promoting the synthesis of carotenoid-derived volatiles. *Foods* 10, 2678. https://doi.org/10.3390/foods10112678

Çolak, N.G., Eken, N.T., Ülger, M., Frary, A., and Doğanlar, S. (2020). Exploring wild alleles from *Solanum pimpinellifolium* with the potential to improve tomato flavor compounds. *Plant Sci.* 298, 110567. https://doi.org/10.1016/j.plantsci.2020.110567

Collins, E.J., Bowyer, C., Tsouza, A., and Chopra, M. (2022a). Tomatoes: An extensive review of the associated health impacts of tomatoes and factors that can affect their cultivation. *Biology (Basel)* 11, 239. https://doi.org/10.3390/biology11020239

Collins, E.J., Bowyer, C., Tsouza, A., and Chopra, M. (2022b). Impacts of tomatoes and factors that can affect their cultivation. *Biology (Basel)* 11, 239. https://doi.org/10.3390/biology11020239

Cortina, P.R., Santiago, A.N., Sance, M.M., Peralta, I.E., Carrari, F., and Asia, R. (2018). Neuronal network analyses reveal novel associations between volatile organic compounds and sensory properties of tomato fruits. *Metabolomics* 14, 57. https://doi.org/10.1007/s11306-018-1355-7

D'Angelo, M., Zanor, M.I., Sance, M., Cortina, P.R., Boggio, S.B., Asprelli, P., et al. (2018). Contrasting metabolic profiles of tasty Andean varieties of tomato fruit in comparison with commercial ones. *J. Sci. Food Agric.* 98, 4128–4134. https://doi.org/10.1002/jsfa.8930

Delgado-Vargas, V.A., Ayala-Garay, O.J., Arévalo-Galarza, M.L., and Gautier, H. (2023). Increased temperature affects tomato fruit physicochemical traits at harvest depending on fruit developmental stage and genotype. *Horticulturae* 9, 212. https://doi.org/10.3390/horticulturae9020212

Duckena, L., Alksnis, R., Erdberga, I., Alsina, I., Dubova, L., and Duma, M. (2023). Non-destructive quality evaluation of 80 tomato varieties using Vis-NIR spectroscopy. *Foods* 12, 1990. https://doi.org/10.3390/foods12101990

Dzakovich, M.P., Goggans, M.L., Thomas-Ahner, J.M., Moran, N.E., Clinton, S.K., Francis, D.M., et al. (2024). Transcriptomics and metabolomics reveal tomato consumption alters hepatic xenobiotic metabolism and induces steroidal alkaloid metabolite accumulation in mice. *Mol. Nutr. Food Res.* 68, e2300239. https://doi.org/10.1002/mnfr.202300239

Ebert, A.W., and Chou, Y.-Y. (2015). The tomato collection maintained by AVRDC—The World Vegetable Center: Composition, germplasm dissemination and use in breeding. *Acta Hortic.* 1101, 169–176. https://doi.org/10.17660/ActaHortic.2015.1101.26

Fan, X., Lu, N., Xu, W., Zhuang, Y., Jin, J., Mao, X., et al. (2023). Response of flavor substances in tomato fruit to light spectrum and daily light integral. *Plants (Basel)* 12, 2832. https://doi.org/10.3390/plants12152832

Gai, W., Yang, F., Yuan, L., Ul Haq, S., Wang, Y., Wang, Y., et al. (2023). Multiple-model GWAS identifies optimal allelic combinations of quantitative trait loci for malic acid in tomato. *Hortic. Res.* 10, uhad021. https://doi.org/10.1093/hr/uhad021

Gai, W., Yuan, L., Yang, F., Ahiakpa, J.K., Li, F., Ge, P., et al. (2024). Genome-wide variants and optimal allelic combinations for citric acid in tomato. *Hortic. Res.* 11, uhae070. https://doi.org/10.1093/hr/uhae070

Gao, L., Gonda, I., Sun, H., Ma, Q., Bao, K., Tieman, D.M., et al. (2019). The tomato pan-genome uncovers new genes and a rare allele regulating fruit flavor. *Nat. Genet.* 51, 1044-1051. https://doi.org/10.1038/s41588-019-0410-2

Gascuel, Q., Diretto, G., Monforte, A.J., Fortes, A.M., and Granell, A. (2017). Use of natural diversity and biotechnology to increase the quality and nutritional content of tomato and grape. *Front. Plant Sci.* 8, 652. https://doi.org/10.3389/fpls.2017.00652

Gramazio, P., Pereira-Dias, L., Vilanova, S., Prohens, J., Soler, S., Esteras, J., et al. (2020). Morphoagronomic characterization and whole-genome resequencing of eight highly diverse wild and weedy *S. pimpinellifolium* and *S. lycopersicum* var. *cerasiforme* accessions used for the first interspecific tomato MAGIC population. *Hortic. Res.* 7, 174. https://doi.org/10.1038/s41438-020-00395-w

Guan, S., Liu, C., Yao, Z., Wan, H., Ruan, M., Wang, R., et al. (2024). Detection and analysis of VOCs in Cherry tomato based on GC-MS and GC×GC-TOF MS techniques. *Foods* 13, 1279. https://doi.org/10.3390/foods13081279

Gupta, P., Dhanya, A.J., Sharma, R., and Sreelakshmi, Y. (2024a). Identification of volatiles in tomato fruit using headspace solid-phase-micro-extraction (HS-SPME) coupled with gas chromatography-mass spectrometry (GC-MS). *Methods Mol. Biol.* 2788, 39–48. https://doi.org/10.1007/978-1-0716-3782-1_3

Gupta, P., Dholaniya, P.S., Princy, K., Madhavan, A.S., Sreelakshmi, Y., and Sharma, R. (2024b). Augmenting tomato functional genomics with a genome-wide induced genetic variation resource. *Front. Plant Sci.* 14, 1290937. https://doi.org/10.3389/fpls.2023.1290937

Gur, A., and Zamir, D. (2004). Unused natural variation can lift yield barriers in plant breeding. *PLoS Biol.* 2, e245. https://doi.org/10.1371/journal.pbio.0020245

Hagenguth, J., Kanski, L., Kahle, H., Becker, H.C., and Horneburg, B. (2024). Flavour improvement in early generations of fresh market tomatoes (*Solanum lycopersicum*): I. Identification of QTL for sensory attributes, physicochemical measurements and volatile compounds. *Plant Breed.* 143, 334–349. https://doi.org/10.1111/pbr.13168

Hagenguth, J., Kanski, L., Kahle, H., Naumann, M., Pawelzik, E., Becker, H.C., et al. (2022). Breeders' sensory test: A new tool for early selection in breeding for tomato (*Solanum lycopersicum*) flavour. *Plant Breed.* 141, 96–107. https://doi.org/10.1111/pbr.12994

Ibáñez, G., Valcárcel, M., Cebolla-Cornejo, J., and Roselló, S. (2019). FT-MIR determination of taste-related compounds in tomato: A high throughput phenotyping analysis for selection programs. *J. Sci. Food Agric.* 99, 5140–5148. https://doi.org/10.1002/jsfa.9760

Jian, W., Cao, H., Yuan, S., Liu, Y., Lu, J., Lu, W., et al. (2019). SlMYB75, an MYB-type transcription factor, promotes anthocyanin accumulation and enhances volatile aroma production in tomato fruits. *Hortic. Res.* 6, 22. https://doi.org/10.1038/s41438-018-0098-y

Khan, A.W., Garg, V., Roorkiwal, M., Golicz, A.A., Edwards, D., and Varhshney, R.K. (2020). Super-pangenome by integrating the wild side of a species for accelerated crop improvement. *Trends Plant Sci.* 25, 148–158. https://doi.org/10.1016/j.tplants.2019.10.012

Kurina, A.B., Solovieva, A.E., Khrapalova, I.A., and Artemyeva, A.M. (2021). Biochemical composition of tomato fruits of various colors. *Vavilovskii Zhurnal Genet Selektsii.* 25, 514–527. https://doi.org/10.18699/VJ21.058

Li, N., He, Q., Wang, J., Wang, B., Zhao, J., Huang, S., et al. (2023a). Super-pangenome analyses highlight genomic diversity and structural variation across wild and cultivated tomato species. *Nat. Genet.* 55, 852–860. https://doi.org/10.1038/s41588-023-01340-y

Li, W., Li, Y., Liang, Y., Ni. L., Huang, H., Wei, Y., et al. (2023b). Generating novel tomato germplasm using the ancestral wild relative of *Solanum pimpinellifolium*. *Horticulturae* 9, 34. https://doi.org/10.3390/horticulturae9010034

Li, X., Tieman, D., Liu, Z., Chen, K., and Klee, H.J. (2020). Identification of a lipase gene with a role in tomato fruit short-chain fatty acid-derived flavor volatiles by genome-wide association. *Plant J.* 104, 631–644. https://doi.org/10.1111/tpj.14951

Liu, Z., Alseekh, S., Brotman, Y., Zheng, Y., Fei, Z., Tieman, D.M., et al. (2016). Identification of a *Solanum pennellii* chromosome 4 fruit flavor and nutritional quality-associated metabolite QTL. *Front. Plant Sci.* 7, 1671. https://doi.org/10.3389/fpls.2016.01671

Liu, Z., Wu, J., Wang, L., Lu, X., Ahammed, G.L., Zhang, X., et al. (2024). Integration of transcriptome and metabolome reveals regulatory mechanisms of volatile flavor formation during tomato fruit ripening. *Hortic. Plant J.* https://doi.org/10.1016/j.hpj.2023.12.010

Nizampatnam, N.R., Sharma, K., Gupta, P., Pamei, I., Sarma, S., Sreelakshmi, and Sharma, R. (2024). Introgression of a dominant phototropin1 mutant enhances carotenoids and boosts flavour-related volatiles in genome-edited tomato *RIN* mutants. *New Phytol.* 241, 2227–2242. https://doi.org/10.1111/nph.19510

Ortega-Salazar, I., Crum, D., Sbodio, A.O., Sugiyama, Y., Adaskaveg, A., Wang, D., et al. (2023). Double CRISPR knockout of pectin degrading enzymes improves tomato shelf-life while ensuring fruit quality. *Plants People Planet* 6, 330–340. https://doi.org/10.1002/ppp3.10445

Pan, F., Zhang, Q., Zhu, H., Li, J., and Wen, Q. (2023). Transcriptome and metabolome provide insights into fruit ripening of cherry tomato (*Solanum lycopersicum* var. *cerasiforme*). *Plants (Basel)* 12, 3505. https://doi.org/10.3390/plants12193505

Panthee, D.R., Labate, J.A., and Robertson, L.D. (2012). Evaluation of tomato accessions for flavour and flavour-contributing components. *Plant Genet. Resour.* 11, 106–113. https://doi.org/10.1017/S1479262112000421

Pereira, L., Sapota, M., Alonge, M., Zheng, Y., Zhang, Y., Razifard, H., et al. (2021). Natural genetic diversity in tomato flavor genes. *Front. Plant Sci.* 12, 642828. https://doi.org/10.3389/fpls.2021.642828

Rambla, J.L., Medina, A., Fernández-Del-Carmen, A., Barrantes, W., Grandillo, S., Cammareri, M., et al. (2017). Identification, introgression, and validation of fruit volatile QTLs from a red-fruited wild tomato species. *J. Exp. Bot.* 68, 429–442. https://doi.org/10.1093/jxb/erw455

Sim, S.-C., Durstewitz, G., Plieske, J., Wieseke, R., Ganal, M.W., Devynze, A.V., et al. (2012). Development of a large SNP genotyping array and generation of high-density genetic maps in tomato. *PLoS ONE* 7, e40563. https://doi.org/10.1371/journal.pone.0040563

Stikić, R., Jovanović, Z., Vucelić-Radović, B., Marjanović, M., and Savić, S. (2015). Tomato: A model species for fruit growth and development studies. *Botanica SERVICA* 39, 95–102.

Tadmor, Y., Fridman, E., Gur, A., Larkov, O., Lastochkin, E., Ravid, U., et al. (2002). Identification of malodorous, a wild species allele affecting tomato aroma that was selected against during domestication. *J. Agric. Food Chem.* 50, 2005–2009. https://doi.org/10.1021/jf011237x

Tanambell, H., Quek, S.Y., and Bishop, K.S. (2019). Screening of in vitro health benefits of tangerine tomatoes. *Antioxidants (Basel)* 8, 230. https://doi.org/10.3390/antiox8070230

Tieman, D., Taylor, M., Schauer, N., Fernie, A.R., Hanson, A.D., and Klee, H.J. (2006). Tomato aromatic amino acid decarboxylases participate in synthesis of the flavor volatiles 2-phenylethanol and 2-phenylacetaldehyde. *Proc. Natl. Acad. Sci. U. S. A.* 103, 8287–8292. https://doi.org/10.1073/pnas.0602469103

Tieman, D., Zhu, G., Resende Jr, M.F.R., Lin, T., Nguyen, C., Bies, D., et al. (2017). A chemical genetic roadmap to improved tomato flavor. *Science* 355, 391–394. https://doi.org/10.1126/science.aal1556

Tikunov, Y.M., Roohanitaziani, R., Meijer-Dekens, F., Molthoff, J., Paulo, J., Finkers, R., et al. (2020). The genetic and functional analysis of flavor in commercial tomato: The *FLORAL4* gene underlies a QTL for floral aroma volatiles in tomato fruit. *Plant J.* 103, 1189–1204. https://doi.org/10.1111/tpj.14795

Vincis, R., and Fontanini, A. (2019). Central taste anatomy and physiology. *Handb. Clin. Neurol.* 164, 187–204. https://doi.org/10.1016/B978-0-444-63855-7.00012-5

Vinha, A.F., Barreira, S.V.P., Costa, A.S.G., Alves, R.C., and Oliveira, M.B.P.P. (2014). Organic *versus* conventional tomatoes: Influence on physicochemical parameters, bioactive compounds and sensorial attributes. *Food Chem. Toxicol.* 67, 139–144. https://doi.org/10.1016/j.fct.2014.02.018

Waltz, E. (2021). GABA-enriched tomato is first CRISPR-edited food to enter market. *Nat. Biotechnol.* 40, 9–11. https://doi.org/10.1038/d41587-021-00026-2

Wang, X., Gao, L., Jiao, C., Stravoravidis, S., Hosmani, P.S., Saha, S., et al. (2020). Genome of *Solanum pimpinellifolium* provides insights into structural variants during tomato breeding. *Nat. Commun.* 11, 5817. https://doi.org/10.1038/s41467-020-19682-0

Yang, J., Liang, B., Zhang, Y., Liu, Y., Wang, S., Yang, Q., et al. (2022). Genome-wide association study of eigenvectors provides genetic insights into selective breeding for tomato metabolites. *BMC Biol.* 20, 120. https://doi.org/10.1186/s12915-022-01327-x

Yang, T., Ali, M., Lin, L., Li, P., He, H., Zhu, Q., et al. (2024). Recoloring tomato fruit by CRISPR/Cas9-mediated multiplex gene editing. *Hortic. Res.* 10, uhac214. https://doi.org/10.1093/hr/uhac214

Zhang, J., Lyu, H., Chen, J., Cao, X., Du, R., Ma, L., et al. (2024a). Releasing a sugar brake generates sweeter tomato without yield penalty. *Nature* 635, 647–656. https://doi.org/10.1038/s41586-024-08186-2

Zhang, Z., Ye, W., Li, C., Zhou, H., Wang, C., Liu, P., et al. (2024b). Volatilomics-based discovery of key volatiles affecting flavor quality in tomato. *Foods* 13, 879. https://doi.org/10.3390/foods13060879

Zhao, J., Sauvage, C., Zhao, J., Bitton, F., Bauchet, G., Liu, D., et al. (2019). Meta-analysis of genome-wide association studies provides insights into genetic control of tomato flavor. *Nat. Commun.* 10, 1534. https://doi.org/10.1038/s41467-019-09462-w

Zhou, Y., Zhang, Z., Bao, Z., Li, H., Lyu, Y., Zan, Y., et al. (2022). Graph pangenome captures missing heritability and empowers tomato breeding. *Nature* 606, 527–534. https://doi.org/10.1038/s41586-022-04808-9

18 Profiling *Brassica* leafy Germplasm for Glucosinolates and Their Compounds

Sangam L. Dwivedi and Autar K. Mattoo

18.1 INTRODUCTION

Brassica diploids (*B. rapa*, AA genome, $2n=2x=20$; *B. nigra*, BB *genome*, $2n=2x=16$; *B. oleracea*, CC genome, $2n=2x=18$) and allotetraploids (*B. juncea*, AABB genome, $2n=4x=36$; *B. napus*, AACC genome, $2n=4x=38$; *B. carinata*, BBCC, $2n=4x=34$) form 'triangle of U' model. U's triangle proposed that allotetraploid species originated from a pairwise cross involving diploid species *B. rapa*, *B. nigra*, and *B. oleracea* (Nagaharu, 1935). They are grown worldwide for seed oil, as condiments, or as culinary vegetables.

Domestication and selection have resulted in extreme morphotypes and extensive axillary branching, with similar morphotypes (specialized enlarged organs as edible products) selected for in two or more *Brassica* species. The examples include leafy heads in Chinese cabbage (*B. rapa pekinensis*) and other cabbages (*B. oleracea* var. *capitata*); the inflorescences of cauliflower (*B. oleracea* var. *botrytis*) and broccoli (*B. oleracea* var. *italica*) form enlarged arrested inflorescences (curds); the brussels sprouts (*B. oleracea* var. *gemmifera*) grow axillary heading buds along their stems; Chinese kale (*B. oleracea* var. *alboglabra*) and broccoletti (*B. rapa* var. *cymosa*) develop into a succulent stem (Cheng et al., 2014).

The Crop Trust (https://zenodo.org/records/7544810; accessed on March 17, 2024) indicated a total of ex situ collection of 85,474 of the six *Brassica* species (*B. olearacea*, *B. napus*, *B. juncea*, *B. carinata*, *B. rapa*, and *B. nigra*), including wild relatives, conserved across gene banks worldwide. In addition, 81,752 accessions of *Brassica* are conserved in 148 institutes spread across 81 countries (Subramanian et al., 2023).

Brassica species are rich sources of glucosinolates (GLSs) and other essential nutrients. GLSs and their hydrolyzed products provide health benefits to humans and also protect plants from stress (Maina et al., 2020). GLSs are sulfur-containing secondary metabolites that are naturally found in *Brassica* species. Both positions of the leaves (outer, middle, or inner) and sections (top, middle, bottom, green/red, or white) within the leaves influence GLS content. Higher GLS content is reported

DOI: 10.1201/9781003467335-18

in the proximal half and white sections and the middle layers of the leaves (Rhee et al., 2020). It has been shown that GLSs and their metabolites are effective in cancer treatment (Orouji et al., 2023; Soundararajan and Kim, 2018), enhance product shelf-life by reducing oxidative processes, and inhibit or retard the microbial growth during processing and storage, with no adverse effect on the organoleptic characteristics of the product (Pateiro et al., 2023).

18.2 VARIABILITY IN LEAF MORPHOTYPES, BIOACTIVE COMPOUNDS, AND MINERALS IN *BRASSICA* LEAFY VEGETABLES

Leaf mustard (*B. juncea*) accessions are characterized by purple, red, and green leaves with a green/white midrib are known for their thick, tender, and spicy leaves with a unique taste and flavor. An assessment of a large collection of leaf mustard germplasms showed huge diversity in leaf characteristics and total and individual GLS content, with many of these compounds being positively correlated with leaf characteristics (Assefa et al., 2023).

Cabbage (*Brassica oleracea* var. *capitata*) germplasm, with diverse origins, has substantial variation in total (3.99 to 23.75 μmol g^{-1} DW) and individual GLS content. Seven aliphatics (glucoiberin, GIB; progoitrin, PRO; epi-progoitrin, EPI; sinigrin, SIN; glucoraphanin, GRA; glucoerucin, GER; and gluconapin, GNA), one aromatic (gluconasturtiin, GNS) and four indolyl (glucobrassicin, GBS; 4-hydroxyglucobrassicin, 4HGBS; 4-methoxyglucobrassicin, 4MGBS; and neoglucobrassicin, NGBS) GLSs are reported in cabbage. The aliphatic GLSs are predominant followed by indolyl and aromatic. They contribute 46%, 51%, and 138% coefficients of variation, respectively. GIB, GBS, and NGBS were the most common GSLs found in all accessions. The hydrolyzed compounds of GBS, SIN, GIB, and GRA are reported to have anticancer properties (Bhandari et al., 2020). The total GLS content of kimchi cabbage (*Brassica rapa* L.) cultivars ranged from 621.15 to 42434.21 μmol kg^{-1} DW, with aliphatic GLSs being predominant (Rhee et al., 2020).

The 12 GLSs among broccoli (*B. oleracea* var. *italica*) floret accessions ranges from 0.467 to 57.156 μmol g^{-1} DW, with highest GLS content ~122-fold higher than the lowest value. Glucobrassicin, neoglucobrassicin, and glucoraphanin GLSs explain 60% of the total variance. Hydroxyglucobrassicin, methoxyglucobrassicin, glucobrassicin, glucoerucin, gluconasturtiin, glucoraphanin, and glucotropaeolin were positively ($P < 0.05$) correlated (Li et al., 2021). Thirteen GLSs, including seven aliphatic, five indole, and one aromatic, among broccoli accessions are known (Yan et al., 2023). The GLS content ranged from 0.1705 to 5.8174 mg g^{-1} DW. Accessions with largest GLS content and other agronomically useful traits are useful resources for broccoli improvement (Yan et al., 2023).

Pak choy (*Brassica rapa chinensis*) accessions are reported to contain 17 individual GLSs, with three distinct groups of the accessions characterized by specific combination of GLSs (Kim et al., 2023). The profiling of GLSs metabolites in a collection of turnip (*B. rapa rapa*) roots and tubers isolated 13 GLSs (8 aliphatic, 4 indolic, and 1 aromatic), and 24 and 16 hydrolyzed (breakdown) products in tubers and leaves, respectively. The 3-butenyl GLS was predominant in both plant organs, and 2-hydroxy-3-butenyl GLS and 2-phenylethyl GLS were significantly higher in tubers (Klopsch et al., 2017).

An assessment of a large collection of *B. napus* accessions revealed 32 phenolics and 36 GLSs. Roots and leaves show striking variations for 1-methylpropyl-GLS (glucocochlearin) and tetra- and penta-glucosyl kaempferol derivatives among accessions. Two main chemotypes (sinapoyl-*O*-hexoside and kaempferol-*O*-trihexoside) may provide useful resources for chemical ecology and breeding of *Brassica* (Missinou et al., 2022). Bróquil (*Brassica oleracea* var. *italica*) landraces show abundant intra- and inter-accession variability for phenolics, flavonoids, GLSs, vitamin C, minerals, and antioxidants (Montaner et al., 2022).

A comparative assessment of nine *Brassica* species revealed that each species has a unique sulfur and phenolic profile, with green mustard (*Brassica juncea*) having the most active antioxidant activity. The species with 25 phytochemicals, including myricetin, quercetin-3-galactoside, quercetin-3-glucoside, pterostilbene, ferulic acid, kaempferol, allyl isothiocyanate, and (-)-epicatechin, in decreasing order, show strongest antioxidant activity. Variation in phenolic compounds were also detected, with *trans*-resveratrol in rocket (*Eruca vesicaria*) leaves being at highest concentration of up to 84 μg g^{-1} DW leaves (Fusari et al., 2020).

Mutants with discrete phenotypes are a useful resource for functional genomic studies. The sequencing of EMS-induced mutant populations of Chinese cabbage (300 M_2 to M_5 mutants) resulted in 750,629 SNPs and 46,272 InDel mutations, covering ~98% of all predicted genes in the A03 genome. Mutations in five genes of the GLS biosynthetic pathway relate to variation in GLS content of the mutant plants. The availability of such large collection of genotyping and phenotyping data of EMS-induced mutant population resource (www.bioinformaticslab.cn/EMSmutation/home) increases its use to facilitate gene mining in Chinese cabbage and possibly in other *Brassica* species (Sun et al., 2022).

18.3 GENETIC AND MOLECULAR BASIS OF BIOACTIVE COMPOUNDS AND MINERALS IN *BRASSICA* LEAFY VEGETABLES

Brassica leaves are the source of nutrients, including minerals, vitamins, soluble fibers, and GLSs, and a key organ of consumption in the form of leafy vegetables. Complex genetic networks regulate cell proliferation and cell expansion of leaves. Understanding the molecular mechanism of leaf size control is a prerequisite for molecular breeding of leaf traits. Multiple genes and QTLs control variation in leaf size that influences leaf growth through different pathways and mechanisms, including phytohormone biosynthesis and signaling, transcription regulation, and small RNAs (Jian et al., 2017; Karamat et al., 2021). A complex locus *RCO*, mapped on chromosome 10, regulates the formation of lobed leaves in *B. juncea*. *BjRCO.1* and *BjRCO.2* within the complex locus are functionally characterized genes involved in the regulation of highly lobed leaves (Chang et al., 2023). *BrrA02.LMI1* is a candidate gene underlying the QTL *qBrrLLA02* controlling leaf lobe formation in *B. rapa*. It is mainly expressed at the leaf margin, with the highest expression at the tip of each lobe. The ectopic expression of *BrrA02.LMI1* in *Arabidopsis* resulted in a deep leaf lobe in the WT, while its downregulation disturbed lobe formation (Li et al., 2023). Thus, it is a valuable genomic resource in regulating leaf morphotypes in breeding *Brassica* crops with distinct leafy characteristics.

Leaf flattening influences yield and quality of Chinese cabbage. A single recessive gene, *Brcwm*, fine-mapped to a 205.66 kb region on chromosome A07 contains 39 genes. A nsSNP within the target interval on exon 4 of *BraA07g021970.3C*, homologous to AT3G55000, results in a proline to serine amino acid substitution. The disrupted leaf flattening trait co-segregated with the nsSNP. *BraA07g021970.3C* expression is significantly higher in flattened leaves compared to wrinkled leaves (cwm mutant), indicating that *BraA07g021970.3C* is the target gene for leaf flattening in Chinese cabbage (Wu et al., 2023). Premature bolting severely reduces harvestable edible products in vegetables. An EMS-induced early-bolting mutant (ebm3) in Chinese cabbage is controlled by a single recessive nuclear gene. *BrEBM3* (*BraA04g017190.3 C*) is the candidate gene of the early bolting mutant. Changes in base substitution from C to T in the 14th exon of *BrEBM3* resulted in an amino acid change from S to F leading to the early-bolting phenotype of emb3. *BrEBM3* is highly expressed in the flower and bud (Tan et al., 2021).

Malformed leaves (ML), such as mushroom type, develop from the leaf veins in some Chinese kale genotypes. Changes in adaxial-abaxial polarity of leaves affect the formation of ML. Two dominant major genes inherit independently and regulate ML phenotype. A major QTL *qML4.1* mapped within 7.4 Mb on chromosome kC4. *BocNGA1* is the key candidate gene regulating ML in Chinese kale. *BocNGA1* contains 15 SNPs in coding sequences and 21 SNPs and 3 InDels in promoter sequences in Boc52 with ML phenotype. *BocNGA1* expression in ML genotypes is significantly lower than in the normal leaves' genotypes, suggesting that *BocNGA1* as a negative regulator for ML genesis in Chinese kale (Feng et al., 2023).

Seedling, rosette, folding, and heading are four vegetative stages in cabbage plants. Rosette leaves, being the largest leaves, provide most of the energy needed to produce the leafy head. GWAS involving over 300 cabbage accessions had significant correlations between morphological traits of rosette leaves and heads. Fifty robust QTLs for rosette leaf and head traits (leaf area with the head area, height and width, leaf width with the head area and head height), identified in multiple years affecting multiple traits, are distributed over all nine chromosomes. Of these, 25 QTLs are associated with both rosette leaf and leafy head traits. Thirteen candidate genes related to auxin and other phytohormones, leaf development, and leaf polarity may regulate leafy head development or rosette leaf expansion in cabbage and Chinese cabbage (Alemán-Báez et al., 2022). An inward curling mutant *icl* in the *OCTOPUS* (*BrOPS*) gene regulates leaf curvature in Chinese cabbage. It positively controls BR signaling by antagonizing *BrBIN2* to promote leaf epinastic growth at the early heading stage in Chinese cabbage (Zhang et al., 2022a).

Brassica oleracea has many diverse morphotypes that are harvested as edible products. *B. oleracea* pangenome, involving 27 high-quality genomes representing diverse morphotypes and their wild relatives, unfolded numerous structural variations (SVs) that exert bidirectional effects on gene expression. SVs either suppress DNA methylation or promote by harboring TF-binding elements. SVs promote

BoPNY and suppress *BoCKX3* in cauliflower and broccoli, respectively. In cabbage, SVs suppress *BoKAN1* and *BoACS4* but promote *BoMYBtf* in ornamental kale. Thus, SVs have a role as dosage regulators of gene expression shaping domestication and diversification in *B. oleracea* genome (Li et al., 2024). Multiple QTLs and candidate genes regulate leaf traits (LL, lamina (leaf) length), LW, lamina width, PL, petiole length, LM, leaf margin) in Chinese cabbage, each trait regulated by a different set of genes (Lee et al., 2019). The auxin-responsive and a CLAVATA3/ESR-related family genes regulate leaf-stem differentiation, while *BoKAN1* is crucial in shaping the leafy heads of cabbage and brussels sprouts (Guo et al., 2024).

The development of curd is a key biological step in cauliflower (*Brassica oleracea* L. var. *botrytis*), influencing the productivity and nutritional quality of curd. The two steps of cauliflower domestication (i.e., curd emergence and curd improvement), initially evolved from broccoli, led to the evolution of nutrient-rich edible inflorescence meristem, i.e., cauliflower. *CAULIFLOWER1* (*CAL1*), *CAL2*, and *FRUITFULL* (*FUL2*), the MDAS-box genes, are the key causal genes regulating curd formation. GWAS analysis detected a RING-type zinc-finger protein-encoding gene, *BOB06G135460*, regulating stem height by influencing cell size (Chen et al., 2024).

Morphological changes associated with cabbage head formation may provide an opportunity to breed new cabbage (*B. oleracea capitata*) cultivars. Unlocking the cellular mechanisms of the leaf curvature is the first essential step to the formation of the leafy head. A comparative assessment of two accessions with distinct head shapes (round and pointed) and one non-heading collard accession unfolded two synergistic cellular mechanisms contributing to the curvature of heading leaves. First, the differential growth across the leaf blade, with increased growth at the leaf's center relative to the margins. Second, the increased expansion of the spongy parenchyma layer compared to the palisade parenchyma layer, leading inward curvature. Both the processes together contribute to the typical leafy heads of cabbage (Liu et al., 2024).

The bitterness in *Brassica* plants is characterized by variation in 2-hydroxy-3-butenyl GLS (progoitrin, PRO) content. *BocODD1* and *BocODD2* regulate biosynthesis of PRO GLS in Chinese kale (Wu et al., 2022). The flowering stalk of Chinese kale is the most used edible part due to its high GLS content and other nutrients. *BocGTR1a* and *BocGTR1c* are highly expressed in leaves and buds, and their expression is enhanced due to wounding and heat stress, suggesting that both regulate GLS accumulation in different tissues, providing scope for selection of specific GLSs in Chinese kale (Jiang et al., 2019). *BoaMYB28* regulates aliphatic GLS content across all developmental leaf stages in Chinese kale, with higher transcript accumulation in mature and inflorescence leaves (Yin et al., 2017).

18.4 DEVELOPING IMPROVED GERMPLASM WITH ENHANCED BIOACTIVE COMPOUNDS AND MINERALS IN *BRASSICA* LEAFY VEGETABLES

The introgression lines (ILs) involving *B. rapa* and *B. juncea* unlocked substantial phenotypic variations for leafy head shapes. SNPs effectively differentiated the two species genomes. ILs on average represented ~59% of the donor genome (*B. juncea*) in the recipient (*B. rapa*) genome. The genetic analysis involving 1,642 SNPs detected significantly enhanced genetic diversity of the new *B. juncea* germplasm (ILs) with introgressed *B. rapa* genome (Zhang et al., 2022b).

The adoption of divergent mass selection resulted in modification of GLS content in the leaves of two *Brassica* species, glucobrassicanapin in *B. napus* and gluconapin in *B. rapa*, through several selection cycles. A linear response to selection was observed with no signs of variability depletion for GLSs modification in either direction. However, response to selection was more effective for reduced GLSs than increasing GLSs. A similar pattern was noticed for other GLSs. Such modifications were also noticed in other plant parts, suggesting GLS translocation in the plant or modification in their synthesis pathway is not organ-specific. Moreover, the populations obtained by selection either yielded at par or excelled better than the original population (Coves et al., 2023). Divergent mass selection for high (SIN, sinigrin; GIB, glucoiberin; GBS, glucobrassicin) and low (SIN, GIB, GBS) GLSs has also been effective in leaves of *B. oleracea* var. *acephala.* SIN and GIB contents, after three cycles of divergent selection, increased by 52% and 78% and decreased by 52% and 45%, respectively. The selection for GBS content was only effective in decreasing concentration, with a reduction of 39%. However, mass selection for GLSs in leaves also affected GLS contents of flower buds and seeds due to the novo synthesis in these organs and/or translocation from leaves (Sotelo et al., 2016).

Microgreens, obtained from the seeds, are living superfood/functional food because of high concentration of carotenoids, phytochemicals, and minerals. An integrated approach involving multiomics, nutriomics, and foodomics should be explored to breed microgreen lines with enhanced multi-nutrient and antioxidant activity, in addition to selecting for fast growth, enhanced nutritional quality, high germination percentage, and prolonged shelf-life (Gupta et al., 2023).

Targeted editing of multiple homologues of *GTR1* and *GTR2* resulted in mustard (*B. juncea*) lines with desired low GLS content in seeds while maintaining high GLS levels in other plant parts, including leaves, with no adverse impact on plant defense and yield parameters (Mann et al., 2023). The introgression of novel alleles of the Myb28^{v} TF from *B. villosa* resulted in broccoli lines with high levels of bioavailable glucoraphanin and sulforaphane GLSs (Sivapalan et al., 2018). Glucoraphanin (methionine-derived 4-methylsulfinylbutyl GLS) is the most abundant GLS in broccoli.

The glucoraphanin in *B. rapa* group of vegetables is undetectable or only present in traces. *BrAOP2* genes convert glucoraphanin to gluconapin. For enriching the beneficial glucoraphanin content in *B. rapa*, the functional *BrAOP2* alleles were replaced by non-functional alleles through marker-assisted backcrossing. The derived progenies with introgression of both non-functional *braop2.2* and *braop2.3* alleles, relative to recurrent parent, significantly increased the glucoraphanin content by 18 times, while introgression of single non-functional *braop2.2* or *braop2.3* had no effect on glucoraphanin content (Liu et al., 2017).

Developing Chinese cabbage with purple color is a new breeding target because of its high nutritional value due to presence of abundant anthocyanidins. Near infrared spectroscopy (NIR) is an ideal screening technique for fast quantitative detection of anthocyanidins in Chinese cabbage breeding populations. The spectral data at later stages of breeding programs should be validated by performing high-performance liquid chromatography. Abundant genetic variability is reported for cyanidin content (93.5 mg kg^{-1} to 12,802.4 mg kg^{-1}), while other anthocyanidins are much lower. NIR prediction models based on partial least square regression with the preprocessing of no-scattering mode and the first-order derivative accurately predicted cyanidin content (Wang et al., 2023). A partial least squares discriminant analysis (PLS-DA) model differentiated subtle leaf color variation, with 96% accuracy and 1–13% higher efficiency than that of the single spectral models. Further validation of Chinese cabbage F_2 population using BSA (bulk segregant analysis) mixing pool identified the candidate genetic locus related to dark leaf color, delimited to 9.76 Mb on chromosome 9 (Xie et al., 2023).

18.5 CONCLUSIONS

Understanding molecular mechanisms of leaf size control is a prerequisite for the molecular breeding of leaf traits. The unlocking of complex genetic networks regulating cell proliferation and cell expansion as well as genes that control specific leaf morphotypes may provide new avenues to tailor *Brassica* germplasm with distinct leaf morphotypes. Targeted genome editing has made it possible to alter GLSs in different plant parts with no adverse effect on plant growth and development.

REFERENCES

Alemán-Báez, J., Qin, J., Cai, C., Zou, C., Bucher, J., Pauolo, M.J., et al. (2022). Genetic dissection of morphological variation in rosette leaves and leafy heads in cabbage (*Brassica oleracea* var. *capitata*). *Theor. Appl. Genet.* 135, 3611–3628. https://doi.org/10.1007/s00122-022-04205-w

Assefa, A.D., Kim, S.-H., Ko, H.C., Ro, N., Subramanian, P., Chung, Y.-J., et al. (2023). Leaf mustard (*Brassica juncea*) germplasm resources showed diverse characteristics in agro-morphological traits and glucosinolate levels. 12, 4374. https://doi.org/10.3390/foods12234374

Bhandari, S.R., Rhee, J., Choi, C.S., Jo, J.S., Shin, Y.K., and Lee, J.G. (2020). Profiling of individual desulfo-glucosinolate content in cabbage head (*Brassica oleracea* var. *capitata*) germplasm. *Molecules* 25, 1860. https://doi.org/10.3390/molecules25081860

Chang, L., Liang, J., Zhang, L., Zhang, Z., Cai, X., Wu, J., et al. (2023). A complex locus regulates highly lobed-leaf formation in *Brassica juncea*. *Theor. Appl. Genet.* 136, 224. https://doi.org/10.1007/s00122-023-04473-0

Chen, R., Chen, K., Yao, X., Zhang, X., Yang, Y., Su, X., et al. (2024). Genomic analyses reveal the stepwise domestication and genetic mechanism of curd biogenesis in cauliflower. *Nat. Genet.* 56, 1235–1244. https://doi.org/10.1038/s41588-024-01744-4

Cheng, F., Wu, J., and Wang, X. (2014). Genome triplication drove the diversification of *Brassica* plants. *Hortic. Res.* 1, 14024. https://doi.org/10.1038/hortres.2014.24

Coves, S., Soengas, P., Velasco, P., Fernández, J.C., and Cartea, M.E. (2023). New vegetable varieties of *Brassica rapa* and *Brassica napus* with modified glucosinolate content obtained by mass selection approach. *Front. Nutr.* 10, 1198121. https://doi.org/10.3389/fnut.2023.1198121

Feng, S., Wu, J., Chen, K., Chen, M., Zhu, Z., Wang, J., et al. (2023). Identification and characterization analysis of candidate genes controlling mushroom leaf development in Chinese kale by BSA-seq. *Mol. Breed.* 43, 17. https://doi.org/10.1007/s11032-023-01364-6

Fusari, C.M., Nazaremo, M.A., Locatelli, D.A., Fontana, A., Beretta, V., and Camargo, A.B. (2020). Phytochemical profile and functionality of *Brassicaceae* species. *Food Biosci.* 36, 100606. https://doi.org/10.1016/j.fbio.2020.100606

Guo, N., Wang, S., Wang, T., Duan, M., Zong, M., Miao, L., et al. (2024). A graph-based pan-genome of *Brassica oleracea* provides new insights into its domestication and morphotype diversification. *Plant Commun.* 5, 100791. https://doi.org/10.1016/j.xplc.2023.100791

Gupta, A., Sharma, T., Singh, S.P., Bhardwaj, A., Srivastava, D., and Kumar, R. (2023). Prospects of microgreens as budding living functional food: Breeding and biofortification through OMICS and other approaches for nutritional security. *Front. Genet.* 14, 1053810. https://doi.org/10.3389/fgene.2023.1053810

Jian, H., Yang, B., Zhang, A., Zhang, L., Xu, X., Li, J., et al. (2017). Screening of candidate leaf morphology genes by integration of QTL mapping and RNA sequencing technologies in oilseed rape (*Brassica napus* L.). *PLoS ONE* 12, e0169641. https://doi.org/10.1371/journal.pone.0169641

Jiang, D., Lei, J., Cao, B., Wu, S., Chen, G., and Chen, C. (2019). Molecular cloning and characterization of three glucosinolate transporter (GTR) genes from Chinese Kale. *Genes (Basel)* 10, 202. https://doi.org/10.3390/genes10030202

Karamat, U., Sun, X., Li, N., and Zhao, J. (2021). Genetic regulators of leaf size in *Brassica* crops. *Hortic Res.* 8, 91. https://doi.org/10.1038/s41438-021-00526-x

Kim, S.H., Ochar, K., Hwang, A., Lee, Y.J., and Kang, H.J. (2023). Variability of glucosinolates in Pak Choy (*Brassica rapa* subsp. *chinensis*) germplasm. *Plants (Basel)* 13, 9. https://doi.org/10.3390/plants13010009

Klopsch, R., Witzel, K., Börner, A., Schreiner, M., and Hanschen, F.S. (2017). Metabolic profiling of glucosinolates and their hydrolysis products in a germplasm collection of *Brassica rapa* turnips. *Food Res. Int.* 100(Pt 3), 392–403. https://doi.org/10.1016/j.foodres.2017.04.016

Lee, S.-M., Song, H., Yi, H., and Hur, Y. (2019). Transcriptomic analysis of contrasting inbred lines and F2 segregant of Chinese cabbage provides valuable information on leaf morphology. *Genes Genom.* 41, 811–829. https://doi.org/10.1007/s13258-019-00809-7

Li, P., Wu, Y., Han, X., Li, H., Wang, L., Chen, B., et al. (2023). *BrrA02.LMI1* encodes a homeobox protein that affects leaf margin development in *Brassica rapa*. *Int. J. Mol. Sci.* 24, 14205. https://doi.org/10.3390/ijms241814205

Li, X., Wang, Y., Cai, C., Ji, J., Han, F., Zhang, L., et al. (2024). Large-scale gene expression alterations introduced by structural variation drive morphotype diversification in *Brassica oleracea*. *Nat. Genet.* 56, 517–529. https://doi.org/10.1038/s41588-024-01655-4

Li, Z., Zheng, S., Liu, Y., Fang, Z., Yang, L., Zhuang, M., et al. (2021). Characterization of glucosinolates in 80 broccoli genotypes and different organs using UHPLC-Triple-TOF-MS method. *Food Chem.* 334, 127519. https://doi.org/10.1016/j.foodchem.2020.127519

Liu, Z., Alemán-Báez, J., Visser, R.G.F., and Bonnema, G. (2024). Cabbage (*Brassica oleracea* var. *capitata*) development in time: How differential parenchyma tissue growth affects leafy head formation. *Plants (Basel)* 13, 656. https://doi.org/10.3390/plants13050656

Liu, Z., Liang, J., Zheng, S., Zhang, J., Wu, J., Cheng, F., et al. (2017). Enriching glucoraphanin in *Brassica rapa* through replacement of *BrAOP2.2/BrAOP2.3* with non-functional genes. *Front. Plant Sci.* 8, 1329. https://doi.org/10.3389/fpls.2017.01329

Maina, S., Misinzo, G., Bakari, G., and Kim, H.Y. (2020). Human, animal and plant health benefits of glucosinolates and strategies for enhanced bioactivity: A systematic review. *Molecules* 25, 3682. https://doi.org/10.3390/molecules25163682

Mann, A., Kumari, J., Kumar, R., Kumar, P., Pradhan, A.K., and Pental, D. (2023). Targeted editing of multiple homologues of *GTR1* and *GTR2* genes provides the ideal low-seed, high-leaf glucosinolate oilseed mustard with uncompromised defence and yield. *Plant Biotechnol. J.* 21, 2182–2195. https://doi.org/10.1111/pbi.14121

Missinou, A.A., de Carvalho, J.F., Marnet, N., Delhaye, T., Hamzaoui, O., Sayed, D.A., et al. (2022). Identification and quantification of glucosinolates and phenolics in a large panel of *Brassica napus* highlight valuable genetic resources for chemical ecology and breeding. *J. Agric. Food Chem.* 70, 5245–5261. https://doi.org/10.1021/acs.jafc.1c08118

Montaner, C., Mallor, C., Laguna, S., and Zufiaurre, R. (2022). Bioactive compounds, antioxidant activity, and mineral content of bróquil: A traditional crop of *Brassica oleracea* var. *italica*. *Front. Nutr.* 9, 1006012. https://doi.org/10.3389/fnut.2022.1006012

Nagaharu, U. (1935). Genome analysis in Brassica with special reference to the experimental formation of *B. Napus* and peculiar mode of fertilization. *Jap. J. Bot.* 7, 389–452.

Orouji, N., Asl, S.K., Toghipour, Z., Habtemariam, S., Nabavi, S.M., and Rahimi, R. (2023). Glucosinolates in cancer prevention and treatment: Experimental and clinical evidence. *Med. Oncol.* 40, 344. https://doi.org/10.1007/s12032-023-02211-6

Pateiro, M., Domínguez, R., Munekata, P.E.S., Nieto, G., Bangar, S.P., Dhama, K., et al. (2023). Bioactive compounds from leaf vegetables as preservatives. *Foods* 12, 637. https://doi.org/10.3390/foods12030637

Rhee, J.-H., Choi, S., Lee, J.-E., Hur, O.-S., Ro, N.-Y., Hwang, A.-J., et al. (2020). Glucosinolate content in *Brassica* genetic resources and their distribution pattern within and between inner, middle, and outer leaves. *Plants (Basel)* 9, 1421. https://doi.org/10.3390/plants9111421

Sivapalan, T., Melchini, A., Saha, S., Needs, P.W., Traka, M.H., Tapp, H., et al. (2018). Bioavailability of glucoraphanin and sulforaphane from high-glucoraphanin broccoli. *Mol. Nutr. Food Res.* 62, e1700911. https://doi.org/10.1002/mnfr.201700911

Sotelo, T., Velasco, P., Soengas, P., Rodríguez, V.M., and Cartea, M.E. (2016). Modification of leaf glucosinolate contents in *Brassica oleracea* by divergent selection and effect on expression of genes controlling glucosinolate pathway. *Front. Plant Sci.* 7, 1012. https://doi.org/10.3389/fpls.2016.01012

Soundararajan, P., and Kim, J.S. (2018). Anti-carcinogenic glucosinolates in Cruciferous vegetables and their antagonistic effects on prevention of cancers. *Molecules* 23, 2983. https://doi.org/10.3390/molecules23112983

Subramanian, P., Kim, S.-H., and Hahn, B.-S. (2023). *Brassica* biodiversity conservation: Prevailing constraints and future avenues for sustainable distribution of plant genetic resources. *Front. Plant Sci.* 14, 1220134. https://doi.org/10.3389/fpls.2023.1220134

Sun, X., Li, X., Lu, Y., Wang, S., Zhang, X., Zhang, K., et al. (2022). Construction of a high-density mutant population of Chinese cabbage facilitates the genetic dissection of agronomic traits. *Mol. Plant* 15, 913–924. https://doi.org/10.1016/j.molp.2022.02.006

Tan, C., Ren, J., Wang, L., Ye, X., Fu, W., Zhang, J., et al. (2021). A single amino acid residue substitution in *BraA04g017190.3C*, a histone methyltransferase, results in premature bolting in Chinese cabbage (*Brassica rapa* L. ssp. *Pekinensis*). *BMC Plant Biol.* 21, 373. https://doi.org/10.1186/s12870-021-03153-9

Wang, Y.Q., Liu, G.-M., Hu, L.-P., Zhao, X.-Z., Zhang, D.-S., and He, H.-J. (2023). Prediction of anthocyanidins content in purple Chinese cabbage based on visible/near infrared spectroscopy. *Foods* 12, 1922. https://doi.org/10.3390/foods12091922

Wu, S., Zhang, T., Wang, Y., Chen, M., Yang, J., Li, F., et al. (2022). *BocODD1* and *BocODD2* regulate the biosynthesis of progoitrin glucosinolate in Chinese Kale. *Int. J. Mol. Sci.* 23, 14781. https://doi.org/10.3390/ijms232314781

Wu, Y., Xin, Y., Zou, J., Huang, S., Wang, C., and Feng, H. (2023). *BrCWM* mutation disrupted leaf flattening in Chinese cabbage (*Brassica rapa* L. ssp. *pekinensis*). *Int. J. Mol. Sci.* 24, 5225. https://doi.org/10.3390/ijms24065225

Xie, Z., Yan, J., Liang, H., Yue, X., Su, X., Wei, H., et al. (2023). PLS-DA model for accurate identification of Chinese cabbage leaf color based on multispectral imaging. *Veget. Res.* 3, 25 https://doi.org/10.48130/VR-2023-0025

Yan, M., Song, C., Su, S., Li, J., Hu, Z., Lin, S., et al. (2023). Quantification and diversity analyses of glucosinolates in 191 broccoli genotypes highlight valuable genetic resources for molecular breeding. *Agronomy* 13, 2928. https://doi.org/10.3390/agronomy13122928

Yin, L., Chen, H., Cao, B., Lei, J., and Chen, G. (2017). Molecular characterization of MYB28 involved in aliphatic glucosinolate biosynthesis in Chinese Kale (*Brassica oleracea* var. *alboglabra* Bailey). *Front. Plant Sci.* 8, 1083. https://doi.org/10.3389/fpls.2017.01083

Zhang, L., Li, X., Chang, L., Wang, T., Liang, J., Lin, R., et al. (2022b). Expanding the genetic variation of *Brassica juncea* by introgression of the *Brassica rapa* genome. *Hortic. Res.* 9, uhab054. https://doi.org/10.1093/hr/uhab054

Zhang, X., Ma, W., Liu, M., Li, X., Li, J., Lu, Y., et al. (2022a). *OCTOPUS* regulates *BIN2* to control leaf curvature in Chinese cabbage. *Proc. Natl. Acad. Sci. U. S. A.* 119, e2208978119. https://doi.org/10.1073/pnas.2208978119

19 Molecular Insights Regulating Glucosinolate Metabolites Among Brassicaceae Germplasm

A Comprehensive Analysis

Rodomiro Ortiz and Sangam L. Dwivedi

19.1 INTRODUCTION

Brassicaceae, commonly known as the mustard family or crucifers, encompasses a diverse group of plants that includes well-known crops such as broccoli, brussels sprouts, cabbage, cauliflower, kale, mustard, rapeseed, and swede, among others, as well as the model plant species thale cress (*Arabidopsis thaliana*). One of the key features that sets Brassicaceae apart is the presence of glucosinolates (GLSs), a class of secondary metabolites known for their diverse biochemical properties and health-promoting benefits. In the past, efforts to enhance seed quality primarily concentrated on achieving optimal fatty acid balance and minimizing seed GLS content (Nesi et al., 2008), because high levels of GLSs in seeds could impart undesirable bitter flavors and limit their palatability. These two factors were paramount due to their significant impact on the nutritional composition and palatability of seeds, particularly in oilseed crops like *Brassica* species. This historical emphasis underscores the importance of balancing nutritional quality and sensory attributes in seed improvement efforts. This chapter explores, therefore, the molecular insights that regulate GLS metabolites among Brassicaceae germplasm, shedding light on the underlying genetic mechanisms, ecological functions, and potential applications in agriculture and human health.

19.2 THE SIGNIFICANCE OF GLUCOSINOLATES

GLSs are sulfur-containing compounds predominantly found in plants of the Brassicaceae family. These compounds play a crucial role in plant defense against herbivores and pathogens, acting as both deterrents and signaling molecules. Beyond their ecological functions, GLSs have garnered attention for their potential health benefits in humans, including anticancer properties, antioxidant activity, and modulation of the immune system (Liang et al., 2018). They represent a category of

DOI: 10.1201/9781003467335-19

plant secondary metabolites capable of undergoing hydrolysis into diverse breakdown products, including isothiocyanates, thiocyanates, and nitriles. These breakdown products play crucial roles in plant defense mechanisms, aiding in warding off potential threats, while also serving as attractants to natural enemies of insect pests, thereby promoting ecological balance. Certain GLS-derived compounds have shown cancer-preventive characteristics, thus underscoring the significance of diets in human health. Consequently, the levels of these metabolites in the edible parts of plants garner considerable interest, reflecting ongoing research efforts aimed at elucidating their health-promoting effects and potential applications in disease prevention.

The burgeoning volume of data on *A. thaliana*, stemming from diverse omics technologies, has greatly facilitated the discovery of new genes or compounds implicated in GLS biosynthesis and metabolism. Over the past two decades, a substantial body of research has characterized a total of 113 GLS genes, encompassing 23 transcription components, 85 enzymes, and 5 protein transporters (Harun et al., 2020). Despite these advancements, ongoing endeavors persist in the quest to unveil all molecules associated with GLS production. To aid this pursuit, a meticulously curated database, SuCCombase (http://plant-scc.org), has been established to serve as a comprehensive repository of GLS-related information.

19.3 STRUCTURAL DIVERSITY OF GLUCOSINOLATES

The Brassicaceae family displays an astonishing diversity of GLSs, with over 120 different structures identified. Each GLS consists of a β-thioglucose moiety, a sulfate group, and an aglycone side chain that imparts structural specificity. The structural diversity arises from variations in the side chain, leading to distinct GLS profiles among different Brassicaceae species and cultivars (Ishida et al., 2014).

Liu et al. (2020) conducted a systematic exploration of GLS, their precursors, and breakdown products in 12 commonly consumed *Brassica* vegetables, employing gas chromatography–quadrupole time-of-flight mass spectrometer (GC-Q-TOF/MS), liquid chromatography–quadrupole time-of-flight mass spectrometer (LC-Q-TOF/MS), and liquid chromatography–triple quadrupole mass spectrometer (LC-QQQ/MS), utilizing both untargeted and targeted approaches. Their investigation integrated findings with existing literature to construct a comprehensive pathway map for GLS and isothiocyanate biosynthesis. Notably, precursor GLS levels exhibited strong correlations with their downstream breakdown products. Moreover, variations in GLS types and abundances among different genera facilitated the classification of plants based on morphological taxonomy. Validation research on three genera, each with distinct growth environments (underground, damp soil, and aboveground), revealed genus-specific biosynthetic pathways and variations in common GLS biosynthesis pathways. The methodology and resultant findings presented by Liu et al. (2020) serve as a valuable foundation for further exploration into specific facets of GLS metabolism within *Brassica* vegetables. Their comprehensive approach sheds light on the intricate interplay between GLS precursors, breakdown products, and the taxonomy-based classification of plants, offering insights that can drive future research endeavors in this field.

19.4 GENETIC REGULATION OF GLUCOSINOLATE BIOSYNTHESIS

Molecular insights into the regulation of GLS biosynthesis reveal a complex network of genes that coordinate the stepwise conversion of amino acids into the final GLS structures (Malhotra and Bisht, 2020). The key enzymes involved in this process include myrosinase, which hydrolyzes GLSs upon tissue damage, and various biosynthetic enzymes, such as cytochrome P450s and sulfotransferases (Hirani et al., 2012). The presence of these genes and their expression levels determine the types and concentrations of GLSs produced in Brassicaceae plants.

Essoh et al. (2020) conducted a comprehensive examination of gene clusters implicated in GLS biosynthesis, leveraging genomic analysis alongside biochemical pathways and chemical diversity assessments. Their approach enabled the identification of 101 Brassicaceae genes associated with GLS biosynthesis, employing a multi-database strategy. Utilizing the unweighted pair group method with arithmetic mean and principal component analysis on these 101 GLS genes, they observed a clear delineation between genes primarily involved in GLS core structure synthesis and those belonging to the CYP450s and MYBs gene families. Furthermore, Essoh et al. (2020) performed detailed phylogenetic analyses focusing on *CYP79F1-F2* and *CYP81F1-F4* genes to elucidate the disjunction between aliphatic and indolic GLS biosynthetic pathways. These findings indicated a recent diversification of aliphatic *CYP79F1* and *F2* genes in *Brassica* crops, contrasting with an earlier diversification observed for indolic genes, particularly *CYP81F1-F4*. Additionally, their chemical diversity assessments revealed distinct GLS chemo-profiles among *Brassica* crops compared to other species in Brassicaceae genera, with notable variations observed among *Diplotaxis* species, which had elevated GLS contents. These results underscore the unique genetic and chemical characteristics of GLS biosynthesis in *Brassica* crops and highlight the significance of understanding gene evolution and chemical diversity in shaping GLS composition across different plant species within the Brassicaceae family.

The mutation of genes encoding GLS transporters (GTRs) in *Arabidopsis* led to the elimination of GLSs from seeds. However, applying similar mutations to polyploid *Brassica* species necessitates navigating the intricate genetic landscapes shaped by ploidy. While a single gene mutation may yield significant phenotypic changes in diploid organisms like *Arabidopsis*, achieving similar outcomes in polyploid *Brassica* crops often requires targeted approaches to effectively disrupt multiple copies of the gene of interest. In this regard, Nour-Eldin et al. (2017) demonstrated a significant breakthrough in reducing seed GLS levels in *Brassica* crops by mutating GLS transporter genes (*GTR*). While previous research in *Arabidopsis thaliana* was able to show that mutation of *GTR* genes led to the elimination of GLSs from seeds, translating this loss-of-function phenotype to *Brassica* crops presented challenges due to their polyploid nature. However, by mutating one of seven and four of 12 GTR orthologs, Nour-Eldin et al. achieved a substantial reduction (60–70%) in GLS levels in seeds of two different *Brassica* species, *B. rapa* and *B. juncea.* Importantly, this reduction in seed GLSs was stably inherited over multiple generations and maintained in field trials across three locations, indicating the robustness and reliability of the mutant populations. The successful translation of the *gtr* loss-of-function phenotype from a model plant to economically important *Brassica* crops suggest the feasibility of employing a transport engineering approach to reduce seed GLS content in other oilseed crops as well, such as *Camelina sativa* or *Crambe abyssinica*. This pioneering study opens new avenues for targeted genetic manipulation to improve the nutritional quality and agronomic traits of oilseed crops, with potential implications for enhancing human and animal nutrition and expanding the utility of these crops in various agricultural contexts.

As indicated by Coves et al. (2023), mass selection is a promising method for modifying the content of specific GLS in *Brassica* crops. Mass selection involves the selection of plants with desirable traits based on their performance in a population. This approach can be particularly effective for improving GLS profiles in *Brassica* crops due to the high degree of natural genetic variation present within these species.

19.5 MAPPING QUANTITATIVE TRAIT LOCI AND GENETIC VARIATION

The mapping of quantitative trait loci has been instrumental in understanding the genetic variation in GLS content in the Brassicaceae germplasm. This approach allows researchers to identify genomic regions associated with variations in GLS profiles. Research has revealed that different QTLs are responsible for the regulation of specific GLS classes, which highlights the genetic complexity underlying the diversity of these metabolites. In this regard, Feng et al. (2012) conducted an in-depth analysis of the genetic and metabolic mechanisms governing GLS accumulation in oilseed rape or canola (*Brassica napus*), by examining total GLS concentration and its individual components in leaves and seeds of a doubled-haploid (DH) mapping population. Their study integrated QTL affecting GLS concentration in these organs, resulting in the identification of 105 metabolite QTLs (mQTLs). By correlating individual GLS with known metabolic pathways involved in GLS biosynthesis, Feng et al. (2012) were able to predict the functions of genes underlying these mQTLs. Furthermore, this comprehensive analysis enabled the construction of an advanced metabolic network, elucidating epistatic interactions responsible for the GLS composition in both leaves and seeds of *B. napus*. Through their research, Feng et al. (2012) uncovered various previously unknown potential regulatory relationships involved in GLS synthesis, highlighting how genetic variation can intricately influence the biochemical pathways governing GLS production. This study significantly advances our understanding of the genetic and metabolic basis of GLS accumulation in rapeseed oil, shedding light on the complex interplay between genetics and metabolism in shaping plant secondary metabolite profiles.

Tang et al. (2023b) conducted a comprehensive genome-wide association study (mGWAS) focusing on six key GLS metabolites in rapeseed, encompassing three aliphatic GLSs (m145 gluconapin, m150 glucobrassicanapin, and m151 progoitrin), one aromatic GLS (m157 gluconasturtiin), and two indole GLSs (m165 indolylmethyl GLS and m172 4-hydroxyglucobrassicin). Their study identified 113 candidate intervals significantly associated with these GLS metabolites. In the genomic regions linked to the mGWAS peaks, Tang et al. (2023b) predicted 187 candidate genes involved in GLS biosynthesis. Notably, known genes such as *BnaMAM1*,

BnaGGP1, *BnaSUR1*, and *BnaMYB51* were found, thus underscoring their roles in GLS metabolism. Moreover, novel genes including *BnaMYB44*, *BnaERF025*, *BnaE2FC*, *BnaNAC102*, and *BnaDREB1D* were also involved in GLS biosynthesis based on these mGWAS findings, supplemented by analysis of differentially expressed genes. These results provide valuable insights into the genetic underpinnings of GLS biosynthesis in rapeseed, which offers a foundation for a marker-based breeding approach aimed at improving seed quality in *Brassica* species. By elucidating the genetic factors influencing GLS metabolism, this study contributes to the development of rapeseed cultivars with enhanced nutritional profiles and agronomic traits, thereby facilitating the farming of improved Brassica crops.

Brassica rapa is a polyploid species with phenotypically diverse cultivated subspecies (field mustard, turnip, komatsuna, napa cabbage, bomdong, bok choy, and rapini). Variations in GLS profiles among these subspecies are significant, yet the molecular mechanisms driving these variances remain largely elusive. Seo et al. (2017) conducted a comprehensive investigation into GLS content and the expression levels of *BrMYB* transcription factors (TFs) across 38 accessions representing eight subspecies of *B. rapa*. Their study revealed the presence of 12 GLS, which were categorized into three chemical groups based on GLS content patterns and the expression profiles of *BrMYB* TFs. Notably, GLS content and *BrMYB* TF expression levels varied significantly among genotypes, including *B. rapa* subspecies *pekinensis*, *chinensis*, and *rapa*. Among the noted *BrMYB* TFs, *BrMYB28.3*, *BrMYB51.1*, and *BrMYB122.2* positively regulated GLS content across the 38 accessions. Additionally, the expression levels of *BrMYB28s* and *BrMYB34.3* exhibited increases under various abiotic and biotic stress conditions. Specifically, all three *BrMYB51* paralogs displayed markedly elevated expression levels following infection with *Pectobacterium carotovorum* that causes soft rot disease. These findings provide valuable insights into the functional diversity of the 13 *BrMYB* TFs during the evolution of polyploid *B. rapa*. By elucidating the regulatory roles of *BrMYB* TFs in GLS biosynthesis and their responses to environmental stresses, this study enhances our understanding of the molecular mechanisms underlying GLS variation and stress responses in *B. rapa*.

Kim et al. (2023b) conducted a study comparing GLS profiles between Chinese cabbage (*B. rapa pekinensis*) line Chiifu and oilseed-type *B. rapa* (*B. rapa trilocularis*) line LP08. They observed significantly lower levels of total GLS in the Chiifu line compared to LP08. Through comparative transcriptome analysis, they identified differential expression of 8,276 and 9,878 genes between Chiifu and LP08 under light and dark conditions, respectively. Among these, 79 GLS pathway genes were implicated in GLS metabolism under light conditions. Furthermore, QTL analysis using a single-nucleotide polymorphism-based linkage map and 151 F5 individuals from a Chiifu and LP08 cross revealed two major QTL peaks on chromosome 3. High-performance liquid chromatography of GLS profiles from 97 F5 recombinant inbred lines enabled the identification of these QTL peaks. The highest peak harbored the MYB-domain TF gene *BrMYB28.1* (Bra012961), while the second highest peak was near the 2-oxoacid-dependent dioxygenase gene *BrGSL-OH.1* (Bra022920). This study elucidated major genes contributing to the divergence in GLS profiles between Chiifu and LP08, providing molecular insights into the differences between vegetative- and oilseed-type *B. rapa* plants. By uncovering key genetic determinants underlying GLS variation, this research enhances our understanding of GLS metabolism and lays the groundwork for future breeding efforts aimed at manipulating GLS profiles in Brassica crops.

In a study by Zhang et al. (2018), a major quantitative trait locus (QTL) governing aliphatic GLS accumulation in *B. rapa* leaves was investigated. This QTL, comprising three tandem *MAM* genes and two *MYB* genes, was identified in two BC_2DH populations. Among the five candidate genes, only the expression level of *BrMAM-3* (Bra013007) showed a significant correlation with aliphatic GLS accumulation in *B. rapa* leaves. A naturally occurring insertion within exon 1 of BrMAM-3 was identified, predicted to be a loss-of-function mutation, a finding confirmed by qRT-PCR. This loss of function was associated with reduced GLS content in *B. rapa* accessions. Additionally, overexpression of the BrMAM-3 gene led to increased total aliphatic GLSs in transgenic *Arabidopsis* lines. These findings shed light on the molecular mechanisms governing aliphatic GLS accumulation in *B. rapa* leaves, thereby offering insights for the manipulation of total aliphatic GLS content in *Brassica* crops.

The comparative transcriptome analysis conducted by Kim et al. (2023a) revealed substantial differences in gene expression between Chiifu (vegetable type, *B. rapa pekinensis*) and LP08 (oil type, *B. rapa trilocularis*) under both light and dark conditions. Specifically, they were able to notice 8,276 genes that were differentially expressed under light conditions and 9,878 genes under dark conditions. Among these genes, 79 out of 162 *B. rapa* GLS pathway genes were related to GLS metabolism specifically under light conditions. Kim et al. (2023a) further conducted QTL analysis using 151 F_5 individuals derived from a cross between inbred lines representing both types. This analysis led to the identification of two major QTL peaks on chromosome 3. High-performance liquid chromatography was employed to obtain GLS profiles from 97 F_5 recombinant inbred lines. In the highest QTL peak region, the MYB-domain transcription factor gene *BrMYB28.1* (Bra012961) was found, thereby indicating its potential role in regulating GLS metabolism. Additionally, the second highest peak was located near the 2-oxoacid-dependent dioxygenase gene *BrGSL-OH.1* (Bra022920), which suggests its involvement in GLS biosynthesis or regulation. These findings provide valuable insights into the genetic basis of GLS metabolism and variation between different types of *B. rapa*. The identification of key genes and QTL associated with GLS metabolism offers potential targets for further research and breeding efforts aimed at modulating GLS content and composition in *Brassica* crops for various purposes.

19.6 ENVIRONMENTAL FACTORS AND GLUCOSINOLATE VARIATION

Environmental conditions, including soil composition, temperature, and light intensity, also influence GLS content in Brassicaceae plants. The intricate interplay between genetic factors and environmental cues contributes to the adaptability of these plants to diverse ecological niches. Understanding how environmental factors modulate GLS profiles provides valuable insights for crop management and breeding programs.

GLSs serve not only as distinctive flavor compounds in leaf mustard (*Brassica juncea*), but they are also significant secondary metabolites elicited in response to various abiotic stresses. Among them, cold stress stands out as one of the most common challenges encountered by leaf mustard. However, the metabolic response pattern of GLS in leaf mustard under cold stress remains largely unexplored. Tang et al. (2023a) investigated the GLS content of leaf mustard subjected to cold stress and observed a biphasic response characterized by an initial increase followed by a subsequent decrease. Through RNA-seq analysis, they were able to identify notable shifts in gene expression patterns associated with GLS synthesis under cold stress. Specifically, genes involved in aliphatic GLS synthesis exhibited significant upregulation after 24 hours of cold stress, while those related to indole GLS synthesis were prominently upregulated following 48 hours of cold exposure. Additionally, key genes such as *BjBGLU25* and *BjBGLU27* were significantly upregulated, indicating their involvement in GLS metabolism under cold stress conditions. Further exploration into the regulatory mechanisms governing GLS response patterns revealed the potential roles of several TFs including MYB, ERF, IQD, and bHLH. Particularly intriguing was the identification of an unreported TF, *BjMYBS3* (BjuVA05G33250), which appears to play a crucial role in the synthesis of aliphatic GSLs in response to cold stress. Moreover, this study by Tang et al. (2023a) demonstrated that the exogenous application of GLS enhanced the cold stress tolerance of leaf mustard, thus highlighting the potential utility of GLSs as a protective strategy against environmental stress. Their findings gave novel insights into the complex regulatory network governing GLS metabolism in leaf mustard under cold stress conditions, with implications for enhancing stress resilience in other *Brassica* crops.

19.7 ROLE OF GLUCOSINOLATES IN PLANT DEFENSE

The defensive role of GLS in Brassicaceae plants is multifaceted. Upon tissue damage, myrosinase enzymes catalyze the hydrolysis of GLSs, leading to the formation of bioactive breakdown products. These products, including isothiocyanates, nitriles, and thiocyanates, have toxic effects on herbivores and pathogens. The specificity of the defensive response is often linked to the composition and concentration of GLSs in a given plant species.

19.7.1 Implications for Agriculture

Molecular insights into GLS regulation have profound implications for agriculture. Plant-breeding programs can leverage this knowledge to develop Brassicaceae cultivars with enhanced resistance to pests and diseases. Furthermore, optimizing GLS profiles in crops like broccoli and cabbage can contribute to the production of vegetables with improved nutritional and health-promoting attributes. Worth mentioning is the proposed integrated roadmap of Qin et al. (2023) whose aims are to guide ideotype development considering GLS hydrolysis-derived products (GHPs). Their roadmap emphasizes the dual objectives of maximizing the beneficial effects and minimizing the detrimental effects of GHPs, while also considering various end uses of the crops. The key components of this integrated road map are (a) comprehensive evaluation of GHP effects, (b) identification of optimal GHP profiles, (c) implementation of genetic and agronomic strategies to manipulate GHP profiles in crops, (d) integration with end uses, and (e) iterative improvement and validation whereby GHP profiles are continuously refined based on feedback from end-users and ongoing research findings. This proposed integrated roadmap provides a systematic framework for the development of crop ideotypes optimized for GHP content. By balancing the beneficial and detrimental effects of GHPs and tailoring GHP profiles to end use requirements, this roadmap aims to maximize the value and utility of GLS-rich crops in various agricultural and commercial contexts.

19.7.2 Human Health and Nutritional Benefits

Beyond their role in plant defense, GLS offers notable health benefits to humans (Zhu et al., 2023). Numerous investigations have suggested that these compounds may have anticancer properties, particularly in relation to cancers affecting the lungs, colon, and breast. Additionally, the breakdown of GLSs exhibit antioxidant activity and may contribute to the overall nutritional quality of Brassicaceae vegetables.

19.8 CONCLUSION

While significant progress has been made in unraveling the molecular regulation of GLS, challenges remain. The complexity of the biosynthetic pathway, the influence of environmental factors, and the intricate genetic networks involved necessitate further research. Advances in high-throughput sequencing technologies and functional genomics approaches will likely propel our understanding of GLS regulation to new heights. The intricate interplay between genetic factors and environmental cues, coupled with the diverse structural array of GLSs, underscores the complexity of these compounds. As research progresses, harnessing this molecular knowledge offers exciting possibilities for developing resilient crops, promoting sustainable agriculture, and enhancing the nutritional quality of Brassicaceae vegetables for the benefit of both plants and humans. Indeed, the molecular insights into the regulation of GLS metabolites in the Brassicaceae germplasm provide a foundation for addressing key questions in plant biology, agriculture, and human health.

REFERENCES

Coves, S., Soengas, P., Velasco, P., Fernández, J.C., and Cartea, M.E. (2023). New vegetable varieties of *Brassica rapa* and *Brassica napus* with modified glucosinolate content obtained by mass selection approach. *Front. Nutr.* 10, 1198121. https://doi.org/10.3389/fnut.2023.119812

Essoh, A.P., Monteiro, F., Pena, A.R., Pais, S.M., Moura, M., and Romeiras, M.M. (2020). Exploring glucosinolates diversity in Brassicaceae: A genomic and chemical assessment for deciphering abiotic stress tolerance. *Plant Physiol. Biochem.* 150, 151–161. https://doi.org/10.1016/j.plaphy.2020.02.032

Feng, J., Long, Y., Shi, L., Shi, J., Barker, G., and Meng, J. (2012). Characterization of metabolite quantitative trait loci and metabolic networks that control glucosinolate concentration in the seeds and leaves of *Brassica napus*. *New Phytol.* 193, 96–108. https://doi.org/10.1111/j.1469-8137.2011.03890.x

Harun, S., Abdullah-Zawawi, M.R., Goh, H.H., and Mohamed-Hussein, Z.A. (2020). A comprehensive gene inventory for glucosinolate biosynthetic pathway in *Arabidopsis thaliana*. *J. Agric. Food Chem.* 68, 7281–7297. https://doi.org/10.1021/acs.jafc.0c01916

Hirani, A.H., Li, G., Zelmer, C.D., Vetty, P.B.E., Asif, M., and Goyal, A. (2012). Molecular genetics of glucosinolate biosynthesis in brassicas: Genetic manipulation and application aspects. In Goyal, A. (ed.) *Crop Plant*. InTech Open Limited, London, pp. 189–216. https://doi.org/10.5772/45646

Ishida, M., Hara, M., Fukino, N., Kakizaki, T., and Morimitsu, Y. (2014). Glucosinolate metabolism, functionality and breeding for the improvement of Brassicaceae vegetables. *Breed. Sci.* 64, 48–59. https://doi.org/10.1270/jsbbs.64.48

Kim, J.A., Moon, H., Kim, H.S., Choi, D., Kim, N.-S., Jang, J., et al. (2023a). Transcriptome and QTL mapping analyses of major QTL genes controlling glucosinolate contents in vegetable- and oilseed-type *Brassica rapa* plants. *Front. Plant Sci.* 13, 1067508. https://doi.org/10.3389/fpls.2022.1067508

Kim, S.H., Subramanian, P., and Hahn, B.S. (2023b). Glucosinolate diversity analysis in choy sum (*Brassica rapa* subsp. *chinensis* var. *parachinensis*) germplasms for functional food breeding. *Foods* 12, 2400. https://doi.org/10.3390/foods12122400

Liang, X., Lee, H.W., Li, Z., Lu, Y., Zou, L., and Ong, C.N. (2018). Simultaneous quantification of 22 glucosinolates in 12 *brassicaceae* vegetables by hydrophilic interaction chromatography-tandem mass spectrometry. *ACS Omega* 3, 15546–15553. https://doi.org/10.1021/acsomega.8b01668

Liu, Y., Rossi, M., Liang, X., Zhang, H., Zou, L., and Ong, C.N. (2020). An integrated metabolomics study of glucosinolate metabolism in different Brassicaceae genera. *Metabolites* 10, 313. https://doi.org/10.3390/metabo10080313

Malhotra, B., and Bisht, N.C. (2020). Editorial: Glucosinolates: Regulation of biosynthesis and hydrolysis. *Front. Plant Sci.* 11, 620965. https://doi.org/10.3389/fpls.2020.620965

Nesi, N., Delourme, R., Brégeon, M., Falentin, C., and Renard, M. (2008). Genetic and molecular approaches to improve nutritional value of *Brassica napus* L. seed. *C. R. Biol.* 331, 763–771. https://doi.org/10.1016/j.crvi.2008.07.018

Nour-Eldin, H.H., Madsen, S.R., Engelen, S., Jørgensen, M.E., Olsen, C.E., Andersen, J.S., et al. (2017). Reduction of antinutritional glucosinolates in *Brassica* oilseeds by mutation of genes encoding transporters. *Nat. Biotechnol.* 35, 377–382. https://doi.org/10.1038/nbt.3823.

Qin, H., King, G.J., Borpatragohain, P., and Zou, J. (2023). Developing multifunctional crops by engineering Brassicaceae glucosinolate pathways. *Plant Commun.* 4, 100565. https://doi.org/10.1016/j.xplc.2023.100565

Seo, M.S., Jin, M., Sohn, S.H., and Kim, J.S. (2017). Expression profiles of *BrMYB* transcription factors related to glucosinolate biosynthesis and stress response in eight subspecies of *Brassica rapa*. *FEBS Open Bio.* 7, 1646–1659. https://doi.org/10.1002/2211-5463.12231

Tang, B., Zhang, B.-H., Mo, C.-Y., Fu, W.-Y., Yang, W., Wang, Q.-Q., et al. (2023a). Comparative transcriptomics reveal the mechanisms underlying the glucosinolate metabolic response in leaf *Brassica juncea* l. under cold stress. *Agronomy* 13, 1922. https://doi.org/10.3390/agronomy13071922

Tang, Y., Zhang, G., Jiang, X., Shen, S., Guan, M., Tang, Y., et al. (2023b). Genome-wide association study of glucosinolate metabolites (mGWAS) in *Brassica napus* L. *Plants (Basel)* 12, 639. https://doi.org/10.3390/plants12030639

Zhang, J., Wang, H., Liu, Z., Liang, J., Wu, J., Cheng, F., et al. (2018). A naturally occurring variation in the *BrMAM-3* gene is associated with aliphatic glucosinolate accumulation in *Brassica rapa* leaves. *Hortic. Res.* 5, 69. https://doi.org/10.1038/s41438-018-0074-6

Zhu, B., Liang, Z., Zang, Y., Zhu, Z., and Yang, J. (2023). Diversity of glucosinolates among common Brassicaceae vegetables in China. *Hortic. Plant J.* 9, 365e38. https://doi.org/10.1016/j.hpj.2022.08.006

20 CRISPR/Cas-Mediated Genome-Editing to Develop Flavonoids Enriched Fruits and Vegetables

Sangam L. Dwivedi and Autar K. Mattoo

20.1 INTRODUCTION

Genome editing allows plant breeders to make precise changes to DNA by targeting a specific location in a gene within the DNA. It is deployed to add, remove, or alter DNA in the plant genome. Clustered randomly interspersed short palindromic repeats (CRISPR), CRISPR-associated protein 9 (Cas9) systems, transcription activator-like effector nucleases (TALENs), and zinc finger nucleases (ZFNs), among others, have been used to effect precise gene edits. Genome-editing technology has great potential in almost all organisms, from plants and microorganisms to humans and other animals (Dimond et al., 2021; Gaj et al., 2016). Both CRISPR-Cas9 and TALENs have revolutionized the adoption of these technologies in agricultural science.

Multiplex genome-editing (MGE) technologies modify two or more specific DNA loci in a genome with high precision, thereby greatly increases the feasibility of introducing desired changes at multiple nucleotide levels into a target genome. CRISPR/Cas9 system-based MGE tools allow for the simultaneous generation of direct mutations precisely at multiple loci in a gene or multiple genes (Abdelrahman et al., 2021).

The traditional crossbreeding or even genome-assisted crop improvement requires first systematic characterization of natural variability for agronomically beneficial traits in gene pool, identifying and hybridizing parental lines with contrasting phenotypes, followed by time-consuming and laborious characterization of the progenies for multiple generations to derive new germplasm with specific characteristics. In contrast, genome-editing technologies rapidly modify genome in a precise and predictable manner, making it possible to introduce improvements directly in the elite lines. It is equally effective for improving single gene and complex traits or may accelerate domestication of wild species or distantly related crops by multiplex editing of the ortholog genes. Multinational seed companies and publicly funded

 DOI: 10.1201/9781003467335-20

institutions are investing a lot to tailor new crops with specific characteristics. The technological challenges include (i) efficient delivery of genome-editing tools in plants, (ii) simultaneous manipulation of multiple traits, (iii) unravelling of genetic networks controlling key agronomic traits, (iv) precise editing, and (v) government policy support for genome-edited crops (Hua et al., 2019).

20.2 FLAVONOIDS IN PLANTS

Flavonoids are naturally occurring secondary metabolites with variable phenolic structures. Colored fruits and vegetables are excellent sources of flavonoids. Flavones, flavonols, flavanones, flavanonols, flavanols (catechins), anthocyanins, and chalcones are the major classes of flavonoids depending on the carbon of the C ring on which the B ring is attached and the degree of unsaturation and oxidation of the C ring. Flavonoids act as antioxidants (Panche et al., 2016). To date over thousands of flavonoid compounds have been isolated. Flavonoid-rich foods include fruits (apples, berries, cherries, citrus), vegetables (including leafy vegetables), grains (cereals, legumes), nuts, tea, and wine. The variation in flavonoid biosynthesis pathways and genes associated with transport led to the differential accumulation of subclass of flavonoids (Ku et al., 2020).

20.3 PLANTS FLAVONOIDS, ABIOTIC STRESS ADAPTATION, AND HUMAN HEALTH

Flavonoids possess several medicinal benefits, including anticancer, antioxidant, anti-inflammatory, and antiviral properties as well as act as neuro- and cardio-protective effects. A higher intake of flavonoid-rich foods provides multiple health benefits to human subjects (Norouzkhani et al., 2024; Ullah et al., 2020). The biological activities depend upon the type of flavonoids, their mode of action and bioavailability of compounds. Regular consumption of flavonoid-rich foods could potentially minimize the risk of noncommunicable diseases because of antioxidant activity of flavonoids in humans (Dwivedi et al., 2022, and references cited therein).

Flavonoids protect plants from abiotic stress. They are involved in antioxidant activity in response to different abiotic stresses, stress signaling networks, and physiological response of plants under stress conditions (Shomali et al., 2022). For example, anthocyanins are strong antioxidants with free radical scavenging potential. They safeguard plants from a wide range of environmental stresses. Genetic manipulation of AN biosynthetic genes can enhance plant stress tolerance (Li and Ahammed, 2023). Anthocyanin-rich tomato confers tolerance to drought and salinity (Napar et al., 2022).

20.4 GENOME EDITING TO ENHANCE FLAVONOIDS IN FRUIT CROPS

The colored fruits and vegetables, if regularly consumed, provide health benefits to humans. R2R3 MYB TF family controls anthocyanins, phlobaphenes, betalains, and carotenoid biosynthesis while also influencing fruit flavor and texture. MYBs alleles are therefore the key targets for driving novelty in fruits and vegetables, including the use of gene-editing technology (Allan and Espley, 2018).

In grapevine (*Vitis vinifera*), *VvbZIP36* acts as a negative regulator of anthocyanin biosynthesis. Knocking out of one allele of *VvbZIP36* in grapevine with CRISPR/Cas9 promotes anthocyanin accumulation. A range of anthocyanin biosynthesis genes are activated in *VvbZIP36* mutant plants, relative to WT. They play a role in balancing the synthesis of stilbenes, lignans, flavonols, and anthocyanins (Tu et al., 2022).

Pomegranate (*Punica granatum*) fruits are rich sources of phenolics and flavonoids. Dual CRISPR/Cas9-edited *PgUGT84A23* and *PgUGT84A24* lines had unique accumulation of gallic acid 3-O- and 4-O-glucosides (galloylglucose ethers) in the ugt84a23 and ugt84a24, double mutant lines compared to control or single edited lines. The biochemical characterization of the candidate genes isolated *PgUGT72BD1*, which specifically glycosylates the 4-O position of gallic acid (Chang et al., 2019).

Editing of *CmACO1* in melon (*Cucumis melo* var. *reticulatus*) and *FaPG1* in strawberry (*Fragaria* × *ananassa*) by CRISPR/Cas9 systems improved fruit firmness and shelf-life (López-Casado et al., 2023; Nonaka et al., 2023). However, editing of *VvDXS1* resulted in muscat (European grapes characterized with subtle sweetness and aroma, mostly associated with wine making) flavor in grape cultivar Scarlet Royal (Yang et al., 2024). Grape berries with muscat flavor also produce high levels of monoterpenes (Gholami et al., 2008).

20.5 ENHANCING FLAVONOIDS IN VEGETABLE CROPS BY GENOME EDITING

The relative content of β-carotene, a precursor of vitamin A, influences the flesh color in cucumber. Cucumber with orange flesh contains a higher amount of β-carotene than the white-flesh cucumber. A single recessive gene (*CsaV3_6G040750*) regulates orange endocarp phenotype in the CS-B line. Such lines have increased β-carotene content in the endocarp tissues that elevate expression of *CsOr* gene at fruit development stages (Kishor et al., 2021).

Flavonol is the major class of flavonoids in Chinese cabbage (*Brassica rapa pekinensis*). CRISPR/Cas9-mediated *BrFLS1*-KO cabbage lines with distinct metabolism (dihydroflavonol) showed that CRISPR/Cas9 system is efficient in modifying Chinese cabbage lines with distinct metabolites (Park et al., 2023).

Metabolically engineered cisgenic herbicide resistant tomatoes containing *mSl-ALS* and *SlMYB12* genes are highly enriched in leucine (21-fold), valine (9-fold), isoleucine (3-fold), kaempferol (64-fold), and quercetin (45-fold) compared to WT (Vazquez-Vilar et al., 2023). CRISPR/Cas9-mediated multiplex gene editing of red-fruited tomato line with three fruit color genes (*PSY1*, *MYB12*, and *SGR1*) resulted in lines with different fruit colors, including yellow, brown, pink, light-yellow, pink-brown, yellow-green, and light green. More importantly, such lines had no adverse impact on fruit yield or quality (Yang et al., 2023). Targeted editing of *GREENFLESH* (*GF*) locus in two tomato cultivars (MoneyMaker and San Marzano) produced *gf* mutants obtained by CRISPR/Cas9 system and accumulated both carotenoids and chlorophylls in the ripe fruits. The phenotypic assessment of two mutants, one carrying a small insertion of 1 nucleotide and the other containing a larger deletion of 123 bp, indicated substantial increase in carotenoid and tocopherol levels during fruit ripening, and had reduced chlorophyll degradation. More importantly, there were no unintended mutations at putative off-targets (Gianoglio et al., 2022).

Polygalacturonase (*SlPG*) regulates tomato fruit firmness. CRISPR/Cas9-induced targeted mutagenesis of the *SlPG* delays the softening of tomato fruit. T_1-generation homozygous tomato fruits, in relation to WT, exhibited late fruit softening under natural conditions. The firmness value of WT fruit was lower in *slpg* mutant fruit, and the physiological loss of water was higher, demonstrating that targeted mutagenesis of the *SlPG* gene could delay fruit softening with no adverse impact on tomato flavor, aroma, taste, and nutritional value (Nie et al., 2022).

Ascorbate is an essential antioxidant compound. Humans lack ascorbate biosynthetic enzymes, therefore, must intake ascorbate from the food source. The elevation of ascorbate content in tomato fruits will improve their nutritional value. *SlAPX4* gene in tomato is specifically induced during fruit ripening. Ascorbate accumulates at the yellow fruit stage but decreases at later stages of fruit ripening. CRISPR/Cas9-induced mutation in *SlAPX4* increases ascorbate content in ripened tomato fruits, without any adverse impact on tomato plant growth and development, including no change in ascorbate content in leaves. Clearly, it shows that *SlAPX4* mediates a decrease of ascorbate content during the later stage of fruit ripening, and mutation of *SlAPX4* can be used to develop genome-edited tomatoes with elevated ascorbate content in fruits (Do et al., 2022).

The introgression of genome-edited ripening inhibitor mutant allele (*rin*^{CR}) into a dominant *phototropin1* (*Nps1*) mutant resulted in *Nps1/rin*^{CR} hybrid with longer shelf-life and 4.4–7.1-fold higher carotenoid levels than the WT tomato. *Nps1/rin*^{CR} fruits were enriched with higher sucrose, malate, and volatiles associated with tomato taste and flavor, preferred by consumers. Thus, *Nps1* mutant gene can be used to enhance carotenoids, taste, and flavor in tomato germplasm (Nizampatnam et al., 2023).

Target-AID base-editing technology, a modified version of CRISPR/Cas9 system, is an efficient tool for targeting editing of multiple genes. *SlDDB1*, *SlDET1*, and *SlCYC-B* regulate carotenoid accumulation in tomatoes. The edited lines produced via Target-AID technology had variations in carotenoid accumulation, with an additive effect of each single mutation on lycopene and β-carotene accumulation in tomatoes (Hunziker et al., 2020, 2022).

Lettuce (*Lactuca sativa*), a widely consumed vegetable in the USA after potato and tomato, is an excellent source of fiber, iron, folic acid, and vitamin C. The fertile and true-breeding lettuce lines enriched with flavonoids from EMS-mutagenized red lettuce population indicates that pink green kaempferol overproducing kfoA and kfoB mutants accumulated kaempferol to 0.6–1% of their dry weight, while the yellow-green naringenin chalcone overproducing mutant (nco) accumulated naringenin chalcone to 1% dry weight. *Kfo* plants carry a mutation in the *FLAVONOID-3′ HYDROXYLASE* (*F3′H*) gene, *nco in CHALCONE ISOMERASE* (*CHI*), indicating that the non-GMO approach can transform a common crop into a functional food with potential health benefits (Gurdon et al., 2019). The field assessment of these mutant lines revealed higher levels of beneficial phytochemicals, and all, except nco mutants, had a fresh plant weight at harvest comparable to commercial leaf cultivars. This may be released for commercial cultivation or used as parental lines to transfer these traits into locally adapted lettuce cultivars (Cho et al., 2023).

Japan became the first country to announce the release of first gene-edited tomato containing a high level of γ-aminobutyric acid (GABA). GABA is known to reduce blood pressure and promote relaxation in humans (Waltz, 2022). It is marketed as Sicilian Rouge tomatoes by the Tokyo-based company Sanatech Seed, Japan (Ezura, 2022). Numerous examples demonstrate the potential of gene editing technology to achieve crops resilience, productivity, and nutrition in agriculture (Karavolias et al., 2021).

20.6 PUBLIC PERCEPTION OF GENOME-EDITED CROPS

Both traditional and modern plant-breeding methods exploit natural or artificially induced genetic variation followed by hybridization and then careful characterization of the progenies of multiple generations to derive advanced lines for multi-season trials to assess adaptability, productivity, and nutritional quality prior to commercial cultivation. Genome-editing systems, in contrast, provide means to rapidly modify genomes in a precise and predictable way, making it possible to introduce improvements directly into elite cultivars. Gene-edited crops offer numerous advantages, such as increased crop yields, improved nutritional value, and enhanced stress tolerance. The biggest obstacle in the adoption of genome-edited crops is public acceptance of the technology and government regulatory policies in many regions. Lack of a clear and consistent regulatory policy limits the acreage of gene-edited crops (Hua et al., 2019). For achieving widespread acceptance, it is necessary to strengthen safety research, harmonize international regulations, and raise public awareness for acceptance of gene-edited crops (Menz et al., 2020; Zhang and Zhu, 2024).

A five-year (January 2018 to December 2022) analysis of data on gene editing in social media platform revealed that gene-edited crops receive consistently higher favorability ratings than GMO, suggesting that as time passes, foods produced by gene editing will be accepted by the public at large to bring much-needed food and nutritional security and environmental sustainability worldwide (Lynas et al., 2023).

The European Union and New Zealand have brought legislations wherein genome-edited plants are regulated as GMOs, while many other countries move to exempt genome-edited crops (Schmidt et al., 2020). The risk associated with gene-edited crops is comparable to those of traditional and modern breeding methods. However, failure to address regulatory, legal, and trade framework and granting of social license are the main hurdles in promoting the cultivation of genome-edited crops (Pixley et al., 2022). As time passes, it is anticipated that GMO crops will be seen to have a good potential by the public for food and nutritional quality of the produce.

20.7 CONCLUSIONS

The proof-of-concept demonstrates no adverse impact on yield or quality of gene-edited fruits and vegetables. However, greater awareness and government regulatory policies should be in place to encourage the consumption of gene-edited crops to bring much-needed food, nutritional security, and environmental sustainability worldwide to feed the growing world population.

REFERENCES

Abdelrahman, M., Wei, Z., Rohila, J.S., and Zhao, K. (2021). Multiplex genome-editing technologies for revolutionizing plant biology and crop improvement. *Front. Plant Sci.* 12, 721203. https://doi.org/10.3389/fpls.2021.721203

Allan, A.C., and Espley, R.V. (2018). MYBs drive novel consumer traits in fruits and vegetables. *Trends Plant Sci.* 23, P693–P705. https://doi.org/10.1016/j.tplants.2018.06.001

Chang, L., Wu, S., and Tian, L. (2019). Effective genome editing and identification of a regiospecific gallic acid 4-O-glycosyltransferase in pomegranate (*Punica granatum* L.). *Hortic. Res.* 6, 123. https://doi.org/10.1038/s41438-019-0206-7

Cho, E., Gurdon, C., Zhao, R., Peng, H., Poulev, A., Raskin, I., et al. (2023). Phytochemical and agronomic characterization of high-flavonoid lettuce lines grown under field conditions. *Plants (Basel)* 12, 3467. https://doi.org/10.3390/plants12193467

Dimond, R., Lewis, J., and Thomas, G. (2021). Themed issue: Understanding the technical and social landscape of gene editing. *New Genet. Soc.* 40, 361–366. https://doi.org/10.1080/14636778.2021.2004032

Do, J.H., Park, S.Y., Park, S.H., Kim, H.M., Ma, S.H., Mai, T.D., et al. (2022). Development of a genome-edited tomato with high ascorbate content during later stage of fruit ripening through mutation of *SlAPX4*. *Front. Plant Sci.* 13, 836916. https://doi.org/10.3389/fpls.2022.836916

Dwivedi, S.L., Mattoo, A.K., Garg, M., Dutt, S., Singh, B., and Ortiz, R. (2022). Developing germplasm and promoting consumption of anthocyanin-rich grains for health benefits. *Front. Sustain. Food Syst.* 6, 867897. https://doi.org/10.3389/fsufs.2022.867897

Ezura, H. (2022). Letter to the editor: The world's first CRISPR tomato launched to a Japanese market: The social-economic impact of its implementation on crop genome editing. *Plant Cell Physiol.* 63, 731–733. https://doi.org/10.1093/pcp/pcac048

Gaj, T., Sirk, S.J., Shui, S.L., and Liu, J. (2016). Genome-editing technologies: Principles and applications. *Cold Spring Harb. Perspect. Biol.* 8, a023754. https://doi.org/10.1101/cshperspect.a023754

Gholami, M., Hayasaka, Y., Coombe, B.G., and Jackson, J.F. (2008). Biosynthesis of flavour compounds in Muscat Gordo Blanco grape berries. *Aust. J. Grape Wine Res.* 1, 19–24. https://doi.org/10.1111/j.1755-0238.1995.tb00073.x

Gianoglio, S., Comino, C., Moglia, A., Acquadro, A., García-Carpintero, V., Diretto, G., et al. (2022). In-depth characterization of greenflesh tomato mutants obtained by CRISPR/Cas9 editing: A case study with implications for breeding and regulation. *Front. Plant Sci.* 13, 936089. https://doi.org/10.3389/fpls.2022.936089

Gurdon, C., Poulev, A., Armas, I., Satorov, S., Tsai, M., and Raskin, I. (2019). Genetic and phytochemical characterization of lettuce flavonoid biosynthesis mutants. *Sci. Rep.* 9, 3305. https://doi.org/10.1038/s41598-019-39287-y

Hua, K., Zhang, J., Botella, J.R., Ma, C., Kong, F., Liu, B., et al. (2019). Perspectives on the application of genome-editing technologies in crop breeding. *Mol. Plant* 12, 1047–1059. https://doi.org/10.1016/j.molp.2019.06.009

Hunziker, J., Nishida, K., Kondo, A., Ariizumi, T., and Ezura, H. (2022). Phenotypic characterization of high carotenoid tomato mutants generated by the Target-AID base-editing technology. *Front. Plant Sci.* 13, 848560. https://doi.org/10.3389/fpls.2022.848560

Hunziker, J., Nishida, K., Kondo, A., Kishimoto, S., and Ariizumi, T. (2020). Multiple gene substitution by target—AID base—editing technology in tomato. *Sci. Rep.* 10, 20471. https://doi.org/10.1038/s41598-020-77379-2

Karavolias, N.G., Horner, W., Abugu, M.N., and Evanega, S.N. (2021). Application of gene editing for climate change in agriculture. *Front. Sustain. Food Syst.* 5, 685801. https://doi.org/10.3389/fsufs.2021.685801

Kishor, D.S., Lee, H.-Y., Alavilli, H., You, C.-R., Kim, J.-G., Lee, S.-Y., et al. (2021). Identification of an allelic variant of the *CsOr* gene controlling fruit endocarp color in cucumber (*Cucumis sativus* L.) using genotyping-by-sequencing (GBS) and whole-genome sequencing. *Front. Plant Sci.* 12, 802864. https://doi.org/10.3389/fpls.2021.802864

Ku, Y.-S., Ng, M.-S., Cheng, S.-S., Lo, A.W.-Y., Xiao, Z., Shin, T.-S., et al. (2020). Understanding the composition, biosynthesis, accumulation, and transport of flavonoids in crops for the promotion of crops as healthy sources of flavonoids for human consumption. *Nutrients* 12, 1717. https://doi.org/10.3390/nu12061717

Li, Z., and Ahammed, G.J. (2023). Plant stress response and adaptation via anthocyanins: A review. *Plant Stress* 10, 100230. https://doi.org/10.1016/j.stress.2023.100230

López-Casado, G., Sánchez-Raya, C., Ric-Varas, P.D., Paniagua, C., Blanco-Portales, R., Muñoz-Blanco, J., et al. (2023). CRISPR/Cas9 editing of the polygalacturonase *FaPG1* gene improves strawberry fruit firmness. *Hortic. Res.* 10, uhad011. https://doi.org/10.1093/hr/uhad011

Lynas, M., Adams, S., and Stockert, K. (2023). Gene editing achieves consistently higher favorability in social and traditional media than GMOs. *GM Crops Food* 14, 1–8. https://doi.org/10.1080/21645698.2023.2226889

Menz, J., Modrzejewski, D., Hartung, F., Wilhelm, R., and Sprink, T. (2020). Genome edited crops touch the market: A view on the global development and regulatory environment. *Front. Plant Sci.* 11, 586027. https://doi.org/10.3389/fpls.2020.586027

Napar, W.P.F., Kaleri, A.R., Ahmed, A., Nabi, F., Sajid, S., Ćosić, T., et al. (2022). The anthocyanin-rich tomato genotype LA-1996 displays superior efficiency of mechanisms of tolerance to salinity and drought. *J. Plant Physiol.* 271, 153662. https://doi.org/10.1016/j.jplph.2022.153662

Nie, H., Shi, Y., Geng, X., and Xing, G. (2022). CRISRP/Cas9-mediated targeted mutagenesis of tomato polygalacturonase gene (*SlPG*) delays fruit softening. *Front. Plant Sci.* 13, 729128. https://doi.org/10.3389/fpls.2022.729128

Nizampatnam, N.R., Sharma, K., Gupta, P., Pamei, I., Sarma, S., Sreelakshmi, Y., et al. (2023). Introgression of a dominant phototropin1 mutant super-enhances carotenoids and boosts flavor-related volatiles in genome-edited tomato *RIN* mutants. https://doi.org/10.1101/2023.05.05.539534

Nonaka, S., Ito, M., and Ezura, H. (2023). Targeted modification of *CmACO1* by CRISPR/Cas9 extends the shelf-life of *Cucumis melo* var. *reticulatus* melon. *Front. Genome. Ed.* 5, 1176125. https://doi.org/10.3389/fgeed.2023.1176125

Norouzkhani, N., Afshari, S., Sadatmadani, S.-F., Mollaqasem, M.M., Mosadeghi, S., Ghadri, H., et al. (2024). Therapeutic potential of berries in age-related neurological disorders. *Front. Pharmacol.* 15, 1348127. https://doi.org/10.3389/fphar.2024.1348127

Panche, A.N., Diwan, A.D., and Chandra, S.R. (2016). Flavonoids: An overview. *J. Nutr. Sci.* 5, e47. https://doi.org/10.1017/jns.2016.41

Park, S., Lee, H., Song, J., Lim, C.J., Oh, J., Lee, S.H., et al. (2023). Gene editing of authentic *Brassica rapa flavonol synthase 1* generates dihydroflavonol-accumulating Chinese cabbage. *Hortic. Res.* 10, uhad239. https://doi.org/10.1093/hr/uhad239

Pixley, K.V., Falck-Zepeda, J.B., Paarlberg, R.L., Phillips, P.W.B., Slamet-Loedin, I.H., Dhugga, K.S., et al. (2022). Genome-edited crops for improved food security of smallholder farmers. *Nat. Genet.* 54, 364–367. https://doi.org/10.1038/s41588-022-01046-7

Schmidt, S.M., Belisle, M., and Frommer, W.B. (2020). The evolving landscape around genome editing in agriculture: Many countries have exempted or move to exempt forms of genome editing from GMO regulation of crop plants. *EMBO Rep.* 21, e50680. https://doi.org/10.15252/embr.202050680

Shomali, A., Das, S., Arif, N., Saraf, M., Zahra, N., Yadav, V., et al. (2022). Diverse physiological roles of flavonoids in plant environmental stress responses and tolerance. *Plants (Basel)* 11, 3158. https://doi.org/10.3390/plants11223158

Tu, M., Fang, J., Zhao, R., Liu, X., Yin, W., Wang, Y., et al. (2022). CRISPR/Cas9-mediated mutagenesis of *VvbZIP36* promotes anthocyanin accumulation in grapevine (*Vitis vinifera*). *Hortic. Res.* 9, uhac022. https://doi.org/10.1093/hr/uhac022

Ullah, A., Munir, S., Badshah, S.L., Khan, N., Ghani, L., Poulson, B.G., et al. (2020). Important flavonoids and their role as a therapeutic agent. *Molecules* 25, 5243. https://doi.org/10.3390/molecules25225243

Vazquez-Vilar, M., Fernandez-Del-Carmen, A., Garcia-Carpintero, V., Drapal, M., Presa, S., Ricci, D., et al. (2023). Dually biofortified cisgenic tomatoes with increased flavonoids and branched-chain amino acids content. *Plant Biotechnol. J.* 21, 2683–2697. https://doi.org/10.1111/pbi.14163

Waltz, E. (2022). GABA-enriched tomato is first CRISPR-edited food to enter market. *Nat. Biotechnol.* 40, 9–11. https://www.nature.com/articles/d41587-021-00026-2

Yang, T., Ali, M., Lin, L., Li, P., He, H., Zhu, Q., et al. (2023). Recoloring tomato fruit by CRISPR/Cas9-mediated multiplex gene editing. *Hortic. Res.* 10, uhac214. https://doi.org/10.1093/hr/uhac214

Yang, Y., Wheatley, M., Meakem, V., Galarneau, E., Gutierrez, B., and Zhong, G.-Y. (2024). Editing *VvDXS1* for the creation of muscat flavour in Vitis vinifera cv. Scarlet Royal. *Plant Biotechnol. J.* https://doi.org/10.1111/pbi.14290

Zhang, S., and Zhu, H. (2024). Development and prospect of gene-edited fruits and vegetables. *Food Qual. Safety* 8, 1–15. https://doi.org/10.1093/fqsafe/fyad045

Index

Note: Page numbers in *italics* indicate a figure and page numbers in **bold** indicate a table on the corresponding page.